LEADING MEGAPROJECTS

A TAILORED APPROACH

MOUNIR A. AJAM

CRC Press
Taylor & Francis Group
Boca Raton London New York

CRC Press is an imprint of the
Taylor & Francis Group, an **informa** business
AN AUERBACH BOOK

CRC Press
Taylor & Francis Group
6000 Broken Sound Parkway NW, Suite 300
Boca Raton, FL 33487-2742

Printed on acid-free paper

International Standard Book Number-13: 978-0-367-34050-6 (Hardback)

Visit the Taylor & Francis Web site at
http://www.taylorandfrancis.com

and the CRC Press Web site at
http://www.crcpress.com

Trademarks Covered in This Book

AACE is a registered trademark of the Association for the Advancement of Cost Engineering.
The Customizable and Adaptable Methodology for Managing Projects (CAMMP), The Four Control Reference Points, The Four Dimensions of Project Success, The Seven Elements of Project Maturity, and The SUKAD Way are trademarks of SUKAD Corp.
IPMA is a registered trademark of the International Project Management Association.
ISO is a trademark of The International Organization for Standards.
A Guide to the Project Management Body of Knowledge (*PMBOK® Guide*), PMI, and PMP are registered trademarks of the Project Management Institute.
PRINCE2 is a registered trademark of AXELOS Limited. All rights reserved.
PRiSM and GPM are trademarks of Green Project Management Global.

Dedication

To my beautiful Nabila,

We had to advance our wedding date to relocate and join my first megaproject. Since that day, you have been my patient partner for 25 years, and I am looking forward to the next 25 years. Maybe, one day, we will see Sumer and Akkad working their megaprojects!

Contents

Dedication *vii*

Contents *ix*

Preface *xxiii*
Relevant History *xxiv*
Book Focus *xxv*

Acknowledgments *xxvii*
General Acknowledgments *xxvii*
Acknowledgment for Publishing this Book *xxvii*

About the Author *xxix*
About SUKAD *xxx*

Part A
Current Reality to Tailored Approach

Section I Setting the Scene 1

Chapter 1 Book Introduction 3
1.1 Something to Think About 3
1.2 Why a Book on Megaprojects? 4
1.3 The Blockers 5
1.4 Human Progress and the Giants 5
1.5 The Title of the Book 6
1.5.1 Leading Megaprojects 6
1.5.2 A Tailored Approach 6
1.5.3 Enabling Executives 7
1.6 Our Approach 7
1.7 Proposed Solution 8

1.8 Depth of Coverage 9
1.8.1 Integration of Various Aspects 9
1.8.2 Practitioner Perspective 9
1.8.3 What Is Not Included? 10
1.9 Book Structure 10
1.9.1 Part A: Current Reality to Tailored Approach 10
1.9.2 Part B: The Simulation 11
1.9.3 Closing 12
1.10 Chapter Closing Comments 12

Chapter 2 Relevant Definitions 13
2.1 Introduction 13
2.2 Definitions 13
2.2.1 Project Versus Program 13
2.2.2 Project Versus Product 15
2.2.3 Life Cycles 16
2.2.4 Phases and Stages 18
2.2.5 Gates 20
2.2.6 Project and Organizational Project Management 22
2.2.7 Capital Project 24
2.2.8 Megaproject 24
2.2.9 Asset Management 25
2.2.10 Project Owner and Service Provider 26
2.2.11 Methodology, Method, Tailored Methods 27
2.2.12 Agile and Agility 28
2.2.13 Project Classification 29
2.2.14 Success and Failure 30
2.2.15 Assessing Performance 31
2.2.16 Organization 31
2.3 Closing Comments 32

Chapter 3 The Literature Review 33
3.1 Introduction 33
3.2 A Summary of the References 33
3.2.1 References from Professional Associations 33
3.2.2 Project Management for Engineering and Construction 33
3.2.3 Industrial Megaprojects and Capital Projects 34
3.2.4 The Oxford Handbook of Megaproject Management 35
3.2.5 The Construction Industry Institute 35
3.2.6 Other References 36
3.2.7 Comparing the References 37
3.3 Closing Comments 37

Chapter 4 Professional Associations **39**
4.1 What Exists? 39
4.2 The Project Management Institute 40
4.3 Chapter Closing 40

Section II The Challenges— Current State of Practice **43**

Chapter 5 Observed Challenges **45**
5.1 Introduction to the Section 45
5.2 "The One Thing" 45
5.3 The Observed Challenges 46
5.3.1 Politics in Infrastructure (Good and Bad) 46
5.3.2 The Complexity of Megaprojects 46
5.3.3 Culture of Fixed Price 47
5.3.4 Culture of Low Bid 47
5.3.5 Overdependence on Contractors and Management Consultants 48
5.3.6 Project Owners' Competence 49
5.3.7 Change Management 50
5.3.8 Lack of Appreciation of Uncertainty 51
5.4 Closing Remarks 51

Chapter 6 Insights from the Global Community **53**
6.1 Introduction 53
6.2 A Venting Moment 53
6.3 Perspectives and Mindsets 54
6.3.1 Forecasting Challenges 54
6.3.2 Do Numbers Lie? 55
6.4 The Statistics 57
6.4.1 Industrial Megaprojects 57
6.4.2 Capital Projects 58
6.4.3 *The Oxford Handbook of Megaproject Management* 58
6.4.4 Study on Australian Projects 59
6.4.5 Other Stats 60
6.4.6 Summary of the Statistics 60
6.5 The Challenges 60
6.5.1 Balancing the Triple Constraints 61
6.5.2 Clarity of Objectives 61
6.5.3 The Front End and Scope Definition 62
6.5.4 Absence of Formal Project Management in the Front End 62
6.5.5 Complexity and Number of Stakeholders 63
6.5.6 Organizational Competence and Capacity 63

6.5.7	Scheduling	63
6.5.8	Cost Reduction	64
6.5.9	The Illusion of Risk Transfer	64
6.5.10	Blame Culture	64
6.5.11	Infrastructure Unique Challenges	65
6.5.12	Other challenges	65
6.6	Chapter Closing	65
Chapter 7	**A Tale of Two Megaprojects**	**67**
7.1	Introduction	67
7.2	Common and Differentiating Features	67
7.3	Performance Results	68
7.3.1	North American Project	68
7.3.2	Asian Project	70
7.4	Contributing Factors	70
7.4.1	NAP Challenges	70
7.4.2	NAP Opportunities	73
7.4.3	AP Challenges	73
7.4.4	AP Opportunities	74
7.5	Performance Conclusions	75
7.6	Chapter Closing	75
Chapter 8	**Other Case Studies**	**77**
8.1	Introduction	77
8.2	The Case Studies	77
8.2.1	The Cart Before the Horse	77
8.2.2	Nice Guys and Leadership	78
8.2.3	Do We Need to Operate This Facility?	79
8.2.4	Optimism Bias	80
8.2.5	Other Case Studies	80
8.3	Chapter Closing	81
Chapter 9	**Section II Summary and Closing**	**83**
9.1	Introduction	83
9.2	Summary of the Challenges	83
9.3	Closing Comments	84
Section III	**Critical Concepts for Megaproject Leadership**	**87**
Chapter 10	**The Critical Concepts, Overview**	**89**
10.1	Introduction to the Section	89
10.2	Challenges to Opportunities	91
10.3	Chapter Closing	92

Chapter 11 Leading Practices and Considerations **95**
11.1 Introduction 95
11.2 Why Do We Need Leading Practices? 95
11.3 Construction Industry Institute 96
11.3.1 Some of the CII Claims 96
11.3.2 What Are Some of the CII Best Practices? 97
11.4 Independent Project Analysis 98
11.4.1 Front-End Loading (FEL) 98
11.4.2 FEL Index 99
11.4.3 Integrated Teams 99
11.4.4 Capital Projects Rules 100
11.5 Other Leading Practices 100
11.5.1 Intermittent Project Reviews 100
11.5.2 Sustainability 101
11.6 Chapter Closing 101

Chapter 12 Organizational Competence and Capacity **103**
12.1 Introduction 103
12.2 What Is the Organizational Project Management Function? 104
12.3 Why Do Organizations Need the OPMF? 105
12.4 How to Transition Out of the Over-Dependence 106
12.5 A Historical Perspective (Side-Tracking a Bit) 107
12.6 Organizational Project Management Systems 107
12.7 Executive and Senior Management 108
12.7.1 Learn the Fundamentals 108
12.7.2 The Responsibility 108
12.7.3 The Consequences 109
12.8 Building and Sustaining the Organizational Project Management Function 110
12.8.1 Overview 110
12.8.2 The Fundamental Elements 111
12.8.3 The Differentiating Elements 112
12.8.4 Strategic and Organizational Aspects 112
12.8.5 Governance 112
12.8.6 Executive Development 113
12.9 Chapter Closing 113

Chapter 13 Mega: Projects, Programs, Assets **115**
13.1 Introduction 115
13.2 Project Versus Program 115
13.3 The Asset Question 116
13.4 Asset Management 117
13.5 Chapter Closing 118

Chapter 14 Section III Closure 119
14.1 Introduction 119
14.2 National Project Management Authority 119
14.3 What Would the NPMA Do? 120
14.4 Closing comments 121

Section IV The Tailored Approach for Megaproject Development 123

Chapter 15 Introduction to the Tailored Approach 125
15.1 A Look Back 125
15.2 This Section 127
15.3 Organizational Project Management 127
15.4 Does CAMMP™ Apply to Megaprojects? 128
15.5 Success One Step at a Time 128
15.6 Why a Stage-Gate Model? 128
15.7 A Bit of Psychology 129

Chapter 16 Critical Success Factors 131
16.1 Introduction 131
16.2 CAMMP™ Critical Success Factors (Principles) 131
16.2.1 Must Link Methods to Organizational System 131
16.2.2 Customize and Adapt to Industry and Project Type 132
16.2.3 Tailor to a Given Project Classification 132
16.2.4 Follow a Stage-Gate Process 133
16.2.5 Integrate a Project Life Cycle with Processes 133
16.2.6 Consider Product and Delivery Processes 134
16.2.7 Implement Project Success Model with Benefits Realization 135
16.2.8 Incorporate Industry Leading Practices 135
16.2.9 Capture Lessons Learned Throughout the Project 135
16.2.10 Defined Roles and Responsibilities 136
16.2.11 Integrate Competence and Sustainability 136
16.3 Megaprojects Critical Success Factors 136
16.3.1 Industrial Megaprojects Recommendations 136
16.3.2 Barshop's Four Principle Practices 137
16.3.3 Balancing the Triple Constraints 137
16.3.4 Shaping/Framing 138
16.3.5 Control the Lock-in Effect 138
16.3.6 Scrutiny of the Need for a Megaproject 139
16.4 Chapter Closing 140

Chapter 17 Integrated and Applied Project Management 141
17.1 CAMMP™ and the SUKAD Way™ 141
17.2 The SUKAD Way™ 141

17.2.1 Organizational Project Management 142
17.2.2 Managing a Project 143
17.2.3 Controlling a Project 143
17.2.4 Other Resources 143
17.3 The CAMMP™ Three Dimensions 143
17.3.1 The First Dimension 144
17.3.2 The Second Dimension 144
17.3.3 The Third Dimension 145
17.3.4 Integrating the Three Dimensions 145
17.4 Projects' Classification 146
17.5 Project Success 148
17.5.1 Non-SUKAD Concepts 148
17.5.2 Perspectives 148
17.5.3 The Four Dimensions of Project Success™ 149
17.5.4 When to Define the Criteria for Success 153
17.5.5 A Project Success Score Card 153
17.6 Closing the Chapter 154

Chapter 18 CAMMP™ for Megaprojects 157
18.1 Introduction 157
18.2 A Capital Project Model 157
18.2.1 An Industry Perspective 157
18.2.2 A SUKAD Perspective 158
18.3 The Four Streams of the Tailored Approach 158
18.4 The Proposed Project Life Cycle 160
18.4.1 Overview 160
18.4.2 The Facilities Stream 160
18.4.3 The Governance (and Business Aspects) Stream 161
18.4.4 The Production Stream 162
18.5 The Phase/Stage Management Process 163
18.6 Gates Relevant Concepts 163
18.6.1 Types of Gates 163
18.6.2 Different Types of Stage Gates? 163
18.6.3 A Common Pitfall 164
18.6.4 Another Pitfall 164
18.6.5 Engineering Stage Gate 164
18.6.6 Stage Gates Criteria 165
18.7 Chapter Summary 165

Chapter 19 CAMMP™ MP: Concept to Success 167
19.1 Introduction 167
19.2 Discovery Phase 170
19.2.1 Phase Overview 170

19.2.2 Concept Stage 171
19.2.3 Feasibility Stage 173
19.2.4 Shaping Stage, Part 1 178
19.3 Development Phase 180
19.3.1 Phase Overview 180
19.3.2 Shaping Stage, Part 2 181
19.3.3 Requirements Stage 182
19.3.4 Strategy Stage 184
19.3.5 Preliminary Engineering Stage 189
19.3.6 Operational Readiness Planning Stage 190
19.3.7 Request for Final Funding 191
19.4 Delivery Phase 191
19.4.1 Phase Overview 191
19.4.2 Engineering Stage 192
19.4.3 Construction Stage 193
19.4.4 Operational Readiness Stage 193
19.4.5 Initial Operations Stage 194
19.4.6 Close Stage 195
19.5 Closing the Chapter 198

Chapter 20 Stage Management Processes and Functions 199
20.1 Navigation Message 199
20.2 The Phase/Stage Management Process 200
20.2.1 The Knowledge Foundation 200
20.2.2 The CAMMP™ Alternative View 200
20.2.3 Stage Management or Phase Management? 203
20.3 The Processes Outputs 203
20.3.1 Authorize 203
20.3.2 Plan Management 204
20.3.3 Plan Details 204
20.3.4 Implement 204
20.3.5 Control 204
20.3.6 Close 205
20.3.7 The Templates 205
20.4 The Project Management Functions 205
20.4.1 Overview 205
20.4.2 What Are the Project Management Functions? 206
20.4.3 Linking the Functions to Processes 206
20.5 Expanding on the Functions 207
20.5.1 Scope Management 207
20.5.2 Cost Management 208

20.5.3 Schedule Management 212
20.5.4 Planning 212
20.5.5 Risk Management 213
20.5.6 Quality Management 214
20.5.7 Procurement Management 216
20.5.8 Project Control 219
20.5.9 Change Management 225
20.6 Chapter Summary 227

Chapter 21 The People Aspects of Megaproject Leadership 229
21.1 Introduction 229
21.2 Essential Recommendations 230
21.2.1 Build and Sustain the Organizational Project Management Function 230
21.2.2 Adequate Staffing 230
21.2.3 Team Competence 231
21.2.4 Team Integration 232
21.2.5 Leadership Continuity 233
21.3 Project Organization 233
21.3.1 Project Governance 233
21.3.2 The Facilities Stream 234
21.3.3 The Other Streams 235
21.4 Chapter Closing 235
21.5 Closing Part A 235

Part B
The Simulation

Section I The Project 237

Chapter 22 The Integrated City Megaproject 239
22.1 Disclaimer 239
22.2 Megaproject Overview 240
22.3 Why a Project, Not a Program 241
22.4 Depth of Coverage 241

Chapter 23 Guide to Part B 243
23.1 Part B 243
23.2 Stage Process Sequence 243
23.3 Process Deliverables and Process Gates 247
23.4 Stage Deliverables and Stage Gates 247
23.5 Chapter Closing 247

Section II The Discovery Phase 249

Chapter 24 Introduction to the Phase 251

Chapter 25 The Concept Stage 253
25.1 Introduction 253
25.2 Process Deliverables and Process Gates 253
25.2.1 Authorize Process 253
25.2.2 Plan Management Process 254
25.2.3 Plan Details Process 254
25.2.4 Implement Process 254
25.2.5 Control Process 254
23.2.6 Close Process 255
25.3 Stage Deliverable and Stage Gate 255
25.3.1 The Project Brief 255
25.3.2 Stage Gate 1: Idea Approval 256

Chapter 26 Feasibility and Shaping Stages 257
26.1 Introduction 257
26.2 Stage Process Sequence 257
26.2.1 Authorize Process 257
26.2.2 Plan Management Process 258
26.2.3 Plan Details Process 258
26.2.4 Implement Process 258
26.2.5 Control Process 258
26.2.6 Close 258
26.3 Process Deliverables and Process Gates 259
26.3.1 Authorize (Stage Work) 259
26.3.2 Plan Management 261
26.3.3 Plan Details 264
26.3.4 Implement Stage Work 271
26.3.5 Control Stage Work 273
26.3.6 Close Stage Work 274
26.4 Stage Deliverables and Stage Gate 275
26.4.1 Feasibility Study Report 275
26.4.2 Stage Gate 2: Project Initial Approval 280
26.4.3 Project Authorization Document 281
26.5 Chapter Summary 283

Section III The Development Phase 285

Chapter 27 Introduction to the Phase 287
27.1 Overview (A Refresher) 287
27.2 Phase Strategy 287

27.3 Revised Model 288
27.4 Closing Remarks 288

Chapter 28 Requirements Stage 289
28.1 Introduction 289
28.2 Process Deliverables and Process Gates 289
28.2.1 Authorize Stage Work 289
28.2.2 Plan Stage Management 290
28.2.3 Plan Stage Details 292
28.2.4 Implement Stage Work 297
28.2.5 Control Stage Work 297
28.2.6 Stage Closing 298
28.3 Stage Deliverable and Stage Gate 298
28.3.1 Project Requirements Document 298
28.3.2 Stage Gate 3: Stakeholders' Alignment 305
28.4 Chapter Summary 305

Chapter 29 Master Plan Stage 307
29.1 Introduction and Stage Purpose 307
29.2 Process Deliverables and Process Gates 307
29.2.1 Authorize Stage Work 307
29.2.2 Plan Stage Management 308
29.2.3 Implement Stage Work 310
29.2.4 Control Stage Work 311
29.2.5 Close Stage 311
29.3 Stage Deliverables and Stage Gate 311
29.3.1 The Project Master Plan 311
29.3.2 The Commercial and Financial Plan 312
29.3.3 Stage Gate 4: Master Plan Approval 313
29.4 Chapter Closing Comments 313

Chapter 30 Strategy Stage 315
30.1 Introduction and Stage Purpose 315
30.2 Special Message 316
30.3 Stage Sequence 316
30.4 Process Deliverables and Process Gates 316
30.4.1 Authorizing the Stage 317
30.4.2 Plan Stage Management 317
30.4.3 Plan Stage Details 317
30.5 Stage Deliverables and Stage Gate 318
30.5.1 Project Management Plan 318
30.5.2 Class 2 Estimate 321
30.5.3 Advance Funding 323

30.5.4 Investment Program 324
30.5.5 Stage Gate 4: Project Management Plan Approval 325
30.6 Chapter Closing Comments 325

Chapter 31 Preliminary Engineering Stage 327
31.1 Introduction and Purpose 327
31.2 Stage Sequence 327
31.3 Process Deliverables and Process Gates 328
31.4 Stage Deliverables and Stage Gate 328
31.4.1 Preliminary Engineering Packages 328
31.4.2 Class 3 Estimate 328
31.4.3 Updated Feasibility and Final Funding 329
31.4.4 Operational Readiness Planning 329
31.4.5 Investment Program 330
31.5 Chapter/Section Closing Comments 330

Section IV The Delivery Phase 331

Chapter 32 Introduction to the Phase 333
32.1 Overview 333
32.2 Purpose of Phase 333
32.3 Author Note 334

Chapter 33 Implementation Stages 335
33.1 Introduction and Stages Purpose 335
33.2 Process Deliverables and Process Gates 336
33.3 Stage Deliverables and Stage Gates 336
33.3.1 Detailed Engineering 336
33.3.2 Construction 336
33.3.3 Operational Readiness 336
33.3.4 Stage Gate: Ready for Handover 336
33.4 Success Factors 337
33.4.1 Chapter Closing Comments 337

Chapter 34 Operational Readiness Stages 339
34.1 Introduction and Stages Purpose 339
34.2 Stage Processes 339
34.3 Deliverables and Gates 339
34.3.1 Operational Readiness Detailed Plan 339
34.3.2 Pre-Handover Work 340
34.3.3 Stage Gate 340
34.3.4 Handover Reviews 340
34.3.5 Start-Up and Commissioning 340
34.3.6 Initial Operations 341

34.3.7 Stage Gate: Final Acceptance 341

Chapter 35 Close Stage 343
35.1 Introduction and Stage Purpose 343
35.2 Stage Sequence 344
35.3 Stage Deliverable and Stage Gate 344
35.3.1 Performance Reconciliation Report 344
35.3.2 Lessons Learned 345
35.3.3 Organizational Records 345
35.3.4 Measuring Project Success 345
35.3.5 Stage Gate: Project Is Closed 345

Chapter 36 Part B Closure 347
36.1 Introduction 347
36.2 Governance 347
36.3 Approvals 348
36.4 Estimating and Reserves 348
36.5 Risk Management 349
36.6 Performance Management 349
36.6.1 General Comments 349
36.6.2 Project versus Organization 350
36.6.3 Sample Project 350
36.6.4 Project Metrics 351
36.7 Chapter Closing Comments 352

Chapter 37 Call for Action 353
37.1 Introduction 353
37.2 Points to Ponder 354
37.2.1 Point 1: Incremental Development 354
37.2.2 Point 2: Are We Using NPV And BCR Properly? 355
37.2.3 Point 3: Are We Assessing Project Success Properly? 356
37.2.4 Point 4: The Role of Education 356
37.2.5 Point 5: How About Shared Services or Authority? 356
37.2.6 Point 6: Contingency and The Management of Risks 357
37.3 Call for Action 357

Appendix A Additional Resources 361

Appendix B Sample Templates 363
Project General Information 363
Stage Management Plan 364
Project Management Plan 365
Strategy Section 365
How to Section 365

People Aspects Section 366
Class 2 Estimate Section 366
Project Manager Comments 366
Stage Gate: Project Management Plan Approval 367
Summary of Prior Work 367
Project Management Plan Approval 367
Management Other Comments or Instructions 367

Acronyms 369

Bibliography 373

Index 377

Preface

Over the last few decades, the field of project management has seen significant growth and acceptance across sectors, industries, and domains. Project management skills are in demand across organizational levels. There are numerous certifications in project management—there may be more than fifty—offered by various professional associations and private companies. It would be safe to say that there are at least a few million practitioners with one project management certificate or another. Yet, we continue to see failed and challenged projects, regardless of industry, domain, size, or complexity.

Failure of low-cost projects in large organizations could be tolerated. However, failure of megaprojects could significantly impact the finances of organizations, private or public—even countries. Just imagine this: a 10% overrun on a $1 billion project—which would be a small project among the mega class—would be $100 million. Can you imagine what companies, governments, or humanity could do with $100 million? Once you are done imagining, can you even visualize if this number were to be 100 or 1,000 times higher?

Megaprojects can often be front page news, and their reputation or legacy will last for decades. Therefore, this is a broad topic with massive global impact. There are many articles, academic papers, and a few books discussing megaprojects. Unfortunately, many of these references focus on failures—maybe a reflection of the current reality. One reference claims that only one project out of 1,000 is successful; another is more optimistic and claims five out of 1,000 (or 0.5%) as successful. However, some of these claims could be designed to be attention grabbers, sensational, or provocative, which would not be right, fair, or even ethical, *if they misrepresent the reality.*

Industrial megaprojects seem to fare better than infrastructure megaprojects, maybe because these projects are taken on by companies with significant scrutiny, whereas infrastructure projects, because they are led by government entities, are affected by numerous factors, including politics. Although industrial megaprojects do better than infrastructure megaprojects, they still do not do well, or as well as they should. Even though their reported success rate is much higher than 0.5%, their performance is still leaving billions of dollars in lost organizational value.

In this book, we will reflect on the sad state of the current practice of megaprojects. However, our intention is not to get stuck in a cycle of doom and sensational reporting, but to learn about the challenges, problems, mistakes, and harmful practices. Knowing the current state of practice (the good, the bad, and the ugly) is a crucial starting point and a current reality. We need to define where we want to go, then figure out how to close the enormous divide between the current reality and a better future. Closing that gap is the primary purpose of this book.

In this Preface, I will think out loud and encourage you to think and ponder these questions:

- What are some of the challenges facing megaprojects, what are the opportunities, and can we improve the state of the current practice?
- Is the organizational culture a factor, and how does project management maturity contribute to the success or failure of projects?
- Should megaprojects be treated as programs, or should they be managed as an integrated component of asset management?
- Have we reached a state of practice at which project owners are overwhelmed or intimidated by project management, which leads them to give up on building their organizational competence and consequently outsource the management of their projects and megaprojects?
- Can project management consultants (management companies), working as representatives of project owners, adequately represent the interests of these project owners to deliver success?
- How about the type of contracts used—do they have a direct or indirect influence on projects?
- We cannot ignore "The Agile Movement" and Agile development; is it the magic wand for capital projects as some Agile advocates are trying to convince us?
- Can we benefit from standardized organizational project management systems, which could help organizations lead their projects from micro to megaprojects? Or is it the perception that standardized processes, such as stage-gate, project life cycle models are perceived as outdated, bureaucratic, and ineffective?
- Some would say, "*Organizational systems, project life cycles, stage-gate processes are already established, so, Mounir, what value can you add?*" Sure, these things have existed for decades, but projects still fail at a high rate. Why? What is missing?
- Are we defining success and failure properly? How do we define success, and what are the differences between success, failure, and the challenged status of projects?

We could be asking many more questions, and we will in due time. However, for now, ten sets of questions are a reasonable limit. Take your time and try to answer the above questions, even speculate if you must. Write down your thoughts and refer to them as you read. Maybe we align, or perhaps we will challenge your conventional wisdom, or you will challenge ours. It is also possible that you might have a book in your head—bring it out.

We would be happy to hear from you, especially if you want to challenge what we present in this book.

Relevant History

We originally wrote a version of this book in early 2015 after a few workshops for the Iraqi Ministry of Transport and its companies. We wrote it to explain The Customizable and Adaptable Methodology for Managing Projects™ (CAMMP™) and how to apply it on various projects. We drafted a simulation of a megaproject to use in support of these workshops. The simulated megaproject scope covers the various areas of operation for the ministry (ports, airports, rail, public transport); however, we never published it. The primary reason we did not release the book is that we were working on updating our project management model, and we preferred to use the latest and greatest before sharing it with the community. Our most recent update to CAMMP™ was published by CRC Press (Taylor & Francis) in December, 2017, as *Project Management beyond Waterfall and Agile*. With the publication of that book, we can revisit the 2015 book on megaprojects, which we do here but with a significant change to the structure and content.

The book in your hands depends on *Project Management beyond Waterfall and Agile*, because it is the primary knowledge source for CAMMP™ and the reason we developed CAMMP™. We urge the reader to consider obtaining our earlier work and reading it before this one. However, we realize that some will not do so. Therefore, we include summaries of the relevant content from that book here, but only the parts that apply to understand vital concepts.

This book is not just an update of the unpublished book. Since 2015, we have conducted many workshops to help us enhance the methodological approach, and we also have performed an intensive literature review, mostly on industrial projects, capital projects, and infrastructure projects. That literature review and research are helping us write a book that we hope will become a leading reference for the leadership of megaprojects. In summary, Part A of this book is mostly new content or re-writing part of the *Project Management beyond Waterfall and Agile* for the context of megaprojects. On the other hand, Part B is the simulation of the model, which is an update of the 2015 book.

Book Focus

This book's core focus is on how to lead megaprojects, concept to success, using a tailored approach.

We will introduce the fundamentals and vital concepts for the tailored approach in the first part of the book, then follow it with a simulation of a realistic project in the second part. However, before presenting the tailored approach and the simulation, the book addresses the *current reality* and *the challenges* of the megaproject management practice.

Our discussions emphasize a project life cycle (PLC) model to lead megaprojects from concepts to closure (concept to success), which makes this book's focus ideal for project owners. This owners' PLC focus is critical because (1) the success and failure of megaprojects start with the project owner in the discovery phase of a project; and (2) there are not enough resources covering projects from concept to closure. Even the numerous postgraduate programs from leading universities are focused on engineering or construction management—that is, the implementation (execution) phase of capital projects. Further, the emphasis on the project owners is that these organizations are the ones that launch the projects and will own their products to generate the benefits from the produced asset. In this context, a project owner could be a private or public entity.

Although this book is ideal for project owners and their executives, senior managers, and project teams, it is highly valuable for all other organizations involved in projects, including engineering, construction, and project/construction management companies. Another message to stress at this time is that this book is also suitable for all industries that handle large and complex projects, so do not let the title of megaprojects intimidate you into believing it is limited to engineering and construction.

—Mounir A. Ajam
9 July 2019

Acknowledgments

General Acknowledgments

We start with acknowledging a special colleague and a previous SUKAD team member, Mr. Luc Bauwmans. Luc has been a significant partner, working with me in delivering many of the SUKAD services, including a postgraduate program on CAMMP™. Luc is a project management expert with considerable experience, including on large and complex projects. Luc had reviewed our past books and this one and offered many useful suggestions. He is currently an Adjunct Professor with the Central Queensland University.

This book has taken me back for many years and brought out many great memories and some bad ones. Therefore, I must acknowledge a few friends and mentors who have left a lasting impression and influenced my career directly or indirectly. These include Ralph C. Spears, John Moore, and Hesham Al-Musaed. Also, I must acknowledge many colleagues, some I have met face to face and others only virtually. This next group includes great supporters such as Professor Christophe Bredillet and Dr. Joel Carboni. In this group we include those whom we exchange messages and feedback almost daily through online platforms like LinkedIn; these include William (Bill) Duncan, Neville Goedhals, Mark Moore, Stephen Maas, Dr. Mihail Sadeanu, Lee Lambert, Jean-Charles Savornin, Dr. Andreas Trautner, and many others (I hope they forgive me for not listing them). Many of these individuals have contributed to this book directly and indirectly, with special thanks to Dr. Mihail and Jean-Charles.

Acknowledgment for Publishing this Book

Finalizing this book for publication could not have been possible without the following:

John Wyzalek from Taylor & Francis, who accepted this book for publication under the Advanced Project Management Series, for his continuous support.

Theron Shreve of DerryField Publishing Services, who guided the transformation of the manuscript into the book that is in your hands.

Susan Culligan of DerryField Publishing Services, who has been patiently editing and typesetting my work, which I know is not easy, my being a non-native speaker of English. She has been wonderful guiding me through the editing process, including some substantial rewrites and re-edits.

About the Author

Mounir A. Ajam is an entrepreneur, author, speaker, coach, advisor, consultant, volunteer leader, and project management thought leader.

He is the author of *The Inheritance (a Story of Friendship, Community, and Project Management)*[*]; *Project Management Foundation*[†]; *Redefining the Basics of Project Management*[‡]; *Applied Project Management*[§]; and *Project Management beyond Waterfall and Agile.*[¶] He is also the author of two e-books series, *Adaptive Project Management and Tailored Methods*[**] series, and *CAMMP™ Case Studies & Simulations Series*[††]; the latter series focuses on micro and small projects.

He is a senior executive with more than three decades of outstanding global and practical experience in capital project industries such as engineering, construction, petroleum, utilities, project management, and management consultancy. He has worked on projects worth billions of U.S. dollars in North America, Europe, Southeast Asia, and West Asia. His experience includes working small and multiple projects and large and complex projects, including megaprojects.

Mounir is a co-founder and the Chief Executive Officer of SUKAD Project Portfolio Management Solutions (SUKAD Corp), founded in the United States in 2019 (previously in the UAE since 2004), a project management provider with registration in the United States, Lebanon, and the United Arab Emirates. Also, he has initiated **the Uruk Initiative** to generate interest among project management professionals from around the world to come together and form a technology startup. This startup is for the development of an online solution and platform for managing projects, with CAMMP™ as the core project management methodological approach. The Uruk PPM Platform is now part of SUKAD Corp. This book will contribute to the Uruk PPM Platform.

Mounir and SUKAD play quite an active role in the project management community through various professional activities that are open to community members at no cost. He is heavily involved with the project management community at the regional and global levels. Globally, he has served in different roles and capacities, including serving on the Global Advisory Group to the Project Management Institute (PMI®) Registered Education Provider program (2005–2007) and as a judge for many PMI® educational awards during the same period. He also served on the 2008 PMI® EMEA (Europe-Middle

[*] Self-published, 2010.
[†] Author House, 2014.
[‡] Author House, 2014.
[§] Author House, 2015.
[¶] CRC Press, 2017.
[**] Bookboon, 2019.
[††] Bookboon, forthcoming.

East-Africa) Congress Project Action Team. Mounir is a graduate of the PMI Leadership Institute Master Class (Class of 2007).

In West Asia, Mounir served on the board of directors for the PMI chapter in the Arabian Gulf, 2004. He led the effort to establish a PMI chapter in the United Arab Emirates. He also led the effort and established the Global Project and Process Management Association (GPPMA) in the UAE. He served as GPPMA board chairperson for three years.

Mounir is an advocate of project management and recognizes its strategic value. He has contributed to project management growth by publishing professional papers and articles on numerous platforms. In recent years, he has been the principal author of the SUKAD Applied Project Management blog site (http://blog.sukad.com) and the Applied Project Management YouTube Channel (https://www.youtube.com/sukad).

For more information about Mounir Ajam, please refer to his personal page at www.mounirajam.com or connect with him on LinkedIn, https://www.linkedin.com/in/mounirajam/.

About SUKAD

SUKAD Vision is ***Project Management for All Aspects of Life!***

SUKAD Mission is ***Be an Agent of Change and a Catalyst for Development!***

SUKAD was established in Dubai, United Arab Emirates, in 2004. In 2012, SUKAD registered in Lebanon, and in early 2019 in the United States. From these legal entities, SUKAD has been providing services, mostly in West Asia but also on a limited basis in Africa, Central, and Southeast Asia. However, recently, with a group of co-founders, we are working on transforming SUKAD into a global integrated project management solutions provider, building on the SUKAD Way™ for Managing Projects. With the SUKAD Way™ as a core element, SUKAD offers learning & professional development services, consultancy & organizational solutions, SUKAD knowledge & innovation products, the Uruk PPM Platform, and open knowledge sharing platform. The Uruk PPM Platform is a pioneering, comprehensive, and versatile cloud-based solution for the management of any project, using the tailored method like the one in this book. Please refer to www.sukad.com for more information.

Part A
Current Reality to Tailored Approach

Section I
Setting the Scene

Chapter 1

Book Introduction

1.1 Something to Think About

I want to start with a provocative statement. Based on my personal experiences and research from various resources with different perspectives, I must ask:

- We know the performance of megaprojects is questionable, and a large percentage of them fail in one dimension or another.
- We know, or maybe we think we know, the challenges that contribute to these failures.
- I believe we do have answers and solutions to minimize, or at least reduce, the percent of failed projects.
- *Then why do these projects still fail at a high rate?*

I want to be provocative but not sensational.

I know that many claims of failures could be questioned as attention grabbers or to sell services. I understand that some of the reports could be misleading, consciously or subconsciously, or at least shared out of context. I have read academic papers complaining about long-standing theories, yet the same author(s) make alarming claims. I have also read papers that dispute these claims as fallacies. I do not want to get into the academic debate, nor do I want to drag you into it. What we know, and almost everyone agrees on, is that *we have a problem,* but we might not have an agreement on the size or the root causes of the problems or how to solve them. It is also possible that some practitioners and academics do not believe a solution is possible; I hope that this book will change your mind and theirs.

I know that one of the challenges relates to how we define success and failure as either/or with nothing in between. When we discuss success, we will offer an alternative view. Our view on the questions raised earlier is this: the answers might be due to the knowing–doing gap syndrome (Pfeffer and Sutton, 1999), or is it Einstein's definition of insanity? I must admit, and agree with Professor Bent Flyvbjerg, that the field of megaproject management is still young, with not enough research or case studies of successful projects to learn from (Flyvbjerg, 2017). I also agree with the professor about the absence of "classics" in this dynamic field of study, based on a variety of reasons. However, I think that one of the biggest challenges is that failure acts as a magnet for media and public attention, whereas success might be overlooked or not shared openly.

In my professional opinion, **we need to share more success stories if we want to break the proprietary barriers.**

1.2 Why a Book on Megaprojects?

I think the last couple of sentences in the previous paragraph directly, if partially, answer this question. There are not enough references or academic programs on megaprojects. I also think that there are not enough educational programs that depend on practitioners to lead the study of megaprojects. Some of what we have seen or read is questioned because some claims are made without significant empirical evidence. In other words, we have read arguments and counter-arguments and even lack of consensus. However, what many academics and practitioners agree on is that we do have a problem, or many issues, and humanity can benefit from closing the performance gaps.

> **Massive Projects, Massive Impact**
>
> From various studies (as quoted by Flyvbjerg in *The Oxford Handbook of Megaproject Management*, page 5), globally, the infrastructure megaprojects market is possibly more than $3 trillion, per year. "If we include the many other fields . . . the market can be $6–9 trillion per year."
>
> Can you even imagine the positive impact if we could help organizations reduce the failure rate and increase the cost savings and benefits by 10%, even 1%?

> What we aspire to here is to provide an excellent reference that presents a model that professionals and organizations can use to increase the chance of success.

The lack of content might exist because managing or leading megaprojects is a humongous endeavor that cannot be covered in a single book or limited to a domain or a degree program. To lead these types of projects successfully would require the vital integration of many fields of studies. These would include governance, engineering, construction, marketing and communication, public relations, project and program management, asset management, and value management—a full list would need more than a page, even without mentioning topics related to the new era of big data, design thinking, and artificial intelligence. Also, the diversity of megaprojects that come from numerous industries prevents or does not facilitate the presence of a "classic" (Pollack et al., 2018).

> To lead these types of projects successfully would require the vital integration of many fields of studies.

My view here is not intended to be discouraging or surrendering to the status quo, but to reflect on the current reality and what would be practical. Therefore, despite our aspiration to provide a comprehensive book, it is physically not possible. However, we cannot let this massive undertaking stop us and will endeavor to provide an excellent resource for organizations, executives, and project teams working on megaprojects.

What we aspire to here is to provide an excellent reference that presents a model that professionals and organizations can use to increase their chances of success. Although our examples and case studies might be leaning toward capital projects, with engineering and construction components, the tailored approach that we will present would apply to all projects, regardless of industry or domain—of course, with the proper tailoring.

1.3 The Blockers

In this book, I offer a logical framework, in a simplified and practical language, and with a positive and optimistic message about serving the project management community. However, in this first chapter, I decided to include this part about some of the adverse blockers to progress: those who might accept the status quo or who do not understand the complexity of megaprojects. I will do this via two stories that could be representative of various scenarios across many organizations. The first story is in the text box, in which this senior professional is arguing that the management of a megaproject is like any other project.

> **Story No 1: Project Management Is the Same**
>
> When I started working this book, I shared the outline online on my LinkedIn page. Here are some of the the responses we get:
>
> "I don't get it—project management is the same, regardless of what we are managing."
>
> "The biggest problem with project management is trying to reinvent the wheel."
>
> "I don't understand, aren't the project management processes universal? You know, applicable to most projects?"

We all know that the basic principles of project management include concepts like the Plan-Do-Check-Act cycle, which are well known in the quality world. We agree that the basics of management do evolve—partially—around processes or process groups that are generic. We understand that some will argue that project management is common sense. However, to believe that the generic guides or processes are all we need to manage capital projects would be uninformed, to say the least.

Think about this brain teaser: *it is common sense not to use common sense on non-common-sense projects.* In other words, megaprojects are a different breed. I like a phrase by Professor Bent Flyvbjerg that says, *"Conventional project managers should not lead megaprojects"* (Flyvbjerg, 2017). Merrow calls megaprojects "fragile" (Merrow, 2011).

The second story is of another senior professional saying, *"Organizational systems, project life cycles, stage-gate process are already established, so, Mounir, what value can you add?"* I am not sure if this person is questioning my personal abilities, or if he is challenging the thinking that we could not improve performance.

My public answer to these two individuals is this: megaprojects are highly complex, their development and execution are the killers or makers of reputations, their successes and failures could make or break companies and countries. In other words, we need a new way of thinking, a cultural mindset that may require shifting paradigm to tackle these projects. Concerning our abilities to make a change, absolutely, we had made an impact in the past, and we will continue to do so, despite the blockers and naysayers.

1.4 Human Progress and the Giants

Human progress is not limited to inventions and inventors; it is about the human capability to grow and build on each other's inventions and contributions. I am not sure what percent of what we present in this reference is our invention. I am willing to accept and admit that we are not reinventing the wheel—there is no need to do so. We believe that the components of what exists are good, but we might need a fresh practitioner's look.* We are proud to say that we aim to build on what came before us, which would empower us to offer a great solution. As Sir Isaac Newton once said, *"If I have seen a little further, it is by standing on the shoulders of Giants."* (The Phrases Finder, n.d.). We must accept that we, all of us, can make a difference, and we need the collective knowledge and experience to deal with an essential and massive topic such as megaprojects. Therefore, we know we have learned from others, and we will share our learnings with the proper credits. We know that some of what we present include a few modifications from the source, and I think the reader can judge whether we are offering something new to the professional community.

> We need a new way of thinking, a culture mindset that may require shifting paradigms to tackle these projects.

1.5 The Title of the Book

Before I continue, I want to reflect on the title, since it helps set the right mindset for understanding this work. We chose a title in two parts: ***Leading Megaprojects: A Tailored Approach***. The reasons:

1.5.1 Leading Megaprojects

The first part of the title is ***Leading Megaprojects,*** with the emphasis on *leading.*

Let us not get into a debate about the differences between manager and leader (management and leadership); I think most of us know the fundamental differences.† I think we all, or most of us, can agree that for any project of some significance, we need both skill sets, among many others. We need leadership *and* management. Which one is needed more, in the project management general context, can be debatable, but not for megaprojects. Due to the massive size, complexity, number of parties and stakeholders involved, and cost, among other vital factors, megaprojects cannot be successful without great leadership. Yes, you read it right, cannot be successful!

Does that mean we only need leaders?

No.

Leaders can define the path for a megaproject, but we still need to support these leaders with seasoned managers and project managers. However, even though we need managers, the emphasis in Leading Megaprojects is on leadership.

1.5.2 A Tailored Approach

The second part of the name is ***A Tailored Approach.***

---

* I personally have had direct exposure to numerous megaprojects and have seen success and failure. I will be sharing a few case studies later.

† If you are interested in my views on this topic, this video https://www.youtube.com/watch?v=a9Kf2rKY9g8 will give you an idea.

After all the references that I have reviewed, and decades in the industry, I have yet to see a practical, comprehensive, and robust model for leading megaprojects, concept to success! In all transparency, I am not saying one does not exist, they do exist, but as proprietary methodologies inside organizations with a high level of project management maturity. I also know that the Independent Project Analysis (IPA) and the Construction Industry Institute (CII) do offer significant knowledge and resources, but they may not be widely known or readily available to practitioners. Both CII and IPA emphasize a stage-gate process, which may seem to be the *de facto* standard. Project life cycle models are in abundance, and so on. However, that does not mean we have a fully integrated product development and delivery approach that is readily available and accessible to the typical practitioner.

Further, we think that maybe there is too much focus on "technical" project management rather than on product development and benefits realization. We also know that many organizations and professional associations publish generic models, especially those that split projects into two major zones: the front end (before final fund authorization) and the execution (after fund authorization). Some references include the business case and feasibility in the project life cycle model, others do not.

Therefore, our view is this: we do not manage projects by following a basic model (or common sense alone); to manage projects effectively, we need a project life cycle model that helps us develop a project from concept to closure (C2C) and beyond; for megaprojects we like to say *concept to success* (C2S). We need a model that would encourage, if not mandate, the integration of the business side with the project side and with a line of sight on production and operation. We must view projects as essential parts of asset management. We realize it is easy to discuss components and the elements of what is required to lead megaprojects, but how do they fit together? Via the ***tailored approach*** that we will be presenting.

1.5.3 Enabling Executives

As you will see throughout this book, megaprojects are massive and are often led by executives who do not fully understand their complexity or who have not been trained in project management. The failure of a megaproject can break companies or significantly affects shareholder values even in the massive conglomerates. Therefore, if executives and organizations do not want to fulfill the theory of insanity, they need help. They need a model—a tailored approach that is suitable for the type of projects that is fit-for-purpose to their industry and application area. We need to build on what the giants have created.

> Executives need a tailored approach that is suitable for the type of projects that is fit-for-purpose to their industry and application area.

1.6 Our Approach

Despite the complexity of megaprojects and the lack of numerous references, as we said before, it is a considerable task, more significant than we can provide in this reference. However, what we offer is a great starting point into the art, science, and psychology of project management and leadership.

What we have done is:

- Review various leading resources on capital projects and megaprojects. We did so to extract from them the prominent challenges, opportunities, and causes of failure and summarize them in a ready reference.
- We also reflect on case studies, mainly those in which we had direct involvement or exposure.
- We reflect on our global experiences and the work that we have been doing at SUKAD, the work on "project and organizational project management," which we have been performing over the last decade or two.

We are offering a great starting point into the art, science, and psychology of project management and leadership.

In other words, we are distilling and refining all the learning from these three vital sources to develop a tailored approach for leading megaprojects. This tailored approach is a customized and adapted methodological approach that would be suitable for these types of projects, or at least as a starting point for organizations to tweak further as they transform the proposed tailored approach into a great fit-for-purpose solution, in line with the organizational context and culture. The launching pad for this approach is the SUKAD's CAMMP™ Model, The Customizable and Adaptable Methodology for Managing Projects™. CAMMP™ is the foundation, but what we offer in this book is **CAMMP™ MP, CAMMP™ for Megaprojects**. (See Figure 1.1)

1.7 Proposed Solution

The result of what we have created is a developed management and leadership solution, a tailored approach to **lead the acquisition phase of any capital investment**—an asset that would generate benefits and goodwill for years and decades to come. We view the leadership of megaprojects as a critical component

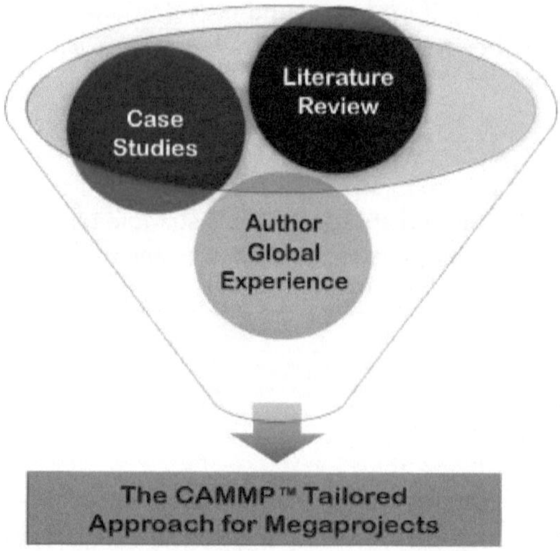

Figure 1.1 The Process for Development of the Tailored Approach for Leading Megaprojects

of asset management, and to deliver a successful and sustainable product we must, collectively, have an excellent understanding of the various topics and field of studies listed earlier.

The tailored approach that we propose is an umbrella approach that would present the model and clarify it via a simulated case study. We say it is an umbrella approach—a big umbrella—because we cover the full spectrum of a megaproject life cycle from needs and concept to closure and into production and operation. In other words, **we provide a model for project owners to cover the breadth and depth of managing and leading a project, concept to success**. We will highlight what is needed as we travel along this exciting journey.

1.8 Depth of Coverage

Several related fields of study have a direct and indirect impact on this dynamic and massive topic. Therefore, we cannot claim that this book is **The Book for Megaprojects**, but it will be unique and offer significant value to you and your organization. Our vision is that this book should be your starting point into the art, science, and psychology of megaprojects, because it gives you an overview beginning to end at a high, but functional level (the big umbrella). Once you see the big picture, then you can go down to the second level of knowledge, or maybe drill deeper if you prefer. Some of the books and references that we have reviewed for this work can help you go to the next level and maybe deeper.

1.8.1 Integration of Various Aspects

We will cover the management and **leadership of megaprojects**, inclusive of discussions on the need to address project management and operations management as vital components of **asset management**. In other words, we are building our solution and the proposed approach on a stage-gate, project life cycle model, which covers projects from **concept to success**. However, we do not limit the model to a purist form of traditional project management. We provide a tailored methodological approach, with an emphasis on leading the work, end-to-end, at the project level, along with the management of every stage of the project. Also, we present the integration of the business, product delivery, and operations management into a cohesive approach. We will discuss all the critical steps from a project management perspective and product delivery perspective. Since we cannot go deeper into all subjects, we will provide references for further study where available. With this approach, we can maintain the essential focus on **the tailored approach for Leading Megaprojects**.

1.8.2 Practitioner Perspective

Another important message to state here, to align expectations, is that the book includes case studies and research. The research is a literature review of many essential references, including leading resources on megaprojects. However, we must stress that the research, or the literature review, is from a practitioner's perspective and not academic, post-graduate type research. The case studies will be presented from the project owner's perspective. The reason we include the literature review and case studies is to extract critical challenges and opportunities, to understand the current state of practice. This understanding helps us to incorporate the learning into the tailored approach and enhance this proposed solution. You will see links and credits referring to these valuable resources throughout the book.

1.8.3 What Is Not Included?

We stated what we would do, and it would be useful to mention what we will *not* do.

We do not intend to cover the technical product-oriented actions; we will list some of what needs to be done, but we cannot include market or facilities analysis, engineering design, or construction and commissioning. Therefore, we will touch on these subjects from the management and leadership perspective but not a technical viewpoint.

We will also mention and present concepts related to scope, cost, time, risk, procurement, and the rest of the organizational project management functions, but we will not go into the details of processes and steps for these topics. We will cover the essentials that would be relevant to megaprojects and the complexity of megaprojects.

1.9 Book Structure

Figure 1.2 presents the high-level structure of the book, showing the main sections. The book consists of two major parts and a closing chapter. Each part has four sections, and each section consists of a few chapters; refer to the table of contents for the detailed outline. The following is a brief review of each section. In simple terms, Part A focuses on megaprojects from the current reality and state of practice to the proposed tailored approach. This part provides the main knowledge content. Part B is a simulation of a megaproject, a practical application of the tailored approach for leading megaprojects. It is not possible to cover all the details and documents of a real project. Therefore, Part B is extremely truncated in term of content, instead focusing on providing clarity on the various steps and components of the tailored approach.

1.9.1 Part A: Current Reality to Tailored Approach

Section I

Section I consists of four chapters, which will help us set the scene and includes definitions of many terms, the introduction of the references we used for the literature review, and discussion of the project management professional association contributions, or lack thereof, to megaprojects.

Section II

Section II consists of five chapters with a focus on the challenges facing megaprojects. We will present our observed challenges (based on decades in the field), insights from the global community (literature review), case studies, and a summary of the current state of practice.

Section III

With Section III, we shift the focus toward the solution. In this section, we present the critical concepts necessary for leading megaprojects. The topics covered here include discussions on leading (best) practices, organizational competence, considerations of program management and asset management, and the critical success factors. All of these are considered, and we build them into the tailored approach.

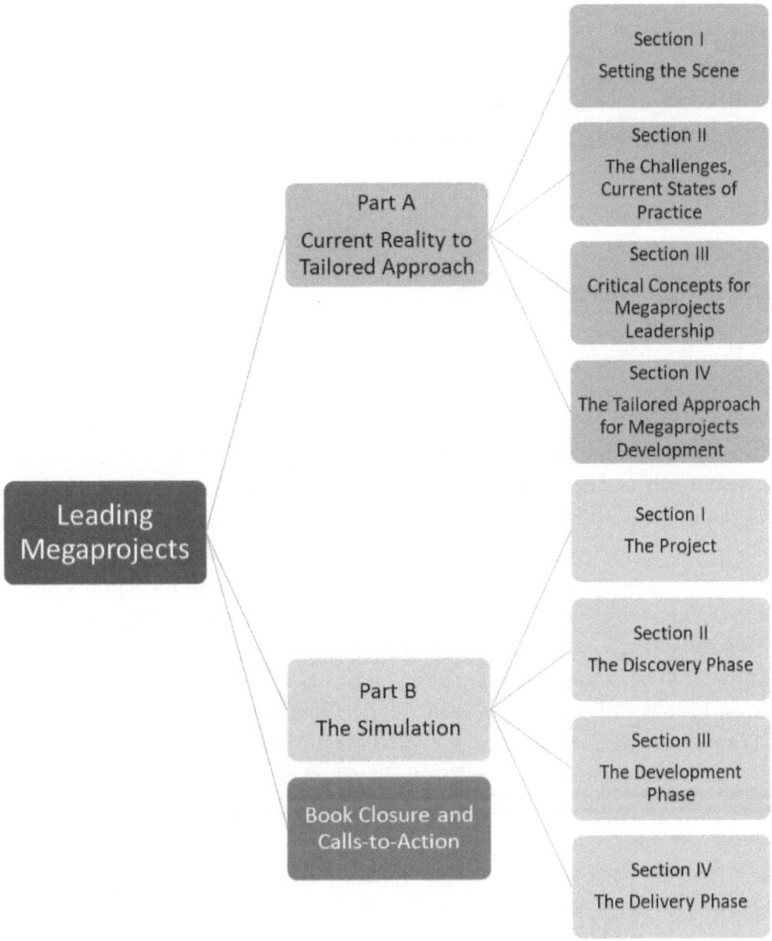

Figure 1.2 The Book Structure

Section IV

It is time to present a tailored approach. In this section, possibly the largest, we start with an overview of CAMMP™, then shift to introducing CAMMP™ for Megaprojects, **CAMMP™ MP.** The section includes details on how to take a megaproject from concept to success, stage by stage. Also, we discuss the core and supporting organizational project management functions in the context of megaprojects. We close this section with a chapter on organizing the project for success—the people aspects.

1.9.2 Part B: The Simulation

This part is dedicated to the simulation, where we take a realistic but hypothetical megaproject and show how to apply the tailored approach step by step. Consequently, what we do not cover in Part A, Section IV, we present here via the simulation. Part B Section I introduces the project, and Sections II to IV each include a phase of the project and its stages.

1.9.3 Closing

The closing is limited to one chapter in which we offer closing comments and a few points to ponder. In this chapter, we will also re-emphasize the call for action and offer a few suggestions for further research and development.

1.10 Chapter Closing Comments

Figure 1.2 and the previous parts of this chapter are overviews of what is or is not included in this book, and the depth of coverage.

What we aspire to is that this book will be a starting point of your journey into the megaproject world. We provide a big umbrella through the tailored approach, and we introduce many topics for which a reader may want to pursue more in-depth understanding. For example, if one wants a more comprehensive knowledge of cost estimating, a useful reference would be AACE International.[‡] Another helpful reference would be the guide from the US Government Accountability Office (GAO).[§] On the other hand, experienced cost engineers might not have experience with megaprojects, and we will provide that context in this book. We present a similar approach for the management of engineering and construction, risk and change management, and other topics.

One last comment before closing this chapter: I write in a direct and simplistic style, to the best of my ability. I do so to make this book a reference for project management professionals at all levels of knowledge and experience. Even senior managers and executives, despite their seniority, might not know enough about project management and its strategic value[¶] or fully appreciate what it takes to lead a megaproject. Therefore, we want to reach them with a practitioner rather than an academic perspective. This is not meant to be a critique of academia or academic research; these are vital for the progress of the field of study. Instead, we wanted a direct language that reflects real-world practice.

With the above understanding, we can move ahead with this book as we continue to set the scene.

‡ https://web.aacei.org/

§ https://www.gao.gov/new.items/d093sp.pdf; suggested by Dr. Mihail Sadeanu.

¶ Unless of course the executive comes from a project management background or has extensive experience working projects.

Chapter 2

Relevant Definitions

2.1 Introduction

Unfortunately, we do not have a universal knowledge source for project management, because there are many professional associations advocating project management. As a result, we might have differences of opinions on some terms that we take for granted in our daily practice. It is important to stress that, sometimes, we do not even agree on the definition of a project; hence, the need for this chapter.

Consequently, we will start with various definitions, which is essential to align on a shared understanding of the context of this book. A reader might not agree with one description or another; that is fine, we do not intend to impose our definition on your work. However, for this book, it is vital to understand our meanings as you approach the various topics.

These definitions could be in a glossary, but we put them here for emphasis and since they are part of the explanations. Further, we include these definitions here knowing that some of them are basic or common and are likely understood by practitioners with excellent expertise in project management, yet we urge even this group of readers to still read them; we will be brief—well, as much as possible.

2.2 Definitions

2.2.1 Project Versus Program

Project

Here are the general points we use to define a project:

- A project could be anything we create from scratch or a significant modification to an existing system, which would require substantial effort in terms of development and delivery.

- A project has a specific product (output), objectives (outcome), defined timeline (schedule), budget, and various other parameters (resources, quality, risks, etc.).
- A project is an investment of time and money to deliver a product or service (output), and it must provide the capabilities to enable the realization of benefits (outcome).
- The project owner can verify and accept the output at completion and close out but often cannot validate the outcome until months or years after the end of the physical work.
- A project can be independent or part of a program. If it is part of a program, it must align with the program objective, which in turn must align with the organizational strategic direction. On the other hand, if the project is independent of a program, it must directly align with the organizational strategic direction.

What Is Not a Project

We need to clarify the definition further by stressing what is *not* a project.

- It is common knowledge that normal, routine, repeated actions (tasks)[*] are not projects. For example, something that requires a typical, repetitive business or manufacturing process to produce a gadget or X item would not be a project.
- A phase of a project, or a stage, is not a project; more on this later.
- Tasks that are part of a project are not projects, regardless of how big the task might be. For example, the graphical design for this book is not a project,[†] it is a piece of writing and publishing this book project, which one could label as a sub-project.

Having said the above, we can still apply the project management process[‡] on a piece of the project, be it a phase, stage, or a sub-project. The processes and process groups, such as the PDCA Cycle,[§] apply for tasks, sub-projects, or stages of a project. It is vital to understand this point: **being able to use these processes on a piece does not make the piece a whole project**. We stress this point because many practitioners confuse the processes and process groups from PMI® (the Project Management Institute) and ISO® (International Standards Organization) as project phases.[¶] Consequently, those who think that the process groups are phases mistake the process groups as the project life cycle, which is a serious problem (except maybe for micro-projects).

Program

Here are the general guidelines that we follow to define a program:

---

* For differentiating tasks from projects, please refer to this blog article http://blog.sukad.com/how-to-differentiate-between-task-and-project/.

† We realize that the graphic designer, especially if an outside agency, might view the graphic design as their project. That is, for the service provider, the graphic design work from initial contact with the client until the successful completion of the work could be viewed as a project—a project for the service provider, not the client; more on this later in the chapter.

‡ Such as those described in ISO 21500 (*Guidance to Project Management*) or the *PMBOK® Guide* (*A Guide to the Project Management Body of Knowledge®*).

§ Plan–Do–Check–Act, also known as the Deming Cycle.

¶ This is possibly one of the most common areas of confusion in the project management community today.

- A program is a group of related projects and may include other operational work.
- The projects are related to a common business objective, often a long-term strategic objective that aligns with the organizational strategic direction, mission, and vision.
- While working on the program, the organization may adjust the projects, such as accelerating or slowing one or more projects. The changes would be driven based on organizational needs and various other factors.*
- The number of projects within a program could also increase or decrease as the organization collects feedback from the ongoing work and completed projects; some might call this *Agile Program Management*, but we call it an adaptive project and program management. It is being nimble or adopting agility (discussed later in this chapter).
- Finally, **each project within the program must deliver some benefits** in addition to the incremental or collective benefits of the entire program. This point might be controversial for some, because we often see posts and discussions insisting that the business value exists only at the program level.

Portfolio Management

In this book, I do not refer to the term *portfolio management* often, because I see it as part of organizational project management and the management of projects and programs in the organization. Portfolio management is the highest level in the context of project management and includes managing the overall portfolio, which consists of various change initiatives, strategic project management, project selection, and termination. In CAMMP™, portfolio management is active in the Discovery Phase, first with the idea and concept, then at Stage Gate 2, in which the initial project authorization is granted through a portfolio management assessment. During the project life cycle, portfolio management is active via the organizational and project governance.

Are Megaprojects Projects or Programs?

We will address this question with a dedicated chapter later in Part A.

2.2.2 Project Versus Product

Product

If one reads the word *product* without the proper context, one could think of products as consumer products, what people buy: a phone, a water bottle, or a car. These products are the result of the manufacturing process of a facility. However, in the project management context, a product is typically the output of a project. In other words, a project scope will lead to final output—a product or a service (PMI, 2017). The project's product would lead to manufacturing and selling the "consumer products" and, consequently, it generates revenues and benefits. Some products, outputs of projects, might be delivered as services or not-for-profit initiatives. Therefore, not all products would produce financial returns, but all should provide benefits. It is vital to differentiate between a product of a project and a consumer product.

* One might consider these adjustments as being driven by the emergent strategy generated from feedback and monitoring actual performance or responses.

Project

We defined "project" earlier. However, there are two points to add, since we are comparing a project to a product:

- As mentioned in the previous point, a project usually leads to the delivery of a product (or service). This could be a new facility or a significant modification to an existing product.
- During the life of a product, there could be numerous projects.

Product Scope versus Project Scope

Since we are talking about a product versus a project, it is also essential to define the term *scope.* Product scope is a description of the final output of the project, the product of the project, such as a three-story office tower, a metro line, an airport terminal, and similar outputs. On the other hand, the project scope is the work that must be done to deliver the product scope. The project life cycle and going through the stages is a form of the project scope.

2.2.3 Life Cycles

Project Life Cycle

We realize that some references, such those from ISO and PMI, define the project life cycle as the time from the beginning to the end of a project. However, in these references, the beginning is with a project charter, or what we[*] call a *Project Authorization Document.* As you will see later, in the CAMMP™ Model, the project starts with the inception of the idea, identification of needs or business objective. CAMMP™ treats that early work—pre-authorization (pre-project charter)—as the Discovery Phase of a project, so it is a phase like any other phase.[†] Therefore, a project life cycle is the period from concept to closure, or what we like to label as C2C. Over this period, we typically break down the project life cycle into phases and stages, each with a focus on some aspects of the product delivery. The duration of a PLC could be weeks or years.

One final comment here: the structure of a project life cycle (phases, stages, stage gates, etc.) would vary from one domain to another and is often a function of the industry, type of project, and maybe its classification.[‡] In Section IV of Part A, we will provide the CAMMP™ project life cycle model and how we would tailor it for megaprojects.

Program Life Cycle

Because a program consists of a few projects and maybe other related work, the program life cycle could look like Figure 2.1. In this image, we show the projects sequentially, but some of them could be in parallel or with a degree of overlap.

---

[*] "We" as in SUKAD and per The Customizable and Adaptable Methodology for Managing Projects™.

[†] This is one of the differences between CAMMP™ and ISO 21500 and the *PMBOK® Guide.*

[‡] We define project classification later in this chapter.

Figure 2.1 A Possible Program Life Cycle

Product Life Cycle

The product life cycle extends across the whole life of the product, regardless of what the product is. As mentioned earlier, the product is the output that the project delivers to the organization. It can be a software application, a book, a new organizational system, a water bottling plant (factory), or a city metro system. Therefore, the product life cycle is a timeframe starting from the moment the idea for the product is identified until the dissolution of the product in one way or another—that is, *the end of its useful life.* This useful life could be a few years for a software application, a car model, or decades for physical facilities such as buildings and factories.

Throughout this useful life, there are shorter periods, called *phases,* of the product life cycle. **These are product life phases and not project phases.** Figure 2.2 is a "typical" product life cycle. Keep in mind that throughout this life cycle, there could be many projects. The biggest project within this product life cycle could be the project that produces the initial facility during the acquisition phase. These are the typical capital projects that could be megaprojects, which we address in this book.

The product life cycle could also be labeled the *asset life cycle.*

Others

The term "____life cycle"* could be used in various contexts in project management. For example, one might hear about the *change* life cycle, the *procurement* life cycle, or the *development* life cycle. The latter

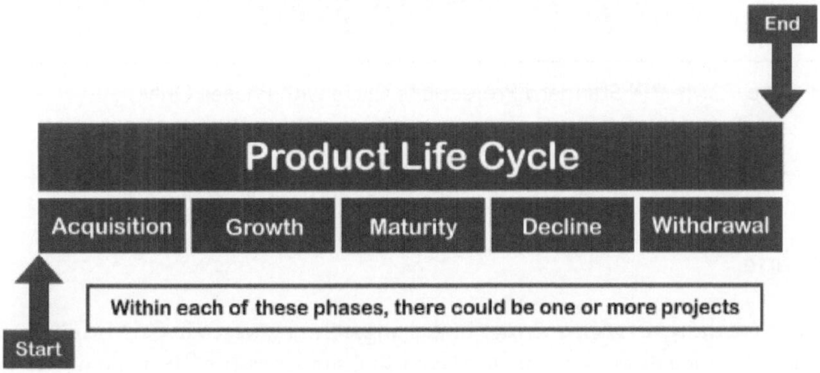

Figure 2.2 A Typical, Generic Product Life Cycle

* Fill in the blank with any words that represent something with a life cycle.

is a common term in technology or software development projects. All these share one common concept: the life cycle is subdivided into steps, phases, or stages.*

2.2.4 Phases and Stages

Product Phase

The term *product phase* is not something we often use in project management, but it is crucial to understand it, especially in relation to the discussion on asset management that we will cover later. Many references on the product life cycle divide it into five phases. The names might be different from one reference to another, but Figure 2.2 is an example. If we want to be immensely simplistic, then one can view the product life as two phases: the acquisition phase (to get the product built and ready for operations) and the operations (and maintenance) phase.

Project Phase

The term *phase* refers to a period of the project life cycle with a specific purpose and a significant output. The CAMMP™ standard model has three phases: *Discovery Phase, Development Phase,* and *Delivery Phase,* as shown in Figure 2.3. These phases would be valid for any project, regardless of size, complexity, domain, or type. They are fixed, and although some organizations may choose different terms for the name of the phases, they need these three phases (Ajam, 2017).† In Section IV of Part A, we will present the tailored CAMMP™ MP (CAMMP™ Megaproject).

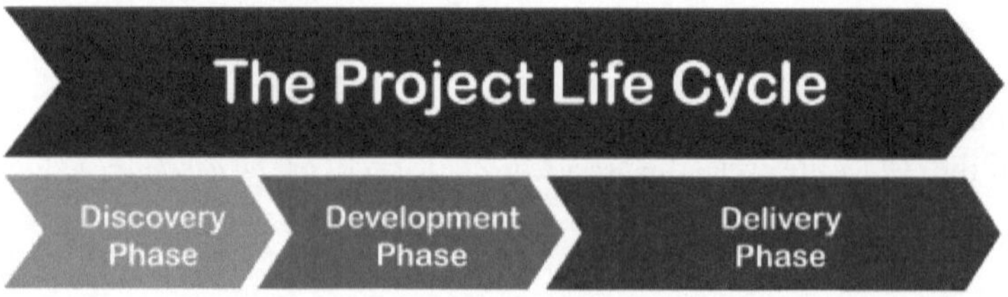

Figure 2.3 The CAMMP™ Standard Project Life Cycle with Phases Only

In this book, if we use the term phase *without context, it would be a project, not a product phase.*

Project Stage

It is common to use the terms *phase* and *stage* interchangeably.‡ However, in CAMMP™, we differentiate between them, even though we could define them using similar words. The main difference is that we

* The right term might vary from one context to another, and generically, these terms could be interchangeable.

† The image and text are from *Project Management beyond Waterfall and Agile*, Chapter 13.

‡ http://blog.sukad.com/what-are-the-differences-or-similarities-between-phases-and-stages/

view the three phases as fixed (although names could change), whereas the stages could vary greatly in names and numbers. The stage level is the level we use for tailoring.

The CAMMP™ standard model has nine stages, as you can see in Figure 2.4 (Ajam, 2017). Please note, the use of the term *standard* does not mean "fixed method" or "set in stone." That rigidity goes against the core principle of CAMMP™, reflected in the name *customizable* and *adaptable*.

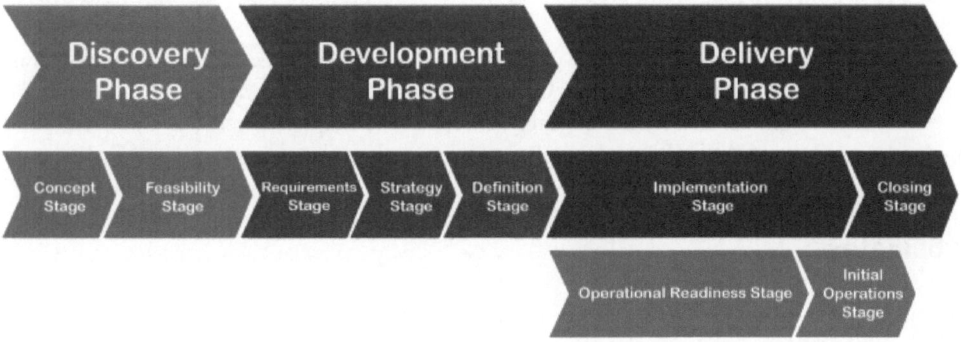

Figure 2.4 The CAMMP™ Standard Project Life Cycle with Phases and Stages

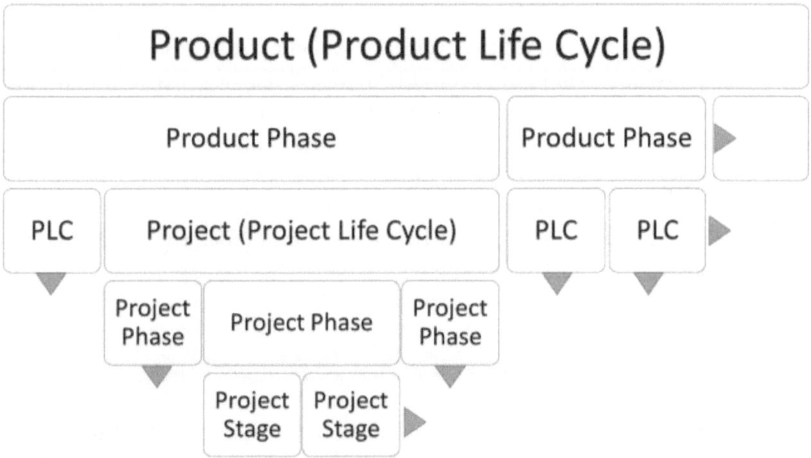

Figure 2.5 The Hierarchy, Product to Project Stage

Putting the Pieces Together

In other words, using the work breakdown analogy, we can visualize via Figure 2.5* and the following text:

* The gray symbols indicate continuity since it would not be practical to show all of the components in this image.

- A product would have a product life cycle, consisting of a few product phases, and each phase could have one or more project.
- For a given project, we use a project life cycle to manage it, concept to closure.
- This project life cycle is subdivided into three defined phases.
- Then each phase is subdivided further into two or more stages.
- Later in this book, you will see that to manage each stage, we go down to one more breakdown level: the project management process—or maybe we should say **stage management process**.

We do the above breakdowns for better governance, aided by the stage gates and process gates that we have not shown yet but will define next. One final comment on stages and phases: throughout this book, these terms might be used interchangeably, except in the chapters specific to CAMMP™ and the tailored approach. In the CAMMP™ context, phases and stages are related, but each has a distinct purpose.

2.2.5 Gates

We're sure you're starting to see the value of these definitions, because some of them are not common knowledge in the project management community. Here is another situation that could be different from your common knowledge: the reader will see two types of gates: *stage gates* and *process gates*.

Stage Gates

Stage gates are the gates along the project life cycle that separate the stages. Other common names for a stage gate would be phase exit, kill point, project review, and maybe other terms. We prefer the term *gate* instead of *exit* because "exit" implies to leave something, so the criteria at these exits would allow people to exit—a one-way flow. On the other hand, a *gate* means an exit *or* entry, and in the context of CAMMP™, it is both. Most stage gates have two purposes—one to confirm that the work of the preceding stage is completed appropriately and satisfactorily (exit a stage), then to decide whether to proceed or not (enter the next stage). Therefore, in the CAMMP™ standard model, there are nine stage gates, one for each of the nine stages (Ajam, 2017, Ch. 12).

Two final comments:

- Remember, these are linked to stages, so if we have fewer or more stages, then we will have fewer or more stage gates.
- An organization, via their organizational project management system (OPMS), can decide to have some of the stage gates be more formal or restrictive than others. For example, funds approval stage gates might require the approval of the highest level of stakeholders in the organization, whereas other stage gates could be approved at lower levels.

Process Gates

Once again, the stage gates happen along the project life cycle, to separate the stages. However, within each stage, the stage (project) management processes apply. In the SUKAD CAMMP™ Model, we have six sets of processes to manage each stage of the project, and for better governance and control we support these processes with process gates. At this time, we illustrate this with Figure 2.6 (Ajam, 2017, Ch. 16),

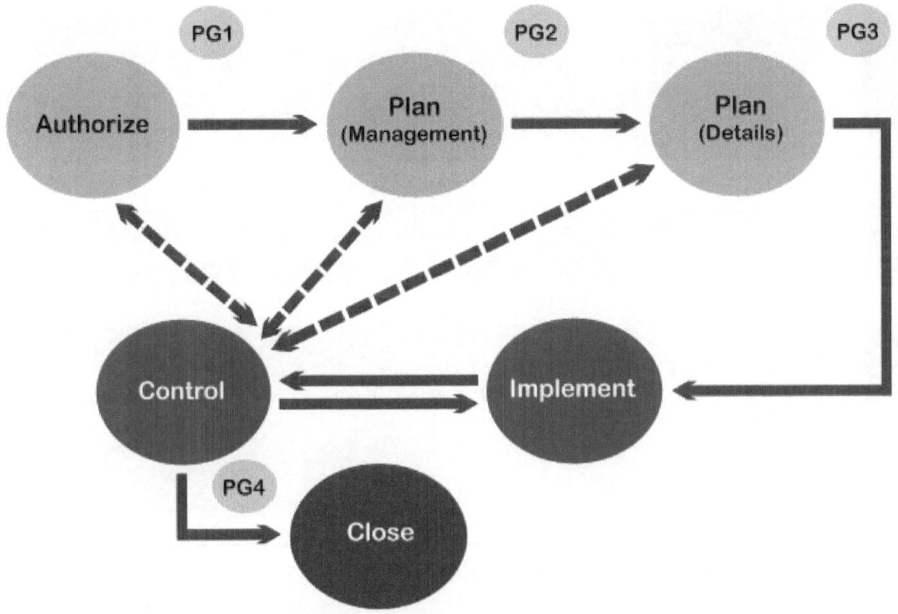

Figure 2.6 The CAMMP™ Six Processes and the Four Process Gates

but we will delay the explanation until later in the book. Just a teaser, for now—the small circles enclosing PG1, PG2, etc. are the process gates, and the large ovals are the six project management processes per CAMMP™. Each of these processes consists of many process steps, which you will see later.*

How formal are these gates?

It would be a function of the organization's culture and maturity, and they could be strict or transparent if they exist.

Stage-Gate Process

The term *stage-gate process* is a common term in the project management community, which refers to a PLC-based methodological approach. Maybe we should clarify that it is a common practice in the capital projects industry but might not be common or even known in technology or the sectors that are relatively new to project management.

Figure 2.7 presents the standard CAMMP™ model, with the phases, stages, and stage gates. If it is not apparent yet, CAMMP™ is a derivative of the stage-gate process.† This image does not show the stage deliverables or the stage management processes for each stage.

* For those familiar with ISO 21500 and the *PMBOK® Guide,* what these references call "a process group," we call "a process." What they call "a process," we call "a process step." You will also notice that we have six processes, since we split planning, whereas these guides have five process groups.

† Which we did not invent; refer to Section 1.3 on page 5. Although we did not invent this concept, we do expand it.

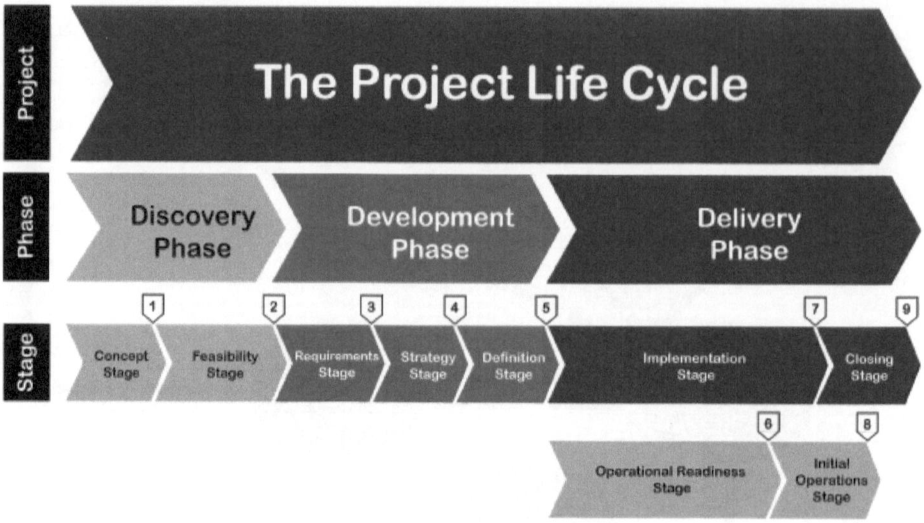

Figure 2.7 The CAMMP™ Standard Project Life Cycle

Here is a sneak peek into the tailored approach that we will be proposing later; can you visualize that CAMMP™ MP integrates the actions (stage management processes) of Figure 2.6 into the project life cycle of Figure 2.7?

2.2.6 Project and Organizational Project Management

Project Management

For us, this is simple; it is about the management actions needed to take a project successfully from concept to closure. Should we say more, especially because this whole book is about project management? I think I should add one point: some (professionals, such as the one in Story No. 1 on page 5) limit their view of project management to the ISO/PMI processes and process groups. We are not saying that is wrong, but it is a purist view that is restricted to the project management "process" actions. Although we agree that we need "pure" project management, we view project management as a holistic process to achieve a result and deliver a product. Therefore, our view of project management is represented by the need of a project life cycle to drive product development, and project management actions (processes and steps) would facilitate the realization of the objectives. Therefore, our view of project management is broader than some of the traditional views, and for megaprojects, we will stretch even our own perspective to a higher level.

We view project management as a process to achieve a result and deliver a product; therefore, we can't view project management in isolation of product delivery.

Organizational Project Management

The most common use of the term *PM* (Project Management) is about managing a single project, whereas *OPM* refers to Organizational Project Management, the organizational approach, for managing the organizational projects. We also use the term OPMS to refer to the OPM System,* which would include governance, policies, procedures, and methods, among other things, for the management of projects. A more comprehensive view of this concept would be the system for managing the organizational projects, programs, and portfolios. Consequently, in the context of this work, PM refers to managing *a* project, and OPM refers to managing *the organization's* projects (Ajam, 2017). We dedicated a whole chapter in *Project Management beyond Waterfall and Agile* for SUKAD's The Seven Elements of Project Management Maturity™, which we show in Figure 2.8.

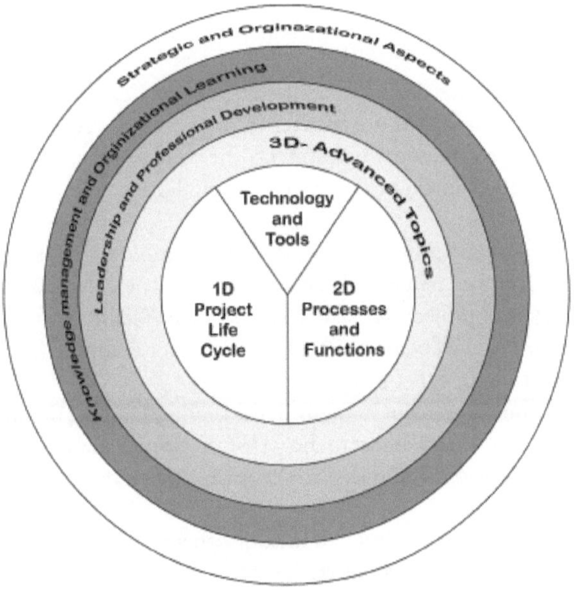

Figure 2.8 The Seven Elements of Project Management Maturity™

Many organizations can and do manage projects without proper OPMS. However, this would be a critical mistake for megaprojects. Unfortunately, many project owners' organizations have not invested in building a robust and sustainable OPMS. In addition to or as part of the OPMS, organizations must have historical records, a knowledge base, organizational records of past projects, metrics, key performance indicators, etc. We call this the Organizational Knowledge Base (OKB). One can consider OKB as an integral part of, or independent of, the OPMS, like two sides of the same coin.

We have also noticed that it is common to use the reference PPM, Project Portfolio Management, as a rough equivalent to OPM. The PPM term is quite common in the software and technology world. Another popular term is EPM, Enterprise Project Management.

* In this context, "system" is not about software, although software tools can be part of this. System is the overall organizational system, inclusive of policies, methodologies, etc.

Organizational Project Management Maturity

This can be a vast topic, and many organizations have published about it and have maturity assessment models. What we will say here, briefly, is that organizational project management maturity (or project management maturity) is about the level of maturity of the project management practice in the organization. In simple English, how good is it? OPMM (Organizational Project Management Maturity) is related to project and organizational governance. It has a direct impact on project performance. A high level of project management maturity means that the organization has adopted a project management *mindset,* as a core element in the organization's culture. What we are saying is that project management maturity and organizational culture have a direct impact on each other and on the organization's abilities to deliver successful projects. *Project Management Maturity is part of the discussion on organizational competence; see Section III of Part A.*

2.2.7 Capital Project

A capital project is a term that we often use to refer to projects that require a capital investment. One might argue that all projects require some investment; the word *investment* is even in our definition of a project. True, however, in the context of this book, a capital project is:

- A project that requires significant capital investment—millions, even billions, of dollars.
- Such a project would require a tremendous effort, years to deliver, and involve numerous stakeholders.
- The **output** of this type of projects is usually a constructed facility, factory, industrial plant, metro system, building, significant software or Enterprise Resource Planning (ERP, usually major software systems) implementation, or other physical entities.
- Due to the construction component in many of these projects, some project management literature (including the Construction Supplement to the *PMBOK® Guide)* labels these as "construction projects."
- The **outcome** of these projects would lead to the realization of benefits, which were expected when this project was authorized.
- The benefits are typically financial but can include other types of benefits, such as for community service or social entrepreneurship projects.

Capital, or facilities projects to be more specific, *can generally be classified into three sectors: buildings, infrastructure, and process* (Oberlender, 2014, Ch. 3, p. 41). Paul Barshop, the author of *Capital Projects,* defines these projects as *"investment of substantial company resources to develop, to improve, or to refurbish an asset that is expected to generate cash flows for more than one year"* (Barshop, 2016, p. 3).

2.2.8 Megaproject

A Common Definition

The megaproject is a term that refers to massive capital projects that would typically cost one or more billion dollars.* However, use this number with some liberty—about a billion dollars will work fine.†

* Dollars in this book means US Dollars.

† This has been a common definition for a while, despite the change in the value of money. This is why it is important not to restrict the definition to a dollar value unless we link it to a year and we adjust as time passes.

They are large-scale and complex ventures and are transformational in nature (Flyvbjerg, 2017).

Here it is important to stress that this $1 billion is not the total cost; it is often a reference to the capital cost of the project (also known as CAPEX, Capital Expenditures*) and exclude the cost associated with setting up the permanent operating organization, which could be significant.† Also, this amount excludes the operating expenditures, also known as OPEX.

How to Write "Megaproject"

We need a couple of lines on how to write this term. We often see three versions for writing megaproject: (a) mega project, (b) megaproject, and (c) mega-project.

We have adopted *megaproject*.

An Alternative Definition

An alternative way of defining megaprojects is to consider projects that could cost as little as $100 million but incorporate complexity into the definition. Levitt and Scott provide four complexity factors to consider; these are (Levitt and Scott, 2014):

- Spatial/technical configuration complexity
- Maturity of involved technologies
- The scale of project's regional economic and political impact
- Cross-institutional complexity: "global megaprojects"

For example, per Levitt and Scott, the $8 billion Trans-Alaska Pipeline project can be broken into many smaller sub-projects, which would make it easier to coordinate. Therefore, it does not fit the first factor listed here. However, because it has a substantial regional economic and political impact (third factor), it would be justified as a megaproject.

Our Definition

Megaprojects are massive capital projects that could be in the billions or hundreds of millions; so, a project estimate of 1 billion UAE dirham (about $270 million), for example, could be classified as a megaproject. The bottom line: do not get hung up on the semantics of the definition; what we are discussing here applies to any significant, large, and complex project.

2.2.9 Asset Management

A few days ago, before writing these words, there was an interesting discussion on LinkedIn® on the definition of an asset, which posed the question, *would every project produce an asset?* I have no intention of repeating the debate here but will provide what is relevant to us. In this book, the term *asset* refers to any facility, system, process, or related things that are tangible and owned by an organization. Building

* CAPEX = Capital Expenditures, the funds that will cover the development and construction of the project leading to initial operations.

† We are not sure if some organizations consider the initial set-up cost as part of the project cost.

a petrochemical plant will produce an asset that would be owned by and generate revenues and profit to a given organization. We can list many other examples, but we think the point is clear.

Why are the "asset" points relevant to this book?

"Industrial corporations create their capital assets primarily through projects" (Merrow, 2011, Ch. 1, p. 11). In other words, megaprojects produce assets, and it would be beneficial to keep this in perspective, because we believe a part of the solution depends on considering asset management, or the asset life cycle, in making decisions on megaprojects.

2.2.10 Project Owner and Service Provider

Project Owner

The term *project owner* refers to the organization that authorizes a project, will operate its output, and will capture the expected benefits. Merrow uses the term *sponsor* to refer to the project owner's organization (Merrow, 2011). It is worth noting here that a project, from the perspective of the **project owner**, does not end with the closeout report. The closeout report might formally close the project management or construction aspects of a project. However, the project itself would have produced a product that would last for years or decades after the closeout report. If we accept a culture of asset management and link project success to the business, then we must go beyond the closeout of the CAPEX work.

Therefore, a product is closed when the product life has reached its natural conclusion and the facilities produced by the project cease to operate. The simulation in this book will not address the operations of such facilities but will discuss how to get ready for their operations and will define the criteria for measuring the success of such initiatives.

The simulation of the project in Part B and the tailored approach presented in Section IV of Part A are from the perspective of the project owner. The project owner could be a private, government, semi-government, or a non-governmental organization.

Service Provider

In the context of this book, a service provider is an organization that is hired to provide services to a project owner. The services could be market research, preliminary engineering, engineering, construction, construction management (CM), project management consultancy (PMC),* and the list goes on. Two important facts to note:

- A service provider, like a project management consultancy, could be working with a client—a project owner—across most of the project life cycle, or over parts of the PLC. On the other hand, other service providers' scope of services might be limited to a stage of the project, such as construction or commissioning.
- A service provider could do all the work with internal resources, or could, in turn, hire other service providers (subcontractors) to do part of the work.

* In this context, a project management consultancy would be a company that helps a project owner manage the project because the owner does not have the necessary project management resources to manage directly. A construction management company is similar to a PMC, but the role could be limited to the construction stage.

Contractor

Depending on where you are in the world, the word *contractor* could mean one thing for you but something else for another person. In other words, the term depends entirely on the context. Some use the term to mean a person working as a freelancer or a supplemental resource under a contract; a contract employee. However, in the context of this book, the term contractor refers to an organization, not an individual. We will do our best not to use the term loosely and out of context. Here are the different usages of the term:

- **General contractor.** This term typically refers to an organization that takes a job from a project owner, and they would lead the work, usually with the help of many subcontractors. This could be limited to engineering, construction, or EPC (engineering–procurement–construction).
- **Engineering contractor.*** These are often companies that would handle the detailed engineering or design work for a project.
- **Construction contractor.** This is easy: the companies that do construction work, either working directly with project owners or under a general contractor.
- **Specialty contractor.** Often, these are contractors that specialize in a given area, such as civil, electrical, mechanical, and other specialties. They typically work as subcontractors.

2.2.11 Methodology, Method, Tailored Methods

Method or Methodology?

It is common that some practitioners confuse the meaning of the terms *method* and *methodology*, using them interchangeably. In general, a method is "a *particular procedure* for accomplishing or approaching something,"† whereas a methodology is "a *system of methods* used in a particular area of study or activity" [my emphasis]. In other words, a method is a specific way that is set or fixed, whereas a methodology is a broader term, "a system of methods."

Furthermore, a method is different from a framework because a method means there is a particular way of doing something—like a systematic process, a step-by-step approach, a defined project life cycle. Note that there may be more than one method of achieving a required outcome. A typical **project management method** may follow a specific project life cycle or a similar approach (Ajam, 2017).

SUKAD developed a methodological approach that is founded on a project life cycle, which is the Customizable and Adaptable Methodology for Managing Projects™ (CAMMP™). Notice the use of the term *methodology* and not *method,* which is intentional. The reason for the use of the term methodology (instead of method) is because **CAMMP™ offers an approach that is not rigid or fixed and must be customized and adapted to the organizational and project context** (Ajam, 2017).

Tailored Method

Continuing with the previous definition, the customized and adapted version of CAMMP™, for a particular project's type, becomes a method, or a tailored method. The simulation that we will present later in the book is based on a tailored approach (method), as the title of this book also indicates.

* In the Arab World, these are called *engineering consultants.*

† *Oxford Online Dictionaries.* Retrieved from http://oxforddictionaries.com/

Potential Confusions

We know that some practitioners might mistake what we are saying to mean a one-size-fits-all or a set-in-stone method. The whole idea behind CAMMP™ is to counter these mistaken approaches. In other words, **one-size-fits-all is anti-CAMMP™**, and vice versa. As we have already stressed, CAMMP™ is a methodological approach that provides a **standard model**.

A standard model, isn't that one-size-fits-all?

No, it is a standardized *approach,* which we customize to industry and organization, then we adapt to a function or project type, and finally, modify to fit a project size or complexity. So **a standard model is the starting point to tailor a method**.

OK, a method, then it must be set in stone.

Nope, a method is a systematic way for doing things that would help an organization and its project management teams to avoid recreating the wheel for every project. We can use a tailored method that is fit-for-purpose, then tweak it further if necessary. For example, a typical capital project, anywhere in the world, would have a similar project life cycle method at the highest level. The details could vary greatly, and that is a function of the given organizational system, maturity, and culture. More on this later in the book.*

2.2.12 Agile and Agility

Agile

In the context of this work, the term *Agile* with a capital letter refers to the Manifesto of Agile Software Development (Beedle et al.), which may include iterative or incremental development of software products; once again, **software products**. Agile, as in the Manifesto, does not work on capital projects, because many of its particulars do not apply. For example, take the Manifesto first principle, *"Our highest priority is to satisfy the customer through early and continuous delivery of valuable software."* (Agile, 2019). Even if we ignore the term *software* and replace it with *product,* how can we deliver a building iteratively or incrementally, *"through early and continuous delivery of valuable* X"? We realize that many Agile advocates keep insisting that Agile works on capital and construction projects. This is nonsense, to say the least. These people are confusing Agile with agility,† or even lean.

Agility

The definition of agility is being flexible, dynamic, and responsive, among various other characteristics. It is a good practice and highly useful in business and in managing projects, regardless of size or complexity. We can definitely use agility concepts on megaprojects—however, not as a project management method. We are sure, as you study our tailored approach for megaprojects, you would notice the embedded agility aspects. Further, megaprojects are not as flexible as we would like them to be—on the contrary, they are fragile.

---

* The following is a link, https://pmquest.sukad.com/catalog/info/id:138, to a sneak peak of one of our online courses that touch on the subject of tailoring.

† If you want a direct view on this topic, please check https://www.youtube.com/playlist?list=PLiXup1lJ-_TNMvGY7q3aix_cZBhMC_lUr.

We can definitely use agility concepts on megaprojects—however, not as a project management method.

Lean

What many Agile advocates miss is that Agile (and its various sisters and brothers) is a subset of lean. Lean, as a concept, is definitely used in the capital projects industry, but again, not as a project management method. Therefore, when people claim that Agile is used in construction, they are likely referring to lean. It is worth noting that there is even a Lean Construction Institute (Lean, 2015).*

2.2.13 Project Classification

A Classification Model

The book *Project Management beyond Waterfall and Agile* (Ajam, 2017) dedicated Chapter 41 to this concept, so we will not repeat that chapter here. In that book, we propose a classification model. Here is the core concept—project classification is a way to categorize projects based on size and complexity. Size and complexity are two categories, and each has a few factors to consider.

Remember, CAMMP™ is about customizing and adapting the methodology to a given context—for example, a capital project. However, some capital projects could be small and straightforward, while others are large and complex. Therefore, and since one-size-fits-all does not work, we need one more step to complete the tailoring of a method.

The CAMMP™ Classes

The CAMMP™ classification model groups projects of the same type based on size and complexity factors. As a result, CAMMP™ proposes three significant classes of projects. These are:

1. Small-simple projects, and we can stretch this class to cover micro projects; we label this CAMMP™ SS.
2. Medium-moderate complexity projects (CAMMP™ MM).
3. Large-complex projects (CAMMP™ LC).

CAMMP™ MP, CAMMP™ for Megaprojects

Although we could stretch the third category to cover megaprojects, in this book, we are expanding on CAMMP™ LC to establish CAMMP™ MP, CAMMP™ for Megaprojects, as an independent project class.

The Potential Pitfalls

This classification system is not generic; it does not apply uniformly across domains and organizations. Each organization or industry can and should create its own classification model. For example, in the capital project industry, some companies classify projects less than $5 million as an operational project; then up to $50 million as small projects, $50–250 million as medium; $250 million to $1 billion as

* www.leanconstruction.org

significant or major (large) projects, and anything over $1 billion as a megaproject. The numbers here are only to demonstrate the concept. Other organizations might have small and large projects; alternatively, they could have three classes, projects, major projects, and megaprojects. Therefore, it is critical that each organization develop its own classification model to fit their context.

2.2.14 *Success and Failure*

How to define success or failure? We need to look at this question from different viewpoints.

The Success of What?

When we say a project is a success or a failure, what do we mean? Did we deliver a bad or a good product? Maybe we finished within the cost and schedule parameters. Perhaps the project met all or some of its objectives. In other words, the definition of success (and failure) might change from one stakeholder to another, depending on what we are assessing and from which perspective. Therefore, SUKAD has developed The Four Dimensions of Project Success™. This is to determine the success of a project using four dimensions*—namely, *product's technical success*, *project management success*, *project delivery success*, and *objectives success* (Ajam, 2017).†

Measurement of Success

The Four Dimensions of Project Success™ tells us when, along the project life cycle, to measure success and from which perspective. However, how do we measure success—that is, quantify it? To assess the success of a project, usually one dimension at a time, we establish the criteria, which would consist of key performance indicators (KPIs) or metrics. These are highly dependent on the organizational system, maturity, and culture. *How many times did we mention system, maturity, and culture in this chapter alone?*

We will discuss some of these KPIs as we discuss the tailored approach and the simulation.

The Judgment

OK, we are performing an assessment, using the metrics or KPIs, so how do we judge and grant a project the label *success* or *failure*? First, we do not believe in these two extremes. Therefore, we recommend four labels or grades. We realize that some prefer an "either/or" scenario, and that is your choice. In our perspective, an either/or view does not provide an adequate assessment that would help us continuously improve effectively and efficiently. One final note: an organization can adapt our four-grade model but modify the definitions, which is fine. The organization can also use a two- or three-grade model. In the end, and regardless of the model, organizations need to understand what went well and what did not.

The four-grade model:

* We will expand on these dimensions in Chapter 17.

† Once again, *Project Management beyond Waterfall and Agile* has a dedicated chapter on this topic, which we will summarize later when we propose the tailored approach.

- **Success**: an assessment of any of the four dimensions would be granted the label, graded **successful if we achieve its criteria within the control limit or tolerance allowed**. For example, if we define project management success* as achieving cost and schedule targets within the allowed tolerance of ±10% of the approved budget, then a final project cost could be 109% of the budget and the project would still be judged a success. Maybe some organizations like to make this allowance ±5% or even zero; that is a function of the organization. (More on this topic in Section II.)
- **Partial Success**: If a project (overall or a specific dimension) missed one of its success criteria and still results in a useful or profitable product, we may label it as partially successful.
- **Partial Failure**: If a project missed more than one of its success criteria and still results in a working/productive facility, we may label it as a partially failed project.
- **Failure**: This would be for projects that are either terminated or missed most (if not all) of its key objectives. We often call these "troubled projects."

2.2.15 Assessing Performance

Concerning the previous discussion on project success, organizations and assessors often use NPV (net present value), BCR (benefit–cost ratio), among other indicators, to select projects from the portfolio (at the project selection time) and to assess the project success upon completion. Most of the literature review that we have conducted refer to these measures. However, later in the book, we would question whether we are using financial metrics properly or not.

Net Present Value

NPV is known in the financial and project world. However, in case someone needs a refresher, here it is, within the context of capital projects. *"NPV is the amount of shareholder wealth created from a capital investment after accounting for the total cost of the investment and the time value of money"* (Barshop, 2016, p.6).

Benefit–Cost Ratio

BCR is also a known term that is determined from a project cost–benefit analysis. BCR is a comparison of the benefits to the cost, and anything over 1.0 means the project is delivering more benefits than the cost of producing those benefits.

2.2.16 Organization

We frequently use the term *organization* and *organizational* throughout this book. We do so for two main reasons. The first reason is to be generic, where we can use this term for public or private organizations, for profit or not for profit. Therefore, an organization could be a company, a ministry, a state or city government, or a non-governmental organization. The second reason is that the term could be a reference to a branch, division, or even a department of a company or a ministry. In other words, as we discuss organizational project management, organizational project management maturity, organizational project management function, and similar uses of the term, we could be referring to a division or the whole entity, and we could be addressing a government entity or a private company.

* Remember, project management success is one of the four dimensions, and we use it to assess the project management function.

2.3 Closing Comments

By now, you would have noticed that our relevant definitions chapter is more comprehensive in scope than a glossary. In addition to defining many of the critical terms that we will use throughout this book, we have given you an indirect view into CAMMP™ and the tailored approach, which will be useful to know as you continue to read.

There are other terms that we will use in this book, but those we might not use frequently; therefore, we will define them when we use them. Further, we have some of these definitions and many others in an online course on the Quest to Mastering Project Management Platform. We would be happy to give the reader one month free access to this relevant online course.*

* The link to the course is https://pmquest.sukad.com/catalog/info/id:128, and to gain access, register on the PM Quest site and send an email with a receipt of purchasing this book.

Chapter 3

The Literature Review

3.1 Introduction

The reader can refer to the bibliography to find the full list of resources we have used for this book. However, we believe it would be appropriate and beneficial if we share a brief on the leading references here. Further, we conducted this research to learn about the challenges and opportunities, the insights from the global professional community, to enhance this book and the solution that we are suggesting. Therefore, it is important to stress that our reviews are not traditional book reviews in which we critique or praise a book or summarize it. The learning will be integrated into the tailored approach for leading megaprojects. Further, for a more in-depth understanding of specific topics, these references are highly recommended.

3.2 A Summary of the References

3.2.1 References from Professional Associations

This is the subject of Chapter 4.

3.2.2 Project Management for Engineering and Construction

Project Management for Engineering and Construction is published by McGraw Hill Education, initially in 1993. The third edition, which we reviewed, was published in 2014. The author is Professor Garold D. Oberlender, PhD, PE, Professor Emeritus of Civil Engineering at Oklahoma State University.

The book presents the principles and techniques of project management, beginning in the conceptual phase by the owner, through coordination of design and construction, to project completion.

(Oberlender, 2014, Ch.1, p. 1). The book structure includes chapters, each addressing a phase of the project (such as construction), whereas other chapters deal with a project management function, such as scheduling, risk, or budgeting. The emphasis is on the earlier phases of the project, because the ability to influence project performance happens early in the project.

Finally, the book presents project management technical aspects (semi-detailed concepts), especially because its primary audience is engineers. In other words, it covers estimating, budgeting, scheduling, and similar topics from a technical perspective, which makes it a useful reference if one wants to go down another level from the level we are discussing in the tailored approach.

3.2.3 Industrial Megaprojects and Capital Projects

Industrial Megaprojects focuses on *"Concepts, Strategies, and Practices for Success"* (Merrow, 2011), as the book subtitle states. The author is Edward W. Merrow, the founder and CEO of Independent Project Analysis, Inc. (IPA), a global leader in the area of project analysis and benchmarking of projects in the process industry.* This is a great reference and is highly recommended, because it contains a great deal of analysis from real projects that IPA analyzed, often with extensive depth.

It is interesting that this book and *Capital Projects* (Barshop, 2016) are both by IPA, and both are highly valuable resources for capital projects in general and megaprojects in particular. *Capital Projects* focuses on executives and what they need to know to lead successful projects. The author is Paul Barshop, who is a director with IPA.

Both of these books depend on data from actual projects analyzed by IPA. The main difference is the focus. *Industrial Megaprojects* focuses on industrial facilities and those that are classified as megaprojects, which means about US$1 billion in 2010 value of money. There are 318 projects in the database, covering the world and from various industries. On the other hand, *Capital Projects* is broader in focus, dealing with all capital projects in the database.† However, many of what we extract from these references are useful for all projects and often for many industries and domains, despite our focus on mega and capital projects.

It is also worth stressing that both of these books share our views and emphasis on the project owners, as the critical link, possibly the weakest in the chain of the management and leadership of capital projects. The significant value that these books bring is the wisdom generated from the information and data from real projects. Benchmarking and projects assessment provides IPA with a wealth of project performance information that is not readily or openly available in other sources.

To reflect on the link between these books and the book you are reading now, *Industrial Megaprojects* identifies crucial mistakes on megaprojects and helps project owners understand how to make the right business decisions and project decisions. Since *Capital Projects* also depends on the data from IPA, it continues the journey of the first book with a focus on the executives or project owners. Both books also offer us a conceptual model of the management of capital projects, with a focus on Front-End Loading (FEL).‡ Using both of these references, and the other references and case studies

* The process industry refers to oil and gas, mining, chemicals, and more.

† *Capital Projects* address smaller projects, but "small" is relative. Each of these "smaller" projects is still large and complex, with costs crossing $100 million.

‡ *Front End* and *Front-End Loading* are the terms commonly used by IPA to refer to the stages of the project that lead to the final fund authorizations, whereas execution would happen after final authorization.

we discuss here, allows us to distill the great insights into a highly practical and tailored model for leading megaprojects.*

Throughout this book, I will refer to challenges collected from these references, so I will not go into more detail at this point. I only want to close this part with another point that supports our focus on project owner as the target audience. I will let Mr. Merrow present the case. *"Most of the big mistakes that the company make in developing and executing these projects stem from a basic lack of being able to pursue a common goal with clarity and good behavior . . . Most big mistakes are made by senior business managers in the sponsoring firms. The reason they make most of the big mistakes is that they have control of the things that matter most; strategy money and people."* (Merrow, 2011, pp. 1–2).

3.2.4 The Oxford Handbook of Megaproject Management

The *Oxford Handbook of Megaproject Management* is edited by Professor Bent Flyvbjerg and was published by Oxford University Press in 2017. The book has many parts, each consisting of a few professional papers, for a total of 25. We did not review all of those papers for the purpose of this book; we selected about half for in-depth review. We judged this was enough, because some topics were similar, and the focus of the other papers was not directly relevant to a tailored model.

The papers in this reference focused mostly on infrastructure projects. If I am not mistaken, all of the academic papers were written by professors or academic researchers. At the end of the book, I will have a point to ponder, maybe a bit controversial, related to this last point.

This is a useful reference, especially for those who want to get more in depth about a topic related to infrastructure megaprojects.

3.2.5 The Construction Industry Institute

We have no specific book that refers to the work of the Construction Industry Institute (CII). For those who are not familiar with CII, it is a consortium of companies (project owners and contractors) that work and support the CII work and research. CII is a center at the University of Texas in Austin, a university in the top 10 in the United States for engineering domains.

Most of my research into CII work comes from personal experience, booklets from past projects, and from their website. CII is well known for a set of best practices in the capital projects industry; they promote 17 specific best practices. They also have an extensive knowledge base that focuses

FEL Index and PDRI

As mentioned above, IPA has Front-End Loading (FEL) and the FEL Index. CII has Front-End Planning and Project Definition Rating Index (PDRI, 2015).

The FEL Index and PDRI are both tools, scoring mechanisms, to help the project management teams and executive management understand the level of development and definition of the front-end work. The score (result) would be used as an indicator to determine whether the project is ready to move to the next stage within the front end.

The final assessment, at the end of the Definition Stage (end of the front end), would be used to determine readiness for final authorization (final funds) and to decide whether to move the project onward to the Implementation Stage.

* I have been following IPA since the early 1990s and have been involved with many projects assessed (benchmarked) by IPA. Once we had a benchmarking meeting with IPA at 4:00 in the morning, because some of the participants from my organization were working night shift on a petrochemical plant shutdown project.

on all functions and stages of projects. CII also puts an extra emphasis on the Front End, they call it *Front-End Planning* and are noted for their PDRI.

CII claims that following their best practices would have significant tangible benefits and have even quantified the benefits. More on this in Chapter 11 on best practices.

3.2.6 Other References

Papers and Reports

We have reviewed numerous other papers and online and media posts on megaprojects. During the writing, news about some of the ongoing megaprojects, such as the UK Cross Rail, the California High-Speed Rail Megaproject, and even a Saudi Rail Project, kept popping up in the news feeds. We also keep hearing about the Chinese Giga Projects (Silk Road) and Saudi Neom Project. Here I want to highlight a couple of the other papers and reports that we have reviewed.

- *Debunking Fake News in a Post-Truth Era. The Plausible Untruths of Cost Underestimation in Transport Infrastructure Projects* (Love and Ahiaga-Dagbui, 2018). This paper offers a few counterpoints to other academic papers, such as are published in *The Oxford Handbook of Megaproject Management* and, most notably, a 2002 paper by Flyvbjerg et al. (Flyvbjerg, Holm, and Buhl, 2002). I will refer to this debate more than once in this book, since the 2002 paper, the counter 2018 paper, and a counter to the counter paper offer us valuable insights. Again, I will not get into the arguments and counter-arguments; I will keep the focus on a few of the core misunderstandings* that are relevant to improving megaproject leadership.
- *Cost overruns in transport infrastructure* (Terrill and Danks, 2016). This report is a study of all transport infrastructure projects in Australia since 2001. One of the aspects of this study is the analysis of cost performance during the development part of the projects (the Front End) and not only the delivery phase.†

Author Experience

I have been working in engineering and project management since 1984, the first three years as coop‡ (assistant engineer) and a couple of years as an engineer upon completing my bachelor's degree. With a master's degree in engineering and construction management, I shifted from engineering to project management, but in the capital projects industry. I had the privilege of working in different project management roles and progressively growing until the time when I co-founded a project management services company, which I have been leading since 2004.

* For example, one of the critical points debated is what should be the reference point for deciding if a project is an overrun situation or not. Another relevant point is one of the positions or beliefs about underestimating cost and overestimating benefit and whether that is an error or a deceptive practice. Once again, our intention here is not to take a side or support one argument or another; rather, we need to understand these challenges and risks so we can offer suggestions on how to solve them or mitigate them.

† This point also relates to the previous note; in particular, what is the reference point for deciding the overrun or the success and failure.

‡ Cooperative education program that blends university learning with real jobs in the industry.

During these decades, I worked on or had been exposed to numerous megaprojects, either as an employee, a contractor, a consultant, or an observer. I have accumulated significant experience from these projects and other business opportunities, often feeling like a sponge, acquiring as much knowledge, skills, and best practices as I can absorb. I will share with you the good and the ugly learning from these decades of operating in the trenches. However, I have to restrict what I share to avoid touching on confidential areas.

3.2.7 Comparing the References

At this time, I want to stress a few vital points. There are many differences between these references.

1. The obvious one is that each of the first three books was written by a single author, whereas most of the papers in *The Oxford Handbook of Megaproject Management* and the other academic papers had more than one author—as many as four would be a typical case.* The multiple authors' scenario is expected in academic research.
2. The second difference is that *Industrial Megaprojects* (Merrow, 2011) and *Capital Projects* (Barshop, 2016) are both written by non-academic professionals and are focused on significant in-depth analysis of the projects in the IPA† database. On the other hand, *The Oxford Handbook of Megaproject Management* and many of the other papers were mostly academic in nature, and although they use projects to make many of the arguments, the research into those projects does not match the depth‡ of what IPA§ would have done.
3. *Industrial Megaprojects* focuses on industrial projects, as the name indicates. *Capital Projects* also leans more toward industrial projects but is not restricted to them. *Project Management for Engineering and Construction* (Oberlander, 2014) is not limited to a type of capital projects; it covers capital projects that the author defines as buildings, infrastructure, and industrial. *The Oxford Handbook of Megaproject Management* and the papers mostly focus on infrastructure projects with a heavier emphasis on transportation projects. Therefore, these references cover many industries that are related to capital projects and megaprojects.
4. There are no technology projects in these references, since not many technology projects would be massive enough to be classified as megaprojects; they would be "large and complex" but not mega. Another challenge on this type of project is the lack of enough publications to fully understand them.

3.3 Closing Comments

We must stress, again, that we are conducting these reviews from a practitioner's perspective and not an academic perspective. Our objective in this book is to produce a tailored approach for leading

* One paper had about 10 authors or contributors.

† Independent Project Analysis, where the authors of these books work and their books depends on the company benchmark analysis.

‡ IPA services include assessment of projects at more than one point along the project life cycle and include face-to-face interviews with the people on the projects in addition to a great deal of data.

§ Although I know IPA, neither I nor any friend or family is associated with them, directly or indirectly. There are no benefits to me from referencing IPA or any of the other sources here. This is important to state since I often reference IPA and CII with a highly favorable view and wanted to ensure that there is no conflict of interest.

megaprojects—a practical approach that can be readily used in the real world. Therefore, as a practitioner and writing for practitioners, our review is to look for challenges, trouble areas, opportunities, new ideas that we can seamlessly integrate into the CAMMP™ tailored approach for megaprojects.

Further, these books and our book are complementary to each other. Our book provides an overview, and the tailored approach for leading megaprojects, concept to success. On the other hand, these various references can provide more technical or specialized project management concepts that would be used in our model or for more in-depth coverage of technical project management. For example, if you want to know what executives should now, Barshop's book is a good choice. If you would like to learn about the various estimating techniques at different phases of the project, Oberlender's book is your source. For understanding the success and failure of megaprojects in different industrial sectors, Merrow provides us with a great book.

I must say, as I was doing the research and literature review, I wanted to do more. The reviews gave me a lot of answers but also generated many more questions. Some of those questions I pursued, and the answers are incorporated into the tailored approach that we will present later. However, some of those questions require extensive research that would be beyond the scope of this book. It is not practical or logical to go after every angle, especially when there are not enough resources or information to reach a conclusive outcome or findings. Therefore, in the final chapter of this book, I will share a few open points to ponder. Although I have my own thoughts on these points, which I will share, I cannot offer answers with certainty or a high degree of confidence, because they need further analysis and research. Consequently, I leave them to the readers, organizations, and research community for further review.

Chapter 4

Professional Associations

4.1 What Exists?

We start by stating that, to our knowledge, most leading professional associations, such as PMI® (Project Management Institute) and IPMA® (International Project Management Associations), do not have any guides, standards, or significant references for capital projects, especially megaprojects. IPMA, as an organization, might be closer to the capital projects industry than PMI—at least this is our perception. However, IPMA has no guides specific to the management or leadership of megaprojects. They have some meetings or conferences around the subject, but I am not aware of any significant reference. We can speculate that the reason for this lack of coverage is likely to stay generic and open to all industries and domains. These, and other organizations, might offer references for programs and program management, but that is a totally different subject.

There are other specialized organizations like those focusing on value management or cost engineering, which have references that are relevant to capital projects. AACE (Association for the Advancement of Cost Engineering) International might be a leading reference here despite the emphasis on cost. However, I am not sure if they have an area of focus addressing the complexity of megaprojects. If they do, they would be doing it from their specialty area of focus—cost—not from an overall management approach. We know that what these organizations offer is of great value to capital projects and megaprojects; we cannot deny that. **Our point is specific to standards or guides dedicated to the overall management of megaprojects, concept to closure.**

We want to be clear and transparent; we have not conducted extensive research on the topic, so our perception could be wrong. The reasons that we did not undertake this extensive research are (1) we already have enough general knowledge about the key concepts from these organizations and how to use them on capital projects and megaprojects, and (2) as we said before, we do not intend to go deep into specialized topics such as value management or cost engineering. For in-depth insights into these topics, we advise you to visit the sites of these various professional associations.

4.2 The Project Management Institute

For many years, PMI has recognized that the *PMBOK® Guide* (PMI, 2017) is a generic, industry-agnostic reference. Therefore, they have produced additional guides, including a series of extensions, labeled "[____] *Extension to the PMBOK® Guide*"; the blank is for a given domain, such as construction. The first edition of the *Construction Extension to the PMBOK® Guide* (PMI, 2016) was released in 2003, and the most recent one published in 2016. Here are a few points about this guide. It covers *"the generally accepted principles for construction projects that are not common to all general project types"* (PMI, 2016, Ch. 1, p. 1).

- Let us start with the title and the term *construction*. Is this guide for the contractors working on the construction phase of a capital project, or is PMI using the word "construction" to refer to a capital project from beginning to end—that is, not limited to the construction phase? The guide did not offer a clear, explicit answer on this question, and our understanding is that it covers capital projects, or projects with a construction component. However, we know PMI, per the *PMBOK® Guide*, considers the discovery phase of a project a "pre-project" and so as outside the project life cycle.
- This guide is quite similar to the *PMBOK® Guide*, in term of structure, in that it focuses on a few general chapters, then a dedicated chapter for each knowledge area.
- The main difference between this guide and the *PMBOK® Guide* is that it offers two more knowledge areas—one for financial management and one for HSSE (health, safety, security, and environmental) management (PMI, 2016).
- In addition to the two knowledge areas, there are additional processes embedded in other knowledge areas.
- In term of process groups, there is no change: we still have five process groups, because these groups are not specific to an industry or domain.
- The guide did not offer clear guidance on a project life cycle model; it just mentioned some concepts that are available in other sources, such as Front-End Loading.

The bottom line, as someone who knows the *PMBOK® Guide* and the construction supplement well, I did not identify any significant knowledge that was not covered in many other references long before this extension was published. However, it is a useful reference for those who have restricted their learning to PMI guides.

4.3 Chapter Closing

In general, the PMI guides (*PMBOK® Guide* and *Construction Extension*) focus on the processes that are useful to manage a stage of a typical capital project.* In any model or approach for managing any type of project, we have always stressed the need for a project life cycle model, which is the appropriate methodological approach for covering a project concept to closure. Besides, we need the project management process to manage each stage of the project. This perspective was lacking emphasis in the PMI guides.

---

* We do not use the term *construction project*, because it is confusing and is not appropriate. We know PMI and other organizations use this term to describe a typical capital project. Out of context, one might think of a construction phase. Therefore, we prefer to use *capital projects*, which would mean covering a project concept to closure, which include a construction phase. *Facilities projects* is another clear term.

It is also unfortunate that other professional associations have not identified megaprojects as a critical topic, despite the high failure rate and the massive consequences to economies and sustainability of the project owners' organization.

Section II
The Challenges—Current State of Practice

Chapter 5

Observed Challenges

5.1 Introduction to the Section

In this section, we address the current state of practice of megaprojects, with a focus on the challenges. We discuss the challenges via the various chapters of this section, starting with the observed challenges—the challenges facing the management of capital projects and megaprojects.

How did we come up with these challenges?

These are general challenges from personal observations and media reports over a period spanning more than three decades of the author's career in project management and the capital project industry. In the next chapter, we will summarize the challenges from the literature review, and Chapters 7 and 8 will be for the challenges gathered from case studies. You will see many common challenges in these chapters, which we will group and summarize in Chapter 9, as we close this section.

5.2 "The One Thing"

I know that some will ask, *"What is the single most important factor that leads to failure?"* or *"Is there such a thing as an ELE (Extinction Level Event)?"** for megaprojects.

I am one of those people who do not believe in "the one thing" syndrome. Every challenge we list here is often a compound of other factors. In other words, a concept on its own, such as fixed-price contracting, might not be the challenge. The challenge would be related to a lack of understanding of the advantages and disadvantages of fixed-price contracting and how we apply this practice in the real world. Another example is complexity. Complexity on its own is not a challenge. However, not knowing how to deal with complexity is the challenge. In other words, we view challenges as risks; there has to be a cause and an effect. Let us review these compounded challenges.

* The reference here is an event that could destroy the chance of success of a megaproject.

5.3 The Observed Challenges

Once again, the following are our observed challenges.* Please note that some of them might be more applicable to one region or industry/sector versus another. Also note that each of these topics deserves a book or further research, which is not feasible here.

5.3.1 Politics in Infrastructure (Good and Bad)

Politics is not a core area of expertise for us. However, since it is an important area, we want to stress it and state that numerous papers and articles have been written about the sad case of infrastructure projects all over the world. The primary concern here is that infrastructure projects are most often, if not exclusively, led by government organizations. As a result, it is common that some government organizations might lack the necessary expertise to lead these projects. It is also common that these projects, at least the front end, may take longer than necessary, may even experience on/off situations resulting from funding issues, elections, or other factors. One more leading factor is the influence of politicians and special interest groups on pushing for projects based on self-interest or gaining popularity. Therefore, some of this segment of stakeholders could actually over-promise benefits and push for lower costs, which could be outdated. This latter situation is complicated when projects are put on hold for a while then re-started, but the politicians (and others) could still use the old numbers, ignoring inflations. More on this in the next chapter.

5.3.2 The Complexity of Megaprojects

We think this is an area that not many can debate or dispute: megaprojects are massive and complex with too many parties involved and numerous stakeholders. Their lengthy durations create added uncertainty, which often increases the complexity. This complexity drives us to seek well-established and highly sophisticated organizational project management systems and competent professionals at all levels. Any organization that wants to obtain value through projects needs an established organizational project management function with adequate individuals and organizational competence.† However, the reality is that not many organizations fully understand the high-level complexity, nor have they built robust systems that would be capable or appropriate to the needed level. It would not be surprising to have organizations using a basic form of project management that might be more suitable for large projects or projects with lower complexity and smaller size than the mega class.

Maybe the issue here is the complexity and in not being able to handle it because of individual and organizational competence, or systems. We realize that this point is a general statement, which we include here to stress, again, **the need for an established organizational project management function**, inclusive of a defined and sustainable organizational project management system.‡

* The reason we are using the term *observed* is because these are observations that have not been proven through formal research or studies. In other words, these are perceptions, although the literature review and case studies do validate them.

† For reference and further reading, IPMA®, The International Project Management Association, has published *Individual Competence Baseline* (IPMA, n.d.a), *Organisational Competence Baseline* (IPMA, n.d.b), and *Project Excellence Baseline* (IPMA, n.d.c).

‡ A link to an article on this point: http://blog.sukad.com/rethink-the-project-management-function/

5.3.3 Culture of Fixed Price

We have observed that a large percentage of capital and megaprojects, at least industrial* projects, are contracted to service providers via a fixed-price contractual arrangement. If the project conditions are suitable for this type of contract, this contracting strategy would be acceptable and expected. However, we observe a few significant issues with this approach (the compounded factors):

- The conditions are not often right for this type of contract. For example, frequently, the project or product scope is not complete; if complete, it is not fixed, nor is it stable. These types of conditions might be suitable for a cost-plus arrangement, but certain clients insist on the fixed price for a variety of reasons, as you will see next.
- The perception by the client organizations (project owners) is that this is a low-risk approach, and some even think it produces the lowest cost. However, what clients miss is that these contracts might be low risk in terms of the contracted cost, because the price is fixed at the beginning; however, once a contractor faces cost challenges, the contractor management might decide to optimize the schedule, possibly slowing the project, along with the possibility of sacrificing quality or even scope. In other words, the initial price might be kept low or fixed, but the project owner ends up paying dearly as a result of delays, quality matters, or production deficiencies. Further, even the initial price might grow because of the next point.
- Fixed price contracts are suitable in a down market and if the scope is set—that is, a fixed, or at least stable, scope, which leads to fixed-price conditions. The reality is elsewhere. In many fixed-price contracts, we have seen that the original scope is *not* entirely defined and is *not* stable. This scenario leads to too many expensive changes that can put pressure on the project owners and even on the contractors. Consequently, even the initial cost grows and produces an overrun† situation.

In Chapters 7 and 8, focusing on case studies, we will share more specific challenges related to fixed price and its consequences on the project outcome.

5.3.4 Culture of Low Bid

We do realize that not all organizations subscribe to a culture of the low bid. Proper procurement practices provide us with the pro and cons of numerous contracting approaches, which is outside the scope of our work. What is in the scope is that the culture of a low bid is widely used and often mandated by policy or unwritten rules.‡ If the context of the project justifies this approach, then go for it. However, we often find the approach is used almost universally, especially in government§ projects, and with disregard to the situation or good practice. This practice is common even where there is a degree of uncertainty in the scope of work, which opens the doors wide for abuse through the change management process.

* Some books, like *Project Management for Engineering and Construction* (Oberlander, 2014), use the term *process* to refer to industrial projects.

† At this time, let us define overrun as actual cost exceeding approved budget. Later, we will offer a slightly different definition.

‡ We have worked (or been in touch) with many projects owners' organizations that refuse to consider any other form of contracting.

§ The term *government* here also includes semi-government—companies owned by government entities.

The consequences are almost always that we hear project owners complain about their contractors' performance, quality, competence, etc., getting into a "blame the contractor" scenario, when most of the trouble started within the project owner's organization and its outdated procurement practices.

One + one = two, but in the context of this challenge, it could be equal to three or even four. What we mean is that if we combine the culture of fixed price with the culture of the low bid, the resulting impact can be significantly higher than the simple addition of their individual consequences. This is also one of the three common causes of megaproject failures: the *"bid low and sue later"* culture (Lenfle and Loch, 2017).*

5.3.5 Overdependence on Contractors and Management Consultants

If you are an investor or organization launching one project only, or maybe you have one capital project once every few years, it would not be worth it for you to invest in building your internal organizational project management function, capacity, and competence. However, if you continually launch projects and manage their products, and you continue to outsource your project management work, then you might be burning money!

Some organizations have decided, consciously or subconsciously, not to build their internal project management function and system.

It is common practice, and logical, that if an organization does not have the capacity or experience to manage a project, it outsources the management of the project indirectly to a general contractor or directly using the help of project management consultant (PMC) or construction management (CM) agency. Again, we can understand if this is happening for one or two projects. However, some organizations have decided, consciously or subconsciously, not to build their internal organizational project management function and system. We think that project management intimidates them. Consequently, they have no choice but to depend on outsourcing. When an organization continues to launch capital projects (including megaprojects) over many decades and always outsource the management of the projects, they are asking for trouble.

The trouble with this approach is that the construction contractors work is done, often, through a fixed-price scenario but the PMC works on a time-and-materials basis (or similar arrangement), often without a price cap, liability, or accountability. In other words, a project is delayed, the PMC revenue (and possibly profit) will increase. We are not saying that PMCs are corrupt or intentionally allowing performance to slack, leading to delays, but what is their incentive to push for a speedy completion? On the other hand, the PMC's ability to manage appropriately might be limited by project owners who want to keep some or most control on decisions, not realizing that delayed decision or procrastination has a direct impact on project performance.†

Executives, project owners, what prevents you from building the project management competence and capacity in your organizations?

* *The Oxford Handbook of Megaproject Management,* p. 23.

† See the next challenge.

We realize that what we are stating in these few sentences would not lead us to gain many friends. However, we have seen these scenarios repeat too many times, and our professional and ethical responsibility mandates us to state what we observe. Notice, we are not blaming one party or another—we are not in the business of blame; it is the system that is broken.

Let us close with a call-to-action!

We believe that research by the Independent Project Analysis (IPA) has concluded that clients, project owners, that manage their project directly achieve better performance.* Therefore, executives, project owners: what prevents you from building the project management competence and capacity in your organizations?

5.3.6 Project Owners' Competence

Here, we might be opening another hot potato as we address a few points related to competence. Before you jump, we have to explain "competent," "not competent," and "incompetent." If we say Mounir is *not competent* in plumbing work, that means Mounir is not trained as a plumber. Therefore, he is not qualified; and should not practice this work. However, if Mounir decides to work as a plumber and he is not qualified, then the incompetent label *plumber* might be appropriate to Mounir since it is likely that he could create a swimming pool in your basement. In this regard, if you want to fix your leaking toilet, hire a ***competent*** plumber.

Back to project management.

Some of the earlier challenges we raised can be complicated when the project owners' staff do not have enough experience in leading or managing capital and megaprojects or even working with PMCs. In other words, they would be "not competent" because, most likely, they have not been trained in project management or have the right level of expertise.† This situation may invite some contractors to take advantage of the situation and go on with low bids, or even manipulate bids to get the contract and hit back with expensive changes. We know this last point is bordering on corruption or unethical behavior, but to think this might not be happening would be naïve, to say the least.

Another point that often affects projects is the misplaced overconfidence of owners' staff, even when they do not have the proper level of competence. This overconfidence, not to say arrogance, often lead to the following:

- A master–slave syndrome—that is, "I am the owner, and you do as I say."‡
- Outsourcing the management of projects, but still wanting to micromanage or interfere in the decision making in areas the owner's staff are not trained for.
- The lack of experience and knowledge lead them away from carrying out effective oversight of the contractor or the PMC/CM.

We know these are harsh words, but please reflect on the following: I have spent most of my career working for or with project owners' organizations, so I know the inside story.§

---

* More on this in the next chapter.

† Project owners, regardless of the industry, might not view project management as a core function for them; therefore, it is ok to outsource.

‡ We have seen young engineers from owner organizations ordering around seasoned construction engineers or managers on how to do something.

§ I will stress again and again, these are generalizations and not applicable to all project owners; these comments are based on observations and limited to some owners. I have had the privilege of working with great project owners and I have seen really bad ones. Since project owners are a focused audience for this group, we must be direct and transparent. If they are not willing to acknowledge this reality—if they have it—then they cannot improve.

An example of the above follows.

When a contractor submits a schedule to the client, a member of the client's staff "approves" the schedule. Listen to this: *"Approved or accepted schedules can produce legal implications to the owner"* (Oberlender, 2014, Ch. 11, p. 323).

Don't the client's staff approve schedule or plans?

The answer might be different depending on the contract laws or standards that you follow. If you approve, do you take on the legal implication as Professor Oberlender stressed in his book? Or, do you have a contract clause that maintains accountability with the contractor? What I have learned as a young project manager working for project owners is to *acknowledge* receipt of the schedule and not *approve* it.*

If you want to approve, then consider this related question: as an owner's project manager or representative, do you have enough experience to judge or accept a schedule, especially by looking at a printout instead of going into the scheduling model, through the software? We are not talking here about target dates, we are talking about the logic and details of the schedule.

One more point, have you heard of "sequestering float" (Oberlender, 2014) and the risk associated with it? In all transparency, I had not heard that term before my research for this book. I knew the concept, but I did not realize it had a name. What is sequestering float? It is the games that a contractor can play† and an owner will miss if the owner's staff does not have the right level of experience. In other words, project owners' teams: understand your limitations, your strengths, and weaknesses, and be humble about it.

The above point is only an example of the severe consequences of the project owners skipping out on building their internal project management system competence and capacity, and as a result, having to over-depend on outsourcing the critical management of their projects.

Should project management be a core function in organizations? The answer is in Part A, Section III.

5.3.7 Change Management

As we solicited opinion from our online network,‡ William (Bill) Duncan§ shared this challenge with us: "Change orders, especially late ones." We agree. Change management, in general, is a common problem area on projects, and the challenge can be magnified owing to:

- Lack of understanding of the topic in general, especially by project owners.
- The absence of change management outside the context of contracts, usually in the early stages of the projects—the foggy zone¶
- Lack of understanding of the cumulative impact of change by service providers and project owners
- Late changes, as Bill suggested, which lead to significantly higher cost in comparison to earlier recognition and resolution of the same changes

Because of the importance of this concept, we had dedicated a chapter for change management in our book *Project Management beyond Waterfall and Agile*, and we will revisit the key concepts in Part A, Section IV.

---

* As I am sure most readers know, legal language is tricky, and often one word, or one misplaced comma, can make a huge difference.

† https://www.enr.com/articles/42405-how-firms-can-avoid-the-construction-schedule-float-game

‡ Link to the post on LinkedIn, https://www.linkedin.com/feed/update/urn:li:activity:6510486121724538880

§ https://www.linkedin.com/in/wrduncan3/

¶ We call project management in these early stages as Management in the Fog; more on this in Section IV of Part A.

5.3.8 Lack of Appreciation of Uncertainty

This is another challenge posted by Bill Duncan and is related to the earlier challenge on competence. What Bill offered is the following*: *"In brief, early estimates ALWAYS have a wide range. The bigger the project, the wider the range. Yet owner management often wants a single number. So a '$2 billion' project should really be quoted as $1.5–3 billion. When the finished product is delivered for $2.8 billion, it's still well within the expected range, but it is perceived as a 40% overrun. Senior management doesn't understand that the uncertainty is there whether they want to acknowledge it or not."*

In other words, project owners, or to be more specific the business executives in project owners' organizations, lack understanding and appreciation of uncertainty. This concept is also related to what we like to call "the dangers of on or ahead of schedule, on or below budget," which we will also discuss later in this book.

This is also one of the three common causes of megaproject failures (Lenfle and Loch, 2017, p. 22).

5.4 Closing Remarks

Table 5.1 summarizes all of these observed challenges.

Table 5.1 The Observed Challenges

No	Challenge
1	Politics in infrastructure
2	The complexity of megaprojects
3	Culture of fixed price
4	Culture of a low bid
5	Overdependence on contractors and management consultants
6	Project owners' competence
7	Change management
8	Lack of appreciation of uncertainty

Remember, these challenges are based on observations. There are many other challenges that we did not list here, since we do not consider them as "top-of-the-list" challenges. Other challenges could be based on research, and those will be in the next chapter. Notice that most of what we listed here are related to the project owners for the reasons we highlighted earlier: owners are a critical focus of this book; did we say that before?

Let us close this chapter by bringing back some questions from the front matter of this book.

- Are megaprojects too complex to manage as projects?

* Shared with Bill's permission.

- Would Agile be the answer to managing megaprojects? This was a question posted or claimed on a LinkedIn post once.
- Instead of Agile, since megaprojects are enormous, could they be managed as programs? We also encounter this view that since megaprojects are significant, they should be treated as programs.
- Have organizations built and are sustaining the proper organizational project management system, methods, knowledge base, etc. to support capital projects in general, and megaprojects in particular?

We could post more questions, and we will do so as we continue with this book. We will also post answers, at least from our viewpoint. In the meantime, continue to ponder these questions.

Chapter 6

Insights from the Global Community

6.1 Introduction

In this chapter, we share insights from the global community. We do so via the literature review that we conducted. There will be two main parts in this chapter. In the first we share general statistical information about project performance from the various resources that we have reviewed. The second part will shift to the challenges from the reviews, and either these challenges were clearly stated by the authors, or we concluded by reading between the lines.

Please note that we are not listing all review comments in this chapter. Some of our findings and observations are spread throughout this book. We are also not getting too deeply into detail to avoid duplicating content. We will continue to highlight the sources for further studies.

6.2 A Venting Moment

Conventional wisdom and common sense would lead one to think that, because megaprojects are massive and complex, organizations would go beyond good practice in managing or leading these types of projects. However, the research tells a different story. For example, the analysis of thousands of capital projects and hundreds of megaprojects by Independent Project Analysis (IPA) has led to the conclusion that *"there is a systematic and undeniable relationship between larger size and poorer preparation"* (Merrow, 2011, p. 223). Therefore, be ready for some shocking statistics.

We cannot avoid stressing our surprise at the state of the current practice. For years, or decades, we have heard about the "hypes" of organizational learning, innovation, artificial intelligence, and organizational project management. We hear about a "profession," professional associations, numerous guides, and generic or specialized certifications, yet we do not see all of that translating into better projects performance.

Despite the billions lost in value on megaprojects, we do not see any of the leading professional associations giving this topic a particular focus. Maybe some of them are too busy hyping Agile, pushing certifications like selling peanuts, and promoting executive playbooks to encourage executives to consider project management.

I have to be direct in what I am saying since this is a critical domain, and we need a dose of reality rather than a series of three letters next to our names. I share my disappointment because despite all of the learning and advancements of science and project management, we are still struggling, with no sign of improvement. We see the bragging about how many certifications one has but nothing about how those certifications or the learning that supposedly went into achieving them is used to improve humanity. Maybe improving humanity would be too much to expect; therefore, we would settle for enhancing the organizational projects' performance. We definitely have the syndrome of the knowing–doing gap (Pfeffer and Sutton, 1999).

Further, we should be careful in thinking that artificial intelligence or machines will take over this domain, since project management is primarily about people, teams, and stakeholders. As we said, we do not see too many signs of improvement. On the contrary, the signs from organizations, even global powers, is that many of these organizations have either eliminated or diluted their internal capacity to manage capital projects and megaprojects (Merrow, 2011) (assuming they had them in the first place).

The challenges are not only about cost or schedule, overruns, or delays. Some of our failures have killed people—whether we reflect on the Titanic disaster,* which was partially blamed on the project and project management, or the recent tragedies of the 737Max.† If we cannot fix what we know, are we capable or ready to meet what will come? Use this chapter as a wakeup call and let us work toward making a real change.

6.3 Perspectives and Mindsets

Although I go more deeply into challenges later in this chapter, I include the following points here because they shed light on our abilities to forecast and avoid manipulative number games. This will be necessary as we review the statistics and try to make sense out of them. We need to have the right mindset and perspective.

6.3.1 Forecasting Challenges

A few days ago, a colleague shared an article on LinkedIn®. The article, *Predictability Limit: Scientists Find Bounds of Weather Forecasting* (Pennsylvania State University, 2019), was not about project management but about the weather. The article documented research on how far in advance scientists can predict the weather and offer a reliable forecast. More than one point of reference and different researchers agree that 10 days is currently the extent of what science can provide us in term of reliable weather predictions. With more scientific advancements, maybe science can help add a few days for advance forecasting.

* https://www.youtube.com/watch?v=wbvfir2x344&t=34s. The case study does not blame project management for the actual accident, but it does blame PM for contributing to fast sinking of the ship and magnifying the loss of life.

† At the time of this writing, it was becoming clear that the tragedies of two airplanes crashes and hundreds killed originated with the projects and related software development.

A few project management professionals in my network were discussing this article and linked it to project management, which raised the question of cross-learning and applicability. The idea is that forecasting the weather requires the study of numerous complicated factors that affect accurate predictability. Could this concept apply to project management, where we are also limited in predicting project performance far in advance? If we cannot predict accurately in advance, then what is the use of estimating and forecasting, especially on projects that could last years? One colleague's view was, should we even budget for the whole project or only do it one stage at a time?

The above scenario sounds logical, and it reminded me of a story of a vice president of a leading oil and gas company being upset with the estimating unit, so he was ready to dismantle it. Let me ask the question, instead of killing the messenger, can we have, or develop, our people to enhance their abilities to forecast projects' performance?

Absolutely.

We can predict and forecast projects in advance.

I say this for two reasons: The first is my own personal experience and what I have seen in some organizations over many years of professional practice. The second one is actual research done at the University of California at Berkeley (UCB). In a study that links project performance with organizational project management maturity, UCB researchers came up with three critical findings, and one of them relates to the question we are addressing here. I quote, *"Project management maturity is strongly correlated with more predictable project management schedule and cost performance"* (Ibbs and Reginato, 2002). In other words, we can predict and forecast projects in advance. We might not be great in the beginning, but we can improve to achieve good predictability, as long as we understand that (1) we are not talking about precision but accuracy, and (2) the accuracy is within a degree of tolerance, as we will discuss next.

6.3.2 Do Numbers Lie?

By now, I am sure you've noticed that I do not mind being direct or provocative at times, especially when we need to be vigilant about a point or a challenge. I know that this point can open a can of worms, so I will try to contain it.

Here is the question: *Do numbers lie?*

We sometimes hear that numbers do not lie, but numbers can be manipulated to make a point, or even to deceive. For example, I was recently discussing and questioning some data that was presented in a LinkedIn post.* I challenged those findings, such as that 0.5% of megaprojects are successful, whereas IPA data shows 35% success.† The response came that the data used for the 0.5% argument is based on 11,000 projects, whereas the IPA data was limited to 318. In other words, we should trust the more extensive database; after all, numbers do not lie. Well, as a professional starting his project management career in

* The post was sharing information from the Project Data Analytics event; speaker: Alexander Budzier, University of Oxford, https://www.linkedin.com/feed/update/urn:li:activity:6521857263655882752.

† To be clear, the 0.5% was quoted without context initially, then we learned it was meant for infrastructure projects. On the other hand, the 35% is for industrial projects. Also, I know how the 35% was generated, since it was documented in *Industrial Megaprojects* (Merrow, 2011), but there was not enough clarity on how the 0.5% was obtained; despite various comments and exchanges, we only got partial answers.

estimating, I know one can manipulate numbers or emphasize the message that one wants to send. I also had to substantiate my numbers to my stakeholders. Therefore, I learned to trust data that is clear and can be verified. The key word here is *trust,* not blind faith.

Consequently, I would trust the data on the 318 because it was based on in-depth analysis, and the criteria of success were evident and explicitly stated. I would also believe the data from the 11,000 projects if the criteria were clear and transparent. Alternatively, I cannot trust it if the presenter cannot provide evidence on how the judgment was made. For example, are all of these 11,000 projects in the database megaprojects or do they include projects of all sizes and complexities? Were the projects judged as a failure if they were a day late or a $1 over budget, or did they have a degree of tolerance? I can raise many questions here, but I believe the point is clear.

> **Overrun or Something Else**
>
> Because estimating is uncertain, it is natural that organizations understand that a cost estimate is just that, an estimate. The final number can vary and could be above or below the estimated value. The range is a function of the organizational culture and system and depends on the level of project definition (where we are on the project life cycle).
>
> Consequently, some organizations consider ±10% as a normal and acceptable range. I also know that, at least in one situation, the organization does not use the term overrun unless the final cost is +110% of budget. Anything between 100% and 109.9% is over-expended.
>
> We do understand that some organizations might use the term overrun even if the project's final cost performance is 100.1%, but is that failure?

Why the lack of trust and why we needed verification and transparency.

Let me be clear: when one says only 0.5% of megaprojects succeed, what was the basis? For example, is a cost performance of 101% (actual cost over estimated cost) considered a failure? In other words, are we confusing or equating an overrun with failure? *See textbox.* Despite the back and forth on the LinkedIn post, and even after getting a copy of the presentation, it was not clear if the author is equating overrun with failure or whether there is a degree of tolerance applied in the statistics of the projects presented. The danger in these situations is that people take these out-of-context numbers and start to use them.

Another concern with the 0.5% relates to defining the starting point of comparison. Was the correlation between the final cost and the feasibility estimate or the final funding estimate?* I know that one aspect of the ongoing debates between Flyvbjerg et al. (2002) and Love et al. (2019) is about the starting point of comparison. So, the question is: should we measure performance from the final funding estimate (sanctioned, approved budget, pre-EPC†) or from the point of the decision to build, which could be the initial estimate? I believe we should know both. Consequently, I would advocate that organizations must assess the performance of the front end (Development Phase in CAMMP™), the implementation (Delivery Phase in CAMMP™), and overall performance. More on this later in the book.

Back to the 0.5% again. Because the presentation and the LinkedIn post did not clarify the basis of the numbers (starting point of comparison), nor did they explain if overrun indicated failure, then I cannot but question that number. My point is that anyone announcing these disastrous levels of performance needs to give us the criteria and context in straightforward language so we can understand and trust. We should not go chasing academic papers or books to find the details. Also, if these numbers are accurate, then we do have a disastrous scenario, much worse than expected. However, if these numbers are not realistic and/or are used out of context, could we not be causing more harm than good? Once

* More on this as we present the tailored approach and cost estimating later in Part A, Section IV.

† Engineering, procurement, construction

again, as a young cost engineer, I learned that forecasts must be accurate and timely, and if they are not (higher or lower), we could be facing many issues and problems on the projects, especially in term of stakeholder management.

Are there opportunities to cross train and share expertise across sectors?

Notice that I did not talk much about the 35% (that comes later in this chapter). I must close this part with the following: (1) if 0.5% is accurate, and its basis is comparable to the basis for the 35% quoted by Merrow (again—sharing similar basis), and (2) knowing that the 0.5% is for infrastructure megaprojects and the 35% is for industrial megaprojects, does this mean there are opportunities to cross-train and share expertise across sectors? Executives, academics, and practitioners should consider this question!

6.4 The Statistics

In this part, we share statistical information from the various references.

6.4.1 *Industrial Megaprojects*

The bottom-line results on project performance from the 318 megaprojects in the IPA database is that 35% of the projects are considered successful and 65% are labeled as failures. These statistics might mean that the failure rate is higher because it is likely that many organizations do not use IPA services, and those with failed projects might not participate in benchmarking exercises. On the other hand, there is a chance that projects not analyzed by IPA could have a higher success rate, but this is doubtful. In other words, failure is often hidden. Obviously, this is an opinion and not substantiated by facts. Another factor to consider: the IPA criteria could be too strict for some organizations, and as we said before, success and failure is not an either-or situation.

The Projects Database

The following is relevant information about the 318 projects that functioned as the dataset for *Industrial Megaprojects* (Merrow, 2011, p. 29).

- Cost range is from $1 to $20 billion, in 2010 value of money.
- Average cost is $3.2 billion.
- Average duration to execute (the span of time from final approval to completion of facilities) is 43 months.
- Average duration from start of scope development through the startup is 66 months.

How does the IPA define success and failure? *Refer to the textbox Defining Success and Failure.*

If we try to better understand the full criteria, we urge you to get the book. I only add a couple more points, specific to the first two cost categories in the success criteria.

- Successful projects, on average, underrun their budget by 2%. Further, they achieve 96% of the industry average, which means they are competitive.
- On the other hand, failed projects averaged about 40% overrun and more than 30% above industry average—not competitive.

One of the exciting findings shared by Merrow is this: *"When we first started examining very large projects as a separate group, we almost immediately noticed an odd phenomenon. The project seems to fall naturally into exceptionally good projects and exceptionally poor projects, with only a very few in the middle, where the bulk of smaller projects would be found"* (Merrow, 2011, p. 37).

6.4.2 Capital Projects

Remember that *Capital Projects* (Barshop, 2016) also depends on the IPA database. However, here the database is much more extensive because it includes all capital projects in the database, even the smaller ones[*] and also non-industrial projects. The size of this database, at the time of this writing, was more than 14,000 projects. Let us start with the critical statistics from the book:

- Only 60% of the finished project meets all of the objectives.[†]
- In term of profitability, the average project is profitable but experiences a reduction in the value[‡] of 22%.
- If organizations have a good process[§] for managing their capital projects—and they use it—they can expect up to a 5% higher NPV than expected. However, if they do not have a good process, or are not using it correctly, they will see up to a 45% reduction in the expected profit. The average is the 22% erosion mentioned in the previous point.
- The breakdown of the 22% is: 10% due to lower sales; 7% to the impact of cost and schedule performance; and 5% to asset performance shortfall and technical production issues. If you scrutinize this statement, you will notice that the challenge is across the project life cycle from the business case all the way to production; hence the need to evaluate projects as assets, which is a point that we discuss later.

Defining Success and Failure

The IPA Model uses five categories for assessing failure and, consequently, success. These are:

- Cost overrun
- Cost competitiveness
- Slip in execution schedule
- Schedule competitiveness
- Production versus plan

The above are the categories and the book also include the limits or thresholds for grading a category as success or failure. For example, for cost overrun the threshold is 25% over budget would be graded as failure (Merrow, 2011, p. 38).

Although the threshold for each category might appear to be generous, missing any of the above categories would classify a project a failure.

A conclusion from both books is that, indeed, megaprojects do not do as well as the full range of capital projects: 35% in comparison to 60% success, respectively. We could also say, these numbers are based on similar analysis and criteria, so we are comparing watermelon to watermelon.

6.4.3 The Oxford Handbook of Megaproject Management

As stated earlier, this book consists of 25 academic papers; it is from these that we have gleaned the following. For starters, it was interesting to note the size of the market, which we mentioned in Chapter 1. The *infrastructure* megaprojects market is possibly more than $3 trillion per year. If we expand this to other fields, the market can be **$6–9 trillion per year**. How much is 1% of that?

Further, research led by Flyvbjerg *"found that on average, a one-year delay or another extension of the implementation phase correlates with an increase in percentage cost overrun of 4.64%"* (Flyvbjerg, 2017, p. 10). The following statistics are about large dam projects (Ansar et al., 2017, p. 175).

---

[*] For capital projects, even smaller projects could still be over $100 million.

[†] Refer to the textbox from the *Industrial Megaprojects* section; IPA has five categories to assess success and failure. Missing any of them will lead to categorizing a project as a failure (Barshop 2016, p.3).

[‡] Actual NPV (net present value) in comparison to promised or expected NPV.

[§] Typically, a good process refers to a well-defined stage-gate process.

- Three out of four projects suffered a cost overrun with the actual costs, on average, 96% higher than estimated.
- Eight out of every 10 projects suffered schedule overrun, and the actual schedule was on average 44% longer than the estimate.
- Although the datasets for benefits is smaller than those used for the cost and schedule performance, these projects suffered a benefits shortfall of at least 11%.
- In other words, with a planned benefit–cost ratio of about 1.4, the cost and schedule overrun and benefits shortfalls lead to failure.

On another subject, in challenging the theory/principle of the Hiding Hand of Albert O. Hirschman (Hirschman, n.d.), the Hiding Hand concept advocates that cost overrun of projects will be outweighed by higher benefits. Flyvbjerg shares with us his findings that refute this principle. Table 8.1 in *The Oxford Handbook of Megaproject Management* provides us with interesting information about infrastructure projects (Flyvbjerg, 2017, p. 175). I am only sharing the bottom line here. For the 1,603 projects in the database with cost information, cost growth averaged 39%.

Similarly, the average benefits shortfall of 786 projects in the database is 10%. In other words, cost overrun is not outweighed by higher benefits. Let us assume that at authorization, these projects were expected to have a benefit–cost ratio (BCR) of 1.4.* With the "average" cost growth and the "average" shortfall in benefits, this BCR of 1.4 drops to about 0.91, which means on "average" these projects fail.

Please note that we are sharing this information as reported, and be careful not to use them as guidelines. The reason we include them here is to show the scale and to create awareness for further research. Also remember the following as you read, which is about the danger of "averages." I always share this with my students: my right hand is in subfreezing temperature, and my left hand is in the oven. On average, my body temperature might be ok, but I just lost two hands. In other words, reporting averages is risky business, since a few projects could have massive overruns, which would skew the average. Refer to the study on Australians projects, below.

The bottom line: dig deeper to understand the numbers. We cannot ignore what all references show us, which is that megaprojects do have problems, but that does not mean it is all doom. Our mandate is to identify the challenges, learn from the successes, and build a robust and sustainable organizational project management system and tailored approach.

Under Estimating Cost, Over Estimating Benefits

A 2002 paper by Flyvbjerg, Holm, and Buhl (2002) emphasizes a common problem in planning where organization underestimate the cost and overestimate the benefits. This situation leads to inflating the benefit–cost ratio of projects. However, the reality is that the performance fails to deliver on these estimates (costs or benefits), which results in significantly lower BCR. Flyvbjerg et al. also claim that this situation could be an error or even deception.

However, Love and Ahiaga-Dagbui (2018) challenge this concept as a planning fallacy. In other words, there is an academic debate on whether underestimating is due to lack of proper project delivery models or is a sign of deceptive practice.

6.4.4 Study on Australian Projects

We mentioned in the previous part that the concept of averages is a risky business, and here is a point to support how averages can skew the stats: *"Ninety percent of Australia's cost overrun problem is explained by 17 percent of projects that*

* Although this number was referenced in some of the papers, do not take this as a basis; in many organizations they want the BCR to be much higher than this to authorize a project.

exceed their promised cost by more than half" (Terrill and Danks, 2016). In other words, the average is skewed heavily owing to a few of the projects in the dataset. This same study shows additional exciting information. We share some:

- One of the leading factors contributing to projects' performance challenges is related to premature announcements by politicians' announcing the projects at a cost per earlier estimates (conceptual estimate). The study finds that most of these projects contribute to most of the overruns.
- In terms of growth over the full life cycle, the projects in the dataset experienced 9% growth before formal funding commitment, 6% between budget commitment and construction, and 9% during construction. In other words, 15% of the increase was in the front end and only 9% during implementation.
- This study also showed that few projects are canceled once announced; two-thirds of the announced projects were completed (even those announced early), which indicates a weak appraisal process (Terrill and Danks, 2016).
- Finally, more significant projects are more likely to overrun and to have higher cost growth (Terrill and Danks, 2016). This finding aligns with the IPA research (Merrow, 2011).

All of these findings support a suggestion, which I will detail later in the book, of **advocating the need to have effective project control during the front-end development.**

6.4.5 Other Stats

We share here information from various sources; here is one snippet. Research from Wood Mackenzie shows the expected capital expenditures on liquified natural gas (LNG)-related projects to be around US$200 billion between 2019 and 2025 (Wood Mackenzie, 2019). They also report that only 10% of the LNG projects are completed under budget, and 60% of projects experienced delays. Unfortunately, the article did not present more detail and context. This is a crucial topic to understand by companies pursuing these types of projects, especially with the boom of LNG projects happening around the time of this writing.

6.4.6 Summary of the Statistics

The bottom-line results from *Industrial Megaprojects,* using the IPA database, is that the success-to-failure split is 35% to 65%, respectively. *Capital Projects,* which also uses the IPA database, expands the search into the smaller capital projects, finding that split change to 60/40, respectively. Concerning infrastructure projects, we have seen the claims shared earlier of 0.5% and the conclusion that on average infrastructure projects fail. However, keep in mind the danger of averages.

The critical message to re-emphasize as we close the statistics part of this chapter is that more significant projects do not do as well as "smaller" projects, yet organizations are not giving these larger projects the attention they deserve.

6.5 The Challenges

The Iron Law of Megaprojects is a term used by Flyvbjerg (Flyvbjerg, 2017, pp. 9–12), which states, *"over budget, over time, under benefits, over and over again."* Maybe it is worth repeating the quotation

"Insanity is doing the same thing over and over again and expecting different results." I know, I have been crediting Albert Einstein with this quotation, like many others. However, the origin of this phrase is disputed, and some believe it is not from Einstein (Bertram, 2019).* Regardless of who defined insanity in this context, it seems that it strongly applies in the megaproject world. More than one source (and our own observations) have listed numerous challenges. In other words, the various references are giving us similar conclusions and somewhat of a common consensus on what ails megaprojects.

Following are some of the challenges we face. We will expand on the answers and solution in Sections III and IV of Part A. Table 6.1 summarizes all of the challenges in this chapter.

Table 6.1 Challenges from Literature Review

No	Challenge
1	Balancing the triple constraints
2	Clarity of objectives
3	The front end and scope definition
4	Absence of formal project management in the front end
5	Complexity and number of stakeholders
6	Organizational competence and capacity
7	Scheduling
8	Cost reduction
9	The illusion of risk transfer
10	Blame culture
11	Infrastructure unique challenges
12	Other challenges: permitting, remote locations, and turnover of project leadership

6.5.1 Balancing the Triple Constraints

There is a need to balance the critical constraints of time, cost, and quality. The main challenge here is that executive stakeholders are often driven by cost or schedule, but less by scope. Oberlender stresses that, *"the source of many problems associated with a project is failure to properly define the project scope"* (Oberlender, 2014, p. 6).

6.5.2 Clarity of Objectives

Clarity of objectives might sound quite basic, and one might say, "Isn't it obvious that we must have clear objectives?" Sure, it might be obvious, but it is more likely that we take this concept for granted, because the project owner's teams do not give clarity of objectives the right focus, as most of the data and analysis show. Barshop stresses that projects without clear objectives are *"50 percent more likely to have a major change or recycle later in the project life cycle . . . On average, you will spend 10 percent more*

* https://www.quora.com/Did-Einstein-really-define-insanity-as-doing-the-same-thing-over-and-over-again-and-expecting-different-results

capital because of inefficiencies created by unclear objectives." Yet only one in three project teams report that they fully understand the project's objectives (Barshop, 2016, p. 77).

6.5.3 The Front End and Scope Definition

This challenge could be combined with the previous two points, but we see them as interrelated but independent matters. Oberlender makes the point that the lack of scope definition in the early stages of the project is the *"root cause of cost overruns, late completion dates, excessive rework, unnecessary disputes, poor team alignment, and other problems"* (Oberlender, 2014, p. 68).

One of the primary root causes for this is rushing through the front end of the project, leading to poor or less than optimal project scope definition. In other words, organizations want to avoid spending significant money before final approval, which often backfires and results in poor scope definition, which ultimately drives costly and late changes.

Although the front-end development is one of the most critical direct contributors to project success, the knowing–doing gap is alive and well in megaprojects. *"Almost every professional agrees that better, more thorough FEL [front-end loading] means better project results"* (Merrow, 2011, p. 222). However, only about 50% of the projects have their FEL classified as "best" or "good," and the other 50% is "fair," "poor," or "screening" quality (Merrow, 2011)*. These numbers are based on projects that have been benchmarked by IPA. We do not have any information on projects outside of the IPA database, but should we imagine they will be any better?

6.5.4 Absence of Formal Project Management in the Front End

Most references put a great deal of focus on the construction phase of a project rather than the overall management of a project (Oberlender, 2014). The various references, especially those from IPA, and my own experience confirm that there is an absence of formal project management in the early stages of the project. For example, we often observe that many organizations do not have any form of change management until they go into the Implementation Stage, and change management is imposed by the contracts. We have also observed, and the literature review confirms this, that many organizations do not put a great deal of emphasis on project management planning; even when an organization might develop a project management plan, they often do it later in the front end, rather than sooner, and would focus on "execution." Some even call this plan the *project execution plan.*

This could be one of the most critical challenges contributing to the lack of clarity of objectives. An interesting point to make here is that the *PMBOK® Guide,* ISO 21500, and other project management references consider the discovery phase of a project (the first part of the front end) as outside the project life cycle, and project management starts with a project charter.

> The *PMBOK® Guide*, ISO 21500, and other project management references consider the discovery phase of a project as outside the project life cycle.

* These ratings are from IPA and the grade they assign in their project analysis or assessment of the front-end development.

6.5.5 Complexity and Number of Stakeholders

We included complexity in Chapter 5, and we include it here again but with the addition of a compounding factor: stakeholders. Capital projects, even small ones, have too many parties involved—project owners, engineering firms, and construction companies, at a minimum. In other words, the number of stakeholders (individuals and organizations) is significant, which adds layers of complexity. However, in addition to what we consider stakeholders, on megaprojects, organizations might have to do a great deal of extra work in terms of infrastructure for remote areas or to satisfy a problematic stakeholder that could be opposing the project. Also, politicians and environmental groups could be detrimental to project success.

6.5.6 Organizational Competence and Capacity

Organizational competence and lack of proper development of the megaprojects's front-end can have a tremendous effect on megaprojects. In Section III of Part A, we dedicate a whole chapter, Chapter 12, for organizational competence and capacity.

Do not outsource the owner team.

The challenge here relates to what we posted in the previous chapter: the overdependence on contractors and consultants, or outsourcing. However, this is not limited to outsourcing project management, but applies also to the technical expertise by industrial firms. Merrow makes the case that from the 1970s to 2000s many industrial firms had decided to dilute or eliminate their technical capacity—the technical and project management expertise necessary to develop and shape the project and lead it to success.

This challenge could be the most critical, because it could be the root cause of many of the other various challenges we share here. It is interesting that when the organization has adequately staffed projects, *"82 percent of the FEL teams were integrated"* (Merrow, 2011, p. 223), and team integration is a critical enabler in megaprojects. *Capital Projects* stresses the following: *"Do not outsource the owner team"* (Barshop, 2016, p. 101).

In properly staffed projects, 82 percent of the FEL teams were integrated, and team integration is a critical enabler in megaprojects (Merrow, 2011).

6.5.7 Scheduling

Scheduling can be a significant issue on projects, and observations from the data indicate that megaprojects are often schedule driven, even if that was not explicitly stated by management. *"Schedule pressure dooms megaprojects more than any other single factor"* (Merrow, 2011, p. 2).

Further, Merrow makes the argument, *"If the economics of the project requires an accelerated schedule, then the appropriate conclusion is that the project is uneconomic and should not be done. Unlike smaller projects, megaprojects cannot be used to fill in a gap in your production or meet a market window"* (p. 3). In other words, projects owners should be careful not to stress the project schedule more than is realistic or possible.

Speed in megaprojects is essential, but we must know where it is useful and where it is dangerous. The research shows that accelerating the front end is risky because it leads to a lack of proper definition, which causes schedule delays. *"All you do if you hit the ground running is fall, in the case of megaprojects . . . Front-end planning needs to be thorough . . ."* (Flyvbjerg, 2017, p. 10).

Ironically, organizations try to shortcut the schedule in the front end, thinking they would be accelerating the schedule. The reality is that the opposite happens. Here is another stat from a different source: CII, the Construction Industry Institute, claims that for every dollar spent on front-end planning the return is $3–$10 (CII, 2019).

6.5.8 Cost Reduction

What we observe often, and this observation is also backed up by the reviews, is that in an ordinary occurrence—the estimate is higher than expected—management (and politicians) push for cost reduction. This executive push for cost reduction invites dangerous games. More on this in Chapter 12.

This executive push for cost reduction invites dangerous games.

Another related point is cost estimation and understanding uncertainty. However, this was covered in Chapter 5. Further, although underestimating cost and overestimating benefits deceptively (Flyvbjerg, Holm, and Buhl, 2002) could exist in practice, we do not think it is a common practice. However, we cannot deny that deception (and the push to ignore certain risks or lack of proper due diligence) does exist, maybe more often than one might think.

6.5.9 The Illusion of Risk Transfer

The illusion of risk transfer to contractors* is a common mistake and another way of highlighting the risks associated with the fixed price culture, which we addressed earlier. Here are the supporting statements from *Industrial Megaprojects*: *"The firms that engineer and construct industrial projects are variable cost firm has very little* [*sic*] *in the way of fixed assets . . . They earn by selling the services of people rather than via the production and sales of products. This simply means they cannot possibly carry the kinds of losses that can and do occur on megaprojects"* (Merrow, 2011, p. 6–7). As a result, they will do whatever they can to survive, which we touched on earlier.

6.5.10 Blame Culture

In a past life, we used to joke that if a project succeeds, then it must be the business-side ingenuity, but if the project failed, then it is project management incompetence; this was the message inside a project owner's organization. Continuing the cycle of scapegoating, owners' project management teams blame the contractors, and in turn, contractors put it on the back of the subcontractors, which ultimately lands on the lowest part of the hierarchical chain.

Blaming the project managers or project management is a risky business and leads to long-term effects that go beyond a single project. We will also address this later in Chapter 12. One more point: we have

* These are our words, rephrased from Introduction, Page 6.

also observed and noted that one of the reasons behind the use of EPC lump-sum contracting is mostly for *blame* transfer rather than *risk* transfer (Merrow, 2011, p. 278).

6.5.11 Infrastructure Unique Challenges

This final challenge is not related to industrial megaprojects, but it was mentioned in *Industrial Megaprojects*, and since it is relevant to our book, we share here.

Merrow touched on the challenge of infrastructure projects and the approach used to get them funded, which involves the question of ethics. *"They are frequently beset by a phenomenon known as 'buy-in and hook,' in which low costs are promised early, knowing full well that the eventual costs would be much higher. Although this sort of deception is not unknown in private-sector ventures, it is not very common simply because there is usually no taxpayer available to foot the bill later"* (Merrow, 2011, p. 20). This point supports the argument by Flyvbjerg et al. (2002) regarding underestimating and whether that is an "error or lie." However, although it is critical to acknowledge that this practice exists, it is not universal. Unfortunately, some practitioners and organizations take this for granted, which is dangerous.

This view might be shared or originated by others in academia. However, since Merrow is not focused on infrastructure projects, we are not sure if his statement is based on IPA benchmark analysis and statistically proven, or if he is adopting the common view advocated and published by Flyvbjerg et al. (2002), Van Wee and Priemus (2017), and Drummond (2017). The argument is that transportation projects are often underestimated due to "optimism bias," "delusional optimism" (Drummond, 2017), or even "deception." That notion has been questioned by Love and Ahiaga-Dagbui (2018). Once again, countries and organizations must understand the problem but must not assume this is the default scenario. Once again, this situation leads to a call for action to build the organizational project management function, competence, and capacity.

6.5.12 Other challenges

There are many other challenges, especially from Merrow (2011), such as permitting problems (p. 232); remote locations (p. 229); and turnover of project leadership, especially in petroleum projects (p. 242). We will touch on some of these challenges in other parts of this book.

6.6 Chapter Closing

There are numerous challenges as we identified from the various references that we have reviewed. Many of them are common and listed in more than one source, as well as being shared by our observations. Although there are some academic disagreements on specific concepts, all of the references agree we do have challenges and significant problems. The disappointment is that many of these problems have been known for decades, and although some remedies are being put in place, we still face numerous failures; and failures in megaprojects have vital if not catastrophic consequences on companies and even countries.

Further, most of the reviews are of recent origin. Oberlender is one of the oldest, but it was updated in 2014; Merrow is circa 2011; Barshop is 2016. Some papers are also old, but most papers that we have reviewed are 2017 to 2019. This is important to stress because the problems are still with us. Murphy has secured a master bedroom in this house; isn't it time to kick him out?

Chapter 7

A Tale of Two Megaprojects

7.1 Introduction

It is time to shift from literature review to actual case studies. In this chapter, I will focus on two case studies of megaprojects that have many similarities but opposite outcomes. Unfortunately, we do not have all of the operational details to discuss the objectives' success of these projects (see textbox), but we know enough to present various aspects related to their management. There are many lessons learned to understand and share via the comparison between these two megaprojects. We will also learn about the power and value of organizational project management (OPM) and the influence of the project management maturity and culture of the organization leading them. In the next chapter, we will touch on a few other case studies, but not to the same level of detail.

For confidentiality reasons, let us call these projects the Asian Project (AP) and the North American Project (NAP).

> **Project Success**
>
> Earlier in the book we touched on project success, and later we will dedicate a part of Chapter 17 to this topic.
>
> If you recall, we listed four dimensions of project success, which are: *product technical success*, *project management success*, *project delivery success*, and *objectives success* (Ajam, 2017). The last two dimensions cannot be assessed until post-project completion, especially objectives success.
>
> In these two case studies, we do not have information post the projects' facilities completion to assess the third and fourth dimensions. Therefore, the focus here is mostly on project management.

7.2 Common and Differentiating Features

Below we compare these two projects in a table for a side by side comparison. Table 7.1 includes the common or similar features, whereas Table 7.2 presents the differentiating features. The notes are related to the tables.

Table 7.1 Case Studies, Similar Features

	The Asian Project	The North American Project
Industry	Petrochemicals	Petrochemicals
The facilities	New facility with a portion inside an existing refinery[a]	New facility with a portion inside an existing refinery
Projects capital cost	About US$2 billion[b]	About US$2 billion
EPC schedule	33 months	36 months
Sponsoring organizations	Three companies: 50/40/10[c]	Two companies: 60/40
Management	*Note 1*	*Note 2*
Procurement	Global	Global

[a] In both cases, the refinery was owned by one of the partners.

[b] Approximate cost in 2019 value.

[c] The 10% partner was a company partially owned by the 40% partner.

Note 1. The overall venture and the project management team (PMT) was led by the 50% partner. In addition, the venture used an integrated team with members from the 40% partner and one or two managers from the 10% partner.

Note 2. The overall venture and PMT for the greenfield facilities were led by the 60% partner, but this project also had an integrated team with resources from both parties. The PMT for the refinery work was by the 40% partner.

7.3 Performance Results

Before I share the performance results (do not peek), reflect on these two tables and the various conditions of each project. Can you speculate on which one had favorable features and factors that would enable success and which one had more challenging conditions and factors that could lead to failure? One hint: one project was successful and one a failure. Keep in mind, in this case, the success and failure are for one dimension—project management—and not the objectives success, project delivery success, or product technical success.

Ready for the answer?

If you are like the vast majority of my audience when I discuss these projects in conferences or events, you would say the North American project had the favorable contribution factors, and the Asian project had the challenging factors; consequently, NAP was a PM success, and AP failed. A few contrarians would say the reverse, and they would be correct. **NAP failed, and AP succeeded, project management wise.**

Here are some of the numbers.

7.3.1 North American Project

The final capital costs were more than 110% of the approved budget. This number did not include the operational readiness work, because that team was part of a different budget; although we do not have the details, they incurred significant challenges owing to schedule. In term of schedule, the project was at least nine months behind the original schedule of 36 months, so about a 25% delay in terms of duration.

(text continues on page 70)

Table 7.2 Case Studies, Differentiating Features

	The Asian Project	The North American Project
Refinery work	*Note 3*	*Note 3*
Project site	Island in Asia, *Note 4*	Mainland North America, *Note 5*
Dispersed teams, *Note 6*	Main partners headquarters in the United States	Main partners headquarters in North America; but parent organizations in Europe
Estimating	All estimates[a] carried out by the project owner team	Not sure on the early estimate but the final estimate was partially based on the bid price.
Preliminary engineering[b]	Process (Onsite): licensor Offsite: internal project owner (different from EPC)	Same company as the EPC Contractor
Contract, commercial aspects	Reimbursable, cost plus incentives[c]	Fixed price, initially open book[d]
Contract, scope	One contract, EPC	One contract, EPC
Contract, responsibilities	Main contractor: EPCM Construction: joint venture (consortium)[e]	Main contractor: EPCM Construction: general contractor (single company)[f]
Engineering	Europe	North America; a few hours by car from the construction site
Construction	Modular, *Note 7*	Stick built on site; mostly

[a] Conceptual/feasibility estimate; development/planning estimate; and final funding/definitive estimate.

[b] Also known as FEED: front-end engineering design.

[c] Incentives for cost, schedule, and safety.

[d] During the bidding process, the project owner team had the right to see the estimated costs details, but upon contract award the books were closed.

[e] Contract issued from EPC contractor.

[f] Same situation, contract issued from EPC contractor.

Note 3. The refinery work for AP was led by the lead partner (the owner of the refinery), the one leading the overall venture, whereas for NAP, the refinery work was led by the second partner (owner of the refinery) with the right of oversight from the lead partner.

Note 4. Since the project was on an island, where the refinery existed as well, it required reclamation work to extend the island to accommodate the new facilities. However, despite the reclamation, the site was still small, with not enough space for all of the required labor (at peak) or laydown yard.

Note 5. The site was large, with open space to accommodate the necessary labor (at peak) and plenty of room for laydown yard.

Note 6. Basically, the AP had owners in North America, engineering in Europe, construction in Asia, whereas the NAP had owners, engineering, and construction in the same country.

Note 7. Due to the site's limitations—island location and labor availability, among other factors—the project depended heavily on modular construction. Offsite facilities were built on site, but most of the onsite facilities (process units) were modular construction at two major offsite locations in neighboring countries. In other words, three construction sites were working in parallel, on the island and the two significant fabrications yards. More on this below.

The performance affected production because of the nine months' delay, and the number floating around was a production opportunity loss of about $2M per day* of delay (resulting in an overall impact of 270+ days' delay).

7.3.2 Asian Project

The final capital cost of the project was at 98% of the budget—an underrun. However, a disclaimer is needed here. The project was done pre-Euro era. The project was funded in US dollars, but all expenses were in seven different currencies, resulting from the team's being dispersed between three countries and procurement using a few of the leading currencies at that time. It happened that during the project, the US dollar experienced heavy pressures and reduced valuation in comparison to the other currencies, which led to a 12% impact on the overall project cost. This risk was owned by executive management and the management reserve and was not allocated to the project management team.

In terms of schedule, the project was completed about one month behind the original date—technically about a 3% time extension, which is considered success since it is within the parameters. As you recall, this project was performed using a cost-plus incentive contract (incentives for cost, safety, and schedule) with a cap on the amount of the incentive. The contractor earned most of the amount of the incentive plan and was also awarded a "replicate project" in a different country.

7.4 Contributing Factors

What contributed to the above performances? I will list the challenges and opportunities for each project first, then offer our assessment.

7.4.1 NAP Challenges

The following are some of the significant challenges facing the North American Project [in no particular order].

PMT Size and Competence

The project team size was too small for this particular project and lacked the necessary competence.† In other words, the project owner team was not adequately staffed. Merrow had stressed the need for adequate staffing and indicated that a lack of proper staffing has negative consequences on the project outcome (Merrow, 2011).‡ Why was the project not adequately staffed? Here are a few reasons:

* Today's value of money.

† Please note an earlier discussion on Mounir the plumber. Saying this team lacked the necessary competence is not a criticism of the people; we are only stating the fact that most of the team members did not have the necessary level of project management experience, and not a single one had megaprojects experience (except a contract employee hired during the engineering stage).

‡ This was one of the challenges identified in the last chapter.

- Lack of resources and lack of a strong OPM function in the owners' organizations (both partners). The organizations did not have enough project professionals to assign to the project.
- Overconfidence or over-dependence on the fixed-price culture, which is the thinking that most project risks have been transferred to the EPC (engineering–procurement–construction) contractor. This point related to the challenge identified earlier as the "illusion of risk transfer."
- Most of the members of this PMT had engineering (not project management) experience on small projects (not large projects or megaprojects). The largest project handled by either of owners* before this project was about 20% the size, so magnitude of scale was a factor.
- There was no dedicated project control staff; owners did not see the need until trouble started to show, at about 35% engineering progress. At that time, the contractor had begun to mobilize the construction team, but too early.

Early Mobilization of Construction

The EPC contractor mobilized the construction team to the site too early, at which point there was not enough engineering work to be done to sustain the team's level of effort. Lack of enough work affects team productivity and morale, especially when these challenges appear at the start of the work when—in a way, almost upon the start of the project—the project would be behind schedule and burning cash, since the construction contractor had already mobilized. This was one of the factors that led to disputes between EPCM and the construction general contractor.

Project Control

In this context, project control refers to scope, cost, and schedule control, which also includes change management, among other related aspects.

Organizations with good project management maturity view project control as a critical function on projects; it is like the navigator role on an airplane or ship: without proper navigation on a continuous basis, the team would not know where they were or how to get where they need to go.

Lack of Integration

Another one of the challenges related to the "EPCM and C" contractors was the lack of full integration in terms of strategy or execution approach. The construction contractor's progress measurement system and construction approach used weekly work packages, and they would not start working on a work package until all materials were available. This way, on a Monday they start work, by Friday evening they should be done, and quality control will come and verify completion. If the work is not 100% done, the contractor will claim zero progress on that work package; so it is 0 or 100.

The EPCM approach was different, and they did not subscribe to this approach, which may seem acceptable, but was not. The scenario exploded due to shoes; not the shoes you put on your feet but piping shoes. Because process projects have a lot of piping on pipe-racks, they use something called "shoes," which the pipes rest on, and which are necessary for thermal expansion and contraction of the pipeline. The EPCM ordered the wrong shoes, which meant the construction contractor either had to delay the work until the shoes were replaced, or install the pipes and later on come back and lift the pipes so they can put the shoes. This is double work and affects the sequence of work and productivity.

---

* The European parent organizations of the project owners must have done large projects, but not their North American subsidiaries.

There were other issues similar to this, but maybe this one was the cutest because the EPCM dropped the shoes, so the C maybe had to throw the shoes at them. The above might be symptoms, whereas the root cause was lack of proper planning and alignment of the implementation approach.

Schedule

This challenge relates to the previous point, but we separate due to criticality. The contract between the project owner and EPC contractor clearly mandated the need to have ONE integrated schedule between the EPCM team and Construction team. However, in reality, they did not integrate their project planning and control functions, including scheduling and schedule control. Consequently, the construction contractor followed their own schedule, which appeared to align at the high level, but did not. Since the project owner did not have any project control or a schedule expert during the contract negotiation and even after engineering and construction started, the owner's team did not discover this crippling error until the damage was done. After that point, it became a recovery plan with re-baselining every six to eight months.

Not Learning the Lessons

The best way to explain this is with another story. At the time of the project, China was booming, and many European and North American manufacturers were shifting some of their manufacturing operations to China. This project was buying valves from one of these companies. By a convenient coincidence, before placing the order, an alert came from a petrochemicals conference warning that flanged valves from this factory had defects. The project owner team passed the message to the EPC contractor, but the contractor chose to dismiss it: "This is a company with a good reputation," was the contractor's position. The owner felt this was a risk; they asked the contractor to reconsider, but since the contract was fixed price, they could not mandate a change unless they were willing to pay extra. To mitigate the risk, the owner requested extra valves for quality testing (destructive tests).

What happened?

Sure enough, all tested valves were defective, which, it goes without saying, had a significant impact on schedule. Sure, the owners did not pay the extra cost of replacing the valves, but they paid dearly in delays.

Change Management

After contract award, the project owner deleted a complete process unit,* with a cost that is about a few percentage points of the total cost. Many other changes added costs. If one assessed the overall cost impact of changes as a percent of the total, the result was about 5% or 6%, which appeared to be normal or acceptable. However, to assess impact, we do not use mathematical assessment (the addition and deletion canceling each other). The real implications take into account the absolute value of the additions and deletion, which in this case would have resulted in more than 10%. A 10% or more of the initial total cost in changes would have a significant impact on productivity, cost, and schedule, as CII

* For those who are not familiar with process industry, a major petrochemical plants would consist of many process units, typically in a train of units.

have shown in their various best practices. The impact was substantial, even without mentioning the cumulative effects of changes. More on this concept in Chapter 11, which deals with leading practices.

In parallel to the above, the project owner added a full cogeneration unit, but to minimize the impact to the main EPC contractor, it was awarded to another contractor. This action led to another layer of coordination and interface, especially with two contractors working on the same site and possibly getting in each other's way.

The Results

All of the above challenges, especially the ones between the main EPCM contractor and the construction contractor, led to dispute and claims, which ended up escalating to the project owner, becoming a blackmail scenario: "Pay us, or we will have to slow down the project." This is what ended up happening, resulting in a nine months' delay. I cannot help but reflect here on the challenge of the fixed price culture, the illusion of risk transfer, and the competence and capacity of the project owner. Almost all of the challenges mentioned in the earlier chapter were invited by Murphy to visit him in his new house. Initially, Murphy had the master bedroom, and now he was claiming the whole house.

7.4.2 NAP Opportunities

As we highlighted in the tables, the project on the mainland had enough space, the site was only two hours by car from the engineering office, the construction contractor was well known and with a good reputation, and the EPCM contractor was part of a global conglomerate. The project did use 3-D modeling to help in the design work. There were operation representatives involved, but here the team was also small, like the PMT. There were other opportunities and positive factors. However, the dominance of the challenges overshadowed the opportunities.

7.4.3 AP Challenges

For the Asian project, the story is almost a 180° from the above. Here are the significant challenges, without expanding on the details:

- The project was put on hold after the discovery phase (feasibility), because the partners had not finalized their agreements (refer to Shaping in Chapter 19) and felt that the economic conditions and down cycle would not justify the project at that time.
- The front-end definition work (definition stage, preliminary engineering) was in two locations: the project owners' home offices working the definition of the offsite and utilities and a licensor site in a different state/province.
- The engineering and construction were continents apart.
- Despite the island reclamation work, the site was still too small to accommodate traditional, stick-built construction.
- The modular construction was an opportunity that included significant secondary risks, such as shipping and construction sequencing risks, among other complexities. For example, the insurance company insisted on having representatives present every time the team loaded or unloaded one of the enormous modules, and there were close to 200 significant modules.

- The currency fluctuations, which we discussed earlier.
- Lack of trust between the EPC contractor and project owner because the budget was finalized by the owner.
- For the onsite work, about half of the labor resided in an accommodation camp on site, but the other half—management team, technicians, etc.—had an hour, two-way commute every day just from the shore to the construction site (not including the commute from their homes).
- Lack of safety culture among the construction labor, yet the project owner subscribed to strict safety management rules and included a safety incentive.
- There were many other challenges, but what we list here would be considered the significant factors.

7.4.4 AP Opportunities

We will list the opportunities here, and I will also be brief.*

- Around the world, many project owners shy away from cost-plus, reimbursable cost contracts. In West Asia,† where I had been working for the last couple of decades, cost-plus is rare, if it exists at all. Merrow states that fixed-price contracting is almost exclusively used in West Asia on megaprojects (Merrow, 2011). The cost-plus incentive contract was indeed a challenge initially; however, the PMT was able to convert it into an opportunity, allowing the team to build relations and work in an integrated way. The primary catalyst to overcoming this challenge was team maturity and transparency.
- Due to the above, the project owner team and contractor team worked in a highly integrated way, below the senior management level. What we mean here is that each party had a project director, engineering manager, construction manager, and project control manager. However, the second tier—the resident engineers, project control, and other staff—worked closely together in total transparency. What helped was the nature of the cost-plus contract, where everything was open. The incentive was another factor: the contractor management promised to pass part of the incentives to their staff.
- Modularization was a significant factor for success, despite the complexity and secondary risks. For starters, instead of cramming thousands of workers onto one site, they were spread onto three locations, and two of the sites were fabrication yards with highly skilled labor and plenty of space. The net result of this action was improved productivity and faster progress. However, a best practice like this does not come easy. The team had to do significant planning, including planning of the installation of the modules, which had to be installed in a certain sequence. The sequence mandated fabrication and delivery of the modules in the right order, owing to lack of laydown space on the island. These requirements kept impacting the work backward, which meant engineering had to progress in the right sequence to support the construction sequence. There were many other actions needed to ensure proper fit-out.
- The project team extensively used best practices, such as constructability reviews, zero incident techniques, team building (internal and external), change management, and front-end planning, among other factors. The owners were members of CII (the Construction Industry Institute) and also utilized the benchmark services of IPA (Independent Project Analysis).

* I put more emphasis on the challenges of the North American project to explain what could go wrong. I also want to put a great deal of emphasis on the good things, to be positive but also to learn. However, Section III is mostly dedicated to what should be done, and Section IV offers the tailored approach. In other words, we are not diluting the focus here.

† What many call Middle East.

- Early end-user involvement. For example, the product quality manager (the one in the operating organization responsible for the petrochemicals products that the plant will sell), the maintenance manager, and other key personnel were in the engineering office from day one of start of engineering.
- Implementing project control and change management even during the front end, before full funding and EPC awards. To enable project control during this phase, the team used the budgetary estimate (the second estimate on the project life cycle). For the post-contract award, the team used a two-step change management process via the use of a change control board. All changes were submitted to the board, and the board either rejected or approved a change, but for assessment only. Once the change assessment was complete, the board made the final recommendation to approve or decline.
- There were numerous other actions, which would take too long to list and discuss.

7.5 Performance Conclusions

I think it is clear now why the Asian Project was a project management success and NAP failed. I will list the vital contribution factors here and close with a conclusion.

- The use of best practices (from CII and IPA). For the AP, the partners were quite familiar with both CII and IPA and capital projects best practices. Further, they actually used many of these practices. On the other hand, the partners of NAP were not as familiar and did not explicitly implement these practices.
- The team size and experience. I had mentioned the team size for the NAP, which was inadequate. On the other hand, the PMT for the AP included many highly skilled project management staff (gathered from many countries where the lead partner operated) and also had adequate numbers. For example, both projects were about the same size, but the PMT for AP was more than three times the size of the NAP team.
- Senior stakeholder involvement: in both projects, there was involvement at the vice president level, but even here there were significant differences.
- Communication and reporting: on one project, the monthly report was an email, on the other project, the report was extensive with highly useful information. I am sure you can guess which report belonged to which organization.
- Flying blind (project control): I touched earlier in a textbox on the role of project control. NAP management did not see the need for this until the damage was done, and at that time, they hired a contract employee (external consultant) to join the team. In contrast, AP had a project control team of three people on the owner side and about 10 on the contractor side, all working together. Also, one of the owner's project control team members was the one leading the estimating effort, so this person knew the project intimately.

I can list many more factors; however, if you analyze them, you would likely come to, or agree with, the conclusion I share below.

7.6 Chapter Closing

As a champion of the need for standardization and project management methodological approach, I hope you do not take what I am about to say here out of context.

Both projects followed a project management method or a project life cycle model with stage gates, which is appropriate to capital projects. The same is true for many of the projects analyzed by IPA, CII, or the various references that we have reviewed. Yet we have seen the different outcomes. In a way, it is like you give a new driver a race car only to have them drive it into a wall. In other words, methods, tools, or cars are not enough to win the race. Putting the race car in the hands of a seasoned and expert driver could win the race. The same thing holds true in project management—project life cycles could look the same, but what is under the hood, and who is driving?

In the project management context, what is under the hood is two things. One is the organizational project management system, which is the governance, policies, methods, procedures, guidelines, and all of the technical components to support the management of the projects and programs. With this organizational project management system (OPMS), we need the organizational knowledge base, competence, and capacity to lead megaprojects. All of these are part of what is under the hood, or maybe the brain of the organizational project management.

All of these could or should lead to a higher level of project management maturity. **Notice I said *could* and *should*; how can we change this to *would* lead?** That is culture! The culture to excel via the proper due diligence, shaping, planning, professional development, competence, leadership, and so on. Culture is the heart of organizational excellence. Or is culture the heart and maturity/process the head?

Project management maturity and organizational culture are the brain, heart, and soul of megaprojects excellence.

Therefore, project management maturity and organizational culture are the brain, heart, and soul of megaprojects excellence. These two factors—starting at the top of organizations and extending throughout the various levels and spread of the organization—are what make the difference. They are our indicators and triggers to ensure that we adequately staff our projects and organize them for success.* Maturity and culture are the results of building the right organizational project management function with the right level of competence and capacity.† They empower organizations to seek the learning and leading practices from organizations like CII and IPA.‡

In the end, as champions of methods and tailored approaches, organizations must build and sustain their organizational project management system and enforce it by building the maturity and the right culture.

* We dedicate a chapter for organizing project for success later.

† There is also a dedicated chapter for this later.

‡ You guessed it, there is a chapter for this as well.

Chapter 8

Other Case Studies

8.1 Introduction

This chapter will touch on a few case studies, each of which focuses on one point or another, rather than a detailed analysis of the project. In some cases, I have to mask the project, because if I mention the type or location, I would be exposing the project.

8.2 The Case Studies

8.2.1 The Cart Before the Horse

This was an airport project.

Once upon a time, there was a major contractor that specialized in roads, highways, and dredging projects—mostly a civil contractor. This contractor saw an opportunity to expand the airport in the capital city in one of the countries in which they were operating. The host country lacked the necessary funds, so the contractor was able to work with a bank (in the contractor's home country) to secure a loan for the project. In this case, the money was acquired for a project without any real scope definition beyond "need to expand the airport." The amount was lower than US$1 billion—in other words, this would have been smaller than a megaproject at the start.

The client (the airport authority, the project owner) hired the construction contractor (the one that secured the loan) and also directly hired a well-known design supervision company specializing in airport design to supervise the technical work. The construction contractor subcontracted engineering work. The owner did not have a "real" project management team, only a few client representatives, mostly for liaison. Therefore, in this situation, the construction contractor secured money to start a project without defined objectives or clear scope—hence the idiom, the cart before the horse.

- **The first problem.** Once the contractor and client started to define the project scope, they realized that the loan amount could not cover the real needs. Therefore, they agreed to split the project into two phases,* and the funds that were supposed to cover the whole project were now limited to phase 1.

- **The second problem.** A few months into phase 1 (which was now the revised project), the project started to fall behind schedule. At that time, the contractor was reporting seven months' delay. My company was asked to perform a quick analysis, and the contractor allowed only a two-day initial assessment on the assumption that there would be follow-up assessments and even hiring supplemental resources to fill the staff gaps that they know they had. On the first day of the evaluation, we discovered that the project was not seven months behind schedule, but more like two years behind. The overall duration would be double the original estimate.

What was happening?

The first challenge was the contractor's competence. Although this was an established international contractor, it was a civil and dredging contractor that had never managed facilities or airport projects, which are highly complex. This project had an added complexity, which was expanding an operating airport. As a result of this situation, the contractor's management team was significantly short-handed, and many team members were not qualified (competence and capacity challenges). The project director was also the human resource manager (wearing more than one hat) and did not have the necessary experience. They had a planning manager part-time, who was unavailable for us to meet. They had a "planning engineer," and it turned out he was a fresh graduate, one year out of college, who learned Primavera in a self-funded class and had limited experience with the tool. In other words; this young engineer knew how to use the tool but was not an experienced scheduler and definitely not a planner.

As an example of the bad situation, the project schedule (engineering and construction schedule) that should have been detailed and developed, based on resource loading and effort required, was developed using top-down estimating techniques—a technique that might be good for concept work but is not suitable for detailed work.

The design supervision contractor realized there were many problems but did not have a mandate for project management and could not interfere. They also did not have any project management staff on the project. The project owner (airport authority) were nowhere to be seen during the assessment; the contractor was supposed to arrange a meeting but did not.

The bottom line: as soon as the project started, it was split into two phases (projects) because the money was not enough for more than half of the work. Then with significant delays and other problems, it was clear that that amount would not even be enough for phase 1. This was a situation in which the original budget was going to triple and quadruple, assuming no other significant challenges. The worst part was that, despite our warnings and assessment report, apparently the contractor and project owner still had no idea how troubled the project was. We are not even sure if they implemented any of our recommendations.

In conclusion, this was a situation in which a project was sanctioned based on an artificial estimate and a low cost promised. This case definitely fit what Flyvbjerg et al. (2002) discussed in terms of estimating "errors or lies."

8.2.2 Nice Guys and Leadership

As a nice guy, I had second thoughts about the title that I chose for this case study and lesson learned. I do not mean to imply that nice guys cannot lead a megaproject. On the contrary, a good leader is, by default, a nice guy (smiley face). Let me get into the story, and you will know the intent of this title. The project was another petrochemical project that would be short a billion dollars in today's value of money.

* Technically, each phase would be a separate project and funded separately.

- **The first challenge.** The complexity of this project was that it had three partners. Further, it included new facilities (greenfield), some modifications inside an existing adjacent refinery (brownfield), and modifications to a megaproject that was still under construction (should we call this part bluefield?). In other words, technically, the project had three different physical areas of work.
- **The second challenge.** In this case, the company that performed the front-end engineering (and the estimate via the bid) was also awarded the EPC on a fixed-price basis. In his book *Industrial Megaprojects,* that was one of the issues or problems highlighted by Merrow: using the FEED* contractor to do the EPC work (Merrow, 2011). Merrow warned against using this practice.
- **The third challenge.** In addition to the previous two complications, the staffing on this project was also an issue (similar to the NAP that we addressed in Chapter 7). A difficulty here was that the project owner team was integrated with personnel from the three partners, but the project director and lead were assigned to the largest-share partner. In other words, the challenge was resources, competence, and capacity.

Here comes the nice guy, the project director (PD).† It was clear that the project director was handling a project that was probably two to three levels above his abilities. The team was not highly experienced either. The PD was a nice guy, but this was not the problem; the problem was that he was a weak manager, which was compounded by lack of experience. The team would meet and agree on something, then the PD would meet with the contractor and agree to something contrary to what the owner (integrated) team agreed on. The owner team would agree to reject a change (with the PD's consent), then the PD would meet with the contractor and approve the change.

Remember, the team included personnel from all three partners. The team recognized the problem but was paralyzed to solve it. The personnel from the two partners' companies (not the lead company) did not want to escalate the problem, because they felt it was an issue for the lead partner to solve. The employees from the leading partner did not escalate the problem to avoid upsetting their boss. Very quickly, the team became dysfunctional. Finally, a contract employee on the team escalated the matter to the executive sponsor, and the issue was opened for assessment, which led to demoting the PD. He was given a choice to stay on the project, go back to the company home office, or take on other work; he decided to stay because he realized his weakness, which he attributed to lack of experience on projects of that size and complexity.

In closing, there is nothing wrong with nice guys; this was a story of a dysfunctional team with a manager who did not have the necessary skills to lead a megaproject.

8.2.3 Do We Need to Operate This Facility?

This is a generic lesson and not a specific case study; it is a pattern in many case studies. This lesson is about many projects and from different domains, including industrial and buildings. The situation is related to the production stream of CAMMP™ MP, in which the actual operation‡ was delayed as a result of lack of proper operational readiness. Some of these projects were high-profile and strategic megaprojects, on country and regional levels.

The challenge here was that the project owner organizations, which would typically be leading the operational readiness activity, were not aligned well with the facilities people (engineering and construction work). Also, the lack of experience might have led to underestimating the work involved to get ready

* FEED, front-end engineering design.

† His actual title was project manager.

‡ In this context operation symbolizes the start of production, occupancy of a building, or opening the "facility" for customers/users.

to close the facilities work and start operations. This scenario is common in organizations that have not done something at the scale or complexity of the projects we're discussing, and did not have the necessary people to handle it (again competence and capacity).

Back to the problem: as a result, for these projects, the facilities would be complete while operational readiness activities were still ongoing or just started. This led to a delay in commissioning and startup and, of course, the ultimate delay of the whole megaproject.

Let me first address what I think is the root cause, or at least one of the root causes. Most of our observations from working with clients and most of the literature on project management do not discuss operational readiness. They have the "pure project management" thinking that we touched on earlier, which is, operational work is not project management work. It is the same thinking that excludes the discovery phase work from the project life cycle as "not led by the project manager." Our views and our CAMMP™ model take a different perspective, which is, our job is to take a project from concept to success. This would encompass the discovery work (concept, feasibility) and the operational readiness work. We also stress the need for project management in these stages, even though the work could be led by a different person than the one managing the engineering and construction of the facility (more on this in Part B, the simulation).

8.2.4 Optimism Bias

This is a similar situation to the previous challenge, where what I am discussing here has been happening on many projects. What we present is from real projects; we just cannot name them or describe them because they are highly unique.

Before the 2008 global financial crisis, there was a boom for capital projects in many worldwide locations, including the region where I had been working, West Asia. It may have appeared that the global crisis caused many megaprojects to be put on hold or canceled altogether. Although the crisis could have been a trigger or a contributing factor, the real issue, the threat, was *optimism bias.* The term optimism bias was used by more than one paper in *The Oxford Handbook of Megaproject Management,* but in this case, it is slightly different; maybe it is *ego* bias.

Let me try to explain. In this country, a project was planned that grabbed media attention at a global level. Instead of waiting for the project to be complete and determine success and failure, the media attention encouraged a similar but bigger project, then another one even much bigger than the second one. Other projects were planned, and many of them were megaprojects and attention magnets. It was like, "build it, and they will come," except in this scenario, it was "build as much and as big as you can, and they will come."

This extreme optimism backfired in a mega way. Billions of dollars were sunk into canceled projects or projects that have been on hold for more than decade and counting. I think most of what contributed to this problem relates to the four sublimes that drive megaproject development (Flyvbjerg, 2017). The four sublimes are technological, political, economic, and aesthetic; an alternative and more blunt term would be the *ego sublime.*

8.2.5 Other Case Studies

There are many other studies to mention here. However, at the request of a reviewer, Dr. Mihail Sadeanu, we are providing a link to a Swiss tunneling project that has received an excellence award.* Further, we

* https://www.iapm.net/en/news-events/iapm-blog/!/show/173/the-gotthard-base-tunnel-project-wins-the-project-excellence-award-/

hear about good projects and successful megaprojects, but we often do not find the case studies. I guess the Heathrow expansion project was highly praised despite the chaos in the luggage handling system upon going into operation.

Further, this is a good place to remind you of the Sydney Opera House. Many projects could be like the Sydney Opera House and receive great reviews and recognition, but project management wise, we might not have the full story. For example, Burj Khalifa in Dubai and Metro Dubai. These are projects that have global or regional legacies, but we do not know the inside story.

8.3 Chapter Closing

In this chapter, we share challenges related to a few cases studies, where we have masked the projects. There are many more case studies and stories that we could share; however, as you will have noticed, many of the stories or causes repeat, confirming that we are not learning. *"Those who do not learn from history are doomed to repeat it,"* a term coined from Professor George Santayana (Clairmont, 2013).

We also shared that following, or the illusion of following, a project life cycle or stage gate process is not enough. We must go deep, build the organizational project management function, which includes the OPM system and organizational knowledge base. Further, a robust organizational project management function must embed the right culture.

Let me close this chapter with a Table 8.1, which summarizes the challenges from the case studies discussed in this chapter and the last one.

Table 8.1 The Challenges from Case Studies

No.	Challenge
1	Project management team size
2	Early mobilization of construction
3	Lack of integration
4	Schedule
5	Not learning the lessons
6	Change management
7	The cart before the horse
8	Team leadership (Dysfunctional team)
9	Operational readiness challenges
10	Optimism, extreme bias

Chapter 9

Section II Summary and Closing

9.1 Introduction

It is time to close Section II, which focused on the challenges and the current state of practice of megaprojects. We must stress that what we shared does not apply on all megaprojects, because, as we have seen, there are successes. However, as we have found and shared already, from observations, case studies, and the literature reviews, the failure and challenges are significantly greater than the successes. This is not a statement to provoke or overstate the failure for attention; it is a reality that we must accept as the first step in searching for the solution. In more than one instance so far, we did imply the solution, which is also the title of the book, a tailored approach that builds on what exists and fills the gaps; remember—standing on the shoulder of giants. There is not much more to say on this topic; what is left is to group the challenges from the various chapters of this section.

9.2 Summary of the Challenges

The observed challenges were summarized in Table 5.1 (page 51). Table 6.1 (page 61) summarizes the challenges from the literature review. Finally, Table 8.1 (page 81) listed the challenges of case studies. Now, Table 9.1 (below) will combine the challenges and group similar challenges. Although we list these challenges as independent challenges, or groups of challenges, each one of them is not totally autonomous. They are part of a system, and often one of them influences or magnify another.

Here we list them. In Chapter 10, we will revisit them and start to link them to the solution. Section III will focus on the critical concepts for megaproject leadership, whereas Section IV will offer the tailored solution, and we will close the book with Part B, which is the simulation.

Table 9.1 The Documented Challenges, Grouped

Group	Challenge
Infrastructure	Politics in infrastructure
	Infrastructure unique challenges
Complexity	The complexity of megaprojects
	Complexity and number of stakeholders
Procurement Practices	Culture of fixed price
	Culture of a low bid
Project Owner Competence and Capacity	Overdependence on contractors and management consultants
	Project owners' competence
	Organizational competence and capacity
	Absence of formal project management in the front end
	Project management team size
Risk and Change Management	Change management
	Lack of appreciation of uncertainty
	The illusion of risk transfer
Planning and Control	Balancing the triple constraints
	Scheduling
	Cost reduction
	Early mobilization of construction
Vision and Objectives	Clarity of objectives
	The front end and scope definition
	Optimism, extreme bias
People Aspects	Blame culture
	Lack of integration
	Team leadership (dysfunctional team)
	Turnover of project leadership
Others	Permitting
	Remote locations
	The cart before the horse
	Not learning the lessons
	Operational readiness challenges

9.3 Closing Comments

If you reflect on this last table, we believe you see a pattern and can conclude that most of these challenges are the one controlled or influenced by the project owner. This is not to say there are no challenges in engineering, construction, or other technical areas of the project. We are not focusing on these challenges for many reasons. The following include some repetitions from earlier in this book.

1. We believe the megaproject success, and failure is driven by the owner. Hence the focus is on project owners; and with an added emphasis on the pre-implementation stages.
2. Challenges and problems in the Implementation Stage, are often driven by issues and deficiencies in the front end. For example, lack of clarity of objective, not enough time in front end development, no proper project management in the early stages, among other factors.
3. More than one reference has shown that most of the cost growth, delays, or production problems are due to the front end; refer to Chapter 6.

We are not ignoring the challenges encountered in implementation. Many of the challenges listed above, change and risk management, project control, leadership, are for the whole project life cycle. However, the focus is on the project owner role and needs for leadership. In conclusion, if you want to get a more in-depth understanding of the implementation (execution) challenges, please refer to Oberlender and Merrow books, and undoubtedly many other references could be specific to engineering or construction.

Section III
Critical Concepts for Megaproject Leadership

Chapter 10

The Critical Concepts, Overview

10.1 Introduction to the Section

We stated in an earlier chapter that methods, procedure, stage-gate process—in other words, standards, processes, and systems—are not enough for success. Many organizations follow some form of process or another, although some organizations do it better than others. If we have the same ingredients, then why are some organizations (or teams) successful, and others are not? Here we reflect on the engines of project management.

> To drive successful project management, we need the seamless integration of people, process, and tools.

To drive successful project management, we need the seamless integration of people, process, and tools (see Figure 10.1). In this context, the process is to encompass the organizational project management system and standardizations; tools would be software or related efficiency enablers, but the people part, although it might be self-explanatory, might not be. Having good people is necessary, but what is more important is to have qualified and competent people who know how to use the tools effectively and follow a mindset of continuously improving the processes. Looking at the engines differently, we believe that the processes and tools components can help an organization go from 0 project management maturity to 2 or 3, on a scale of 5. To go beyond maturity level 3, organizations need the differentiation factors; they need the people engine; they need the right culture to drive excellence.

Please pause and take a moment to study Figure 10.2, repeated from Chapter 2. Notice the differentiating elements—the circles around the core—which include leadership, professional development, knowledge management, and organizational learning. These are the differentiating elements in building and sustaining the organizational project management function through a robust system.

The bottom line is this: we must have an integrated solution that would encompass people, process, and tools. We need to build the process, enable the people, and select the appropriate tools. In other words, **build the organizational project management function, maturity, and culture**. Maybe I should

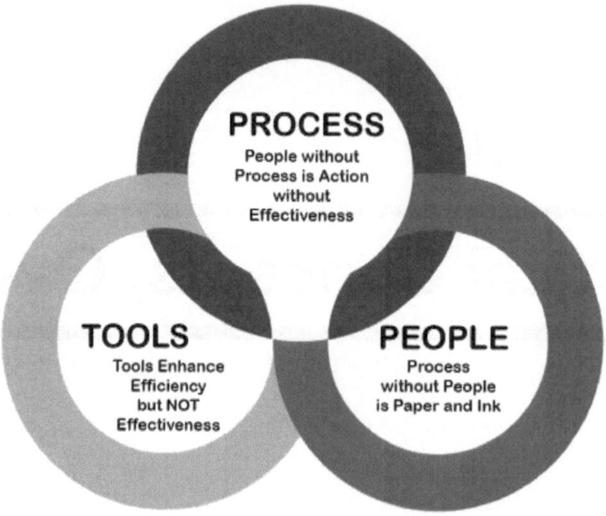

Figure 10.1 The Interactions Between People, Process, and Tools

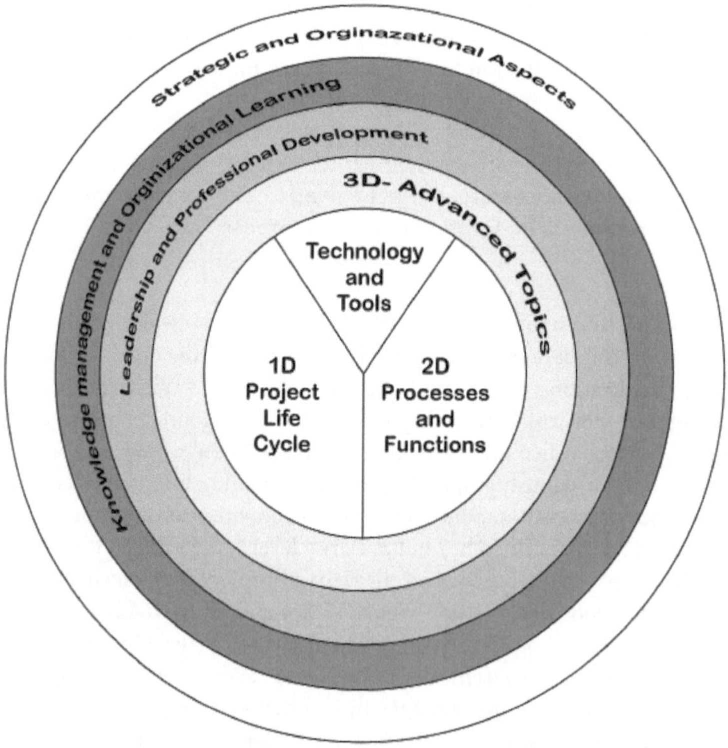

Figure 10.2 The Seven Elements of Project Management Maturity™

end this part with Merrow's reflection on failure in megaprojects: *". . . these results are not inherent like the activities. They are instead, caused by human decisions, ignorance, and uncontrolled, but controllable, human failings. These projects can be fixed"* (Merrow, 2011, p. 19).

10.2 Challenges to Opportunities

In Chapter 9, we listed all of the challenges and grouped them. What we will do next is to go back to Table 9.1 and update it to Table 10.1, in an attempt to link the group of challenges to the rest of the book. You will notice in Table 10.1 the addition of the column on the right. In this column, we define where in this book we will address a given challenge or a group of challenges. The first two sections of Part A focused on setting the scene and the challenges, then Sections III and IV of Part A will focus on the solution. Finally, Part B will use a hypothetical project to simulate the solution in a realistic project setting.

Table 10.1 The Documented Challenges, Grouped, and Where We Address Them in This Book

Group	Challenge	Where We Address This Topic
Infrastructure	Politics in infrastructure	No dedicated chapters for these challenges because we cover them throughout the book. We believe that Section III and IV, especially the tailored approach with phases and stages, allow us to improve the organizational abilities to deal with complexity.
	Infrastructure unique challenges	
Complexity	The complexity of megaprojects	
	Complexity and number of stakeholders	
Procurement Practices	Culture of fixed price	This topic will be part of Chapter 20 in the tailored approach Section IV.
	Culture of a low bid	
Project Owner Competence and Capacity	Overdependence on contractors and management consultants	We will address most of the challenges related to the project owner throughout this book. We also dedicate Chapter 12, Organizational Competence, and Capacity, to this topic.
	Project owners' competence	
	Organizational competence and capacity	
	Absence of formal project management in the front end	
	Project management team size	
Risk and Change Management	Change management	These two groups of challenges are critical. We address them throughout the book, but we will also have dedicated parts in the chapters in Section IV (Tailored Approach). Some of these topics are also part of the Chapter 11, Leading Practices.
	Lack of appreciation of uncertainty	
	The illusion of risk transfer	
Planning and Control	Balancing the triple constraints	
	Scheduling	
	Cost reduction	
	Early mobilization of construction	

(continues on next page)

Table 10.1 The Documented Challenges, Grouped, and Where We Address Them in This Book (cont.)

Group	Challenge	Where We Address This Topic
Vision and Objectives	Clarity of objectives	This is the focus of the discovery and development phases in CAMMP™ in Section IV, Chapter 19.
	The front end and scope definition	
	Optimism, extreme bias	Related to shaping, Section IV, Chapter 19.
People Aspects	Blame culture	We do address many of these challenges in the chapter on organizational competence (Chapter 12), but also in the tailored approach section (Section IV), especially the chapter on the people aspects (Chapter 21).
	Lack of integration	
	Team leadership (dysfunctional team)	
	Turnover of project leadership	
Others	Permitting	These are related to shaping, which is the business theme in the tailored approach (see Section IV).
	Remote locations	
	The cart before the horse	
	Not learning the lessons	Will be addressed in Chapters 19 and 20 in Section IV, Tailored Approach.
	Operational readiness challenges	CAMMP™ has a dedicated stage for operational readiness and another for initial operations; part of the tailored approach in Section IV (also the simulation in Part B).

10.3 Chapter Closing

In closing this chapter, I want to prepare the reader and introduce you to the upcoming chapters in this section. Notice that the title of the section is ***Critical Concepts for Megaproject Leadership***. Therefore, we have dedicated Chapter 11 to leading practices and considerations and Chapter 12 to organizational competence and capacity (building the organizational project management function). In this section, we also believe there is a need to address the question of program management and considering megaprojects (or mega programs) as an integral part of asset management (Chapter 13). We end this section with a closing chapter, Chapter 14, as we transition to the tailored solution section.

Let me close with this piece of information on the engines discussed earlier in this chapter and culture. I came across a timely blog article by the UK Infrastructure and Projects Authority published just two days before writing this chapter (Vickerstaff, 2019). In that article, the author stresses the challenges facing infrastructure projects with delays and cost challenges on some of the high-profile megaprojects in the UK. I will share a couple of quotations from the article:

- *"In order for us to improve future delivery, we need to make sure we are learning the right lessons from projects that have gone before, and we are applying them to projects that come next."*
- *"It will not be possible to simply flick a switch and fix some of these issues, it will require a change in behaviour and culture from all parties, both public and private."*

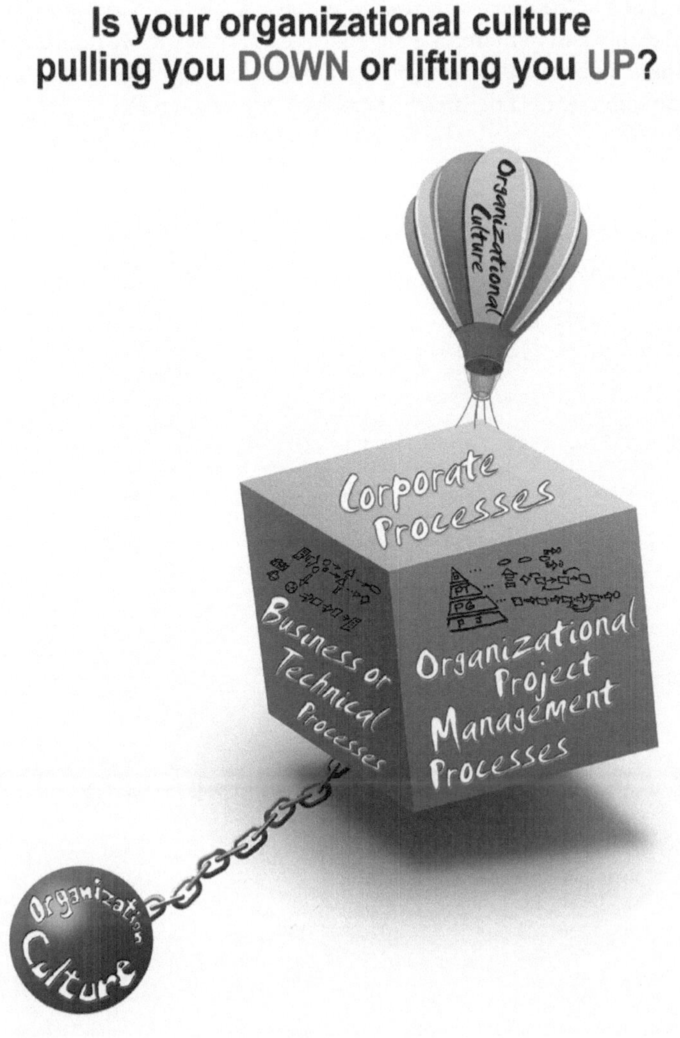

Figure 10.3 Is the Organizational Culture Holding You Down or Lifting You Up?

> **Credit for the Image**
>
> During a workshop in Singapore a few years back, as I was presenting on the importance of culture, my host, friend, and colleague Nah Wee Yang drew a sketch based on a sketch I had on the white board. I had sketched the box you see in the image. Wee Yang added the chain and weight to symbolize that the organizational culture could be holding the organization and project performance down.
>
> Although his illustration reflects a common reality, as a realistic-optimist, I accept reality but I believe in changing it, hence the revised image you see now with the balloon. This book is part of what goes into the balloon; it is the executives' decision to hook their systems to the balloon or the chain.
>
> Reality says, both balloon and chain would be connected—which one will control?

The first lesson stressed is *"Behaviours and culture are more important than process,"* which emphasized what we have stated earlier and reinforce the message to carry forward: we need a process, the tailored approach, we need the organizational systems; however, the underlying enabler is culture—organizational culture, which we can represent via the attached graphical representation.

Chapter 11

Leading Practices and Considerations

11.1 Introduction

I need to start this chapter by explaining what we mean by leading practices. We use the term to refer to what CII (Construction Industry Institute) labels as *best practices.* I believe IPA (Independent Project Analysis) and others call them *value improving practices.* We like the term *leading practices* because these are the practices that enable organizations to elevate their projects' performance beyond what could be achieved through good or common practices.* In other words, we view good or common practices as the general practices that are commonly used in the industry and should lead to good results. However, leading practices require significantly increased effort and discipline, which should lead to higher performance. This is why a reference such as the *PMBOK® Guide* offers 49 processes and hundreds of tools and techniques of *good* practice, whereas CII offers only 17 specific *best* practices.

11.2 Why Do We Need Leading Practices?

I think it was clear from the previous paragraph that best (or leading) practices would be specific practices that would lead to higher performance.

Do they?

Yes, absolutely.

* Many practitioners and training companies that want to promote certifications, such as PMI's PMP® (Project Management Professional®), refer to the *PMBOK® Guide* as the PMI Best Practices. Technically that is an error, because the *PMBOK® Guide* clearly states it is about generally accepted and common practices, not best practices.

As we discuss these leading practices, you will notice that CII has actually quantified the value of these practices versus their absence. For example, CII claims that every dollar spent on front-end planning results in $3 to $10 return (CII, 2019). Barshop (2016) also stated that following a good stage-gate process (implied here including the IPA value-improving practices) would result in an increase of the net present value (NPV) of 5% over what was predicted at project launch, while not following a proper process could result in up to 45% reduction in the NPV value.

For capital projects and megaprojects, CII and IPA are leading organizations in term of references. However, what they offer is not readily available to the average practitioner, because most of what they offer is restricted to their organizational members or those who buy their services. It is good to note that both offer training programs that are open to the public.*

11.3 Construction Industry Institute

As we have mentioned, CII offers a set of best practices for capital projects. These practices are the result of years of research and analysis of real projects from CII member organizations, which include the Who's Who of capital projects from both sides of the fence—project owners and contractors. CII is also part of the University of Texas in Austin, a top-10 engineering university in the United States and among the top globally. In addition to its publications and conferences, CII has a rich knowledge base that covers various aspects of capital projects.

11.3.1 Some of the CII Claims

From a quick review of the CII website,† one will notice the following performance claims from CII membership (CII, 2019):

- Benefits of the materials management program result in a 16% schedule improvement. Materials management is one of the core problems in the construction phase of the project.
- Project planning is critical, and every $1 spent on the Project Definition Rating Index (PDRI) process can save $25 in return. We mentioned PDRI in Chapter 3, and will also touch on it again below.
- Here is something for our techies: adoption of proven technology can enhance construction productivity by 30–45%, among other benefits.
- We already mentioned the $3–$10 payback on every dollar spent in front-end planning.
- The CII Knowledge Base can add value to change management on the order of 11.4% cost growth improvement.

Again, the above are some of the headlines claimed by CII. However, these are not unsubstantiated claims; they are backed by significant research and analysis of real projects. I have also stressed in Chapter 7, "A Tale of Two Megaprojects," that the successful project used the CII practices extensively.

---

* I stress again that I am not involved with either organization, and I do not have any interest or the potential of a conflict of interest, directly or indirectly in CII or IPA.

† https://www.construction-institute.org/

11.3.2 What Are Some of the CII Best Practices?

I will not be able to talk about all of the practices or expand on the details. I will only list a few, especially some for which I do have direct knowledge and experience. The headlines are from CII, but the text and stories are mine.

- **Partnering.** I had the opportunity to work for three years in a partnering arrangement in which my employer, a project owner organization, had partnered with a leading contractor organization to carry out all engineering, procurement, and construction (EPC) for all projects below $50 million (1990s value of money). This was a four-year contract. The first contract with a leading contractor did not go as well as the organization hoped and did not renew the contract. However, the project owner still believed in the benefits of partnering and entered into another partnering relationship with another leading contractor. This was on a reimbursable cost basis, with the budget set by the owner with input from the contractor. I joined the project owner team about two years into this contract to establish a project control unit within the project management office.* Most of the projects were considered small—$2 to $50 million. The portfolio office was responsible for the EPC work but worked closely with the teams handling the front end. When I joined the team, the relationship was rocky, if not chaotic; trust was absent, and many projects were experiencing cost and schedule challenges. However, we were able to work through the challenges, with painful decisions and actions at times. In the end, the owner decided to renew the contract for another four years. In other words, partnering can be challenging and painful and require a great deal of focus, transparency, patience, and integrated teams to earn mutual trust and get over the storming phase.
- **Modularization.** I think the case study that I shared earlier, the AP case in Chapter 7, clearly showed the benefit of modularization. It is interesting that at the time that project was implementing this practice, modularization was not yet listed by CII as a best practice.
- **Constructability.** This is another practice used extensively on the Asia Project. In the old days, constructability started with the question, "Can we build it?" about a given design. However, later, the question was expanded to include, "If yes, then how can we build it efficiently?" To answer these questions, one of the things organizations do to fulfill this practice is to include construction experts in the engineering design, and even in the front end.
- **Change Management.** One of the CII studies, Publication 43-2 (CII, 1994), presents the Quantitative Effects of Project Changes and clearly shows how engineering productivity and construction productivity decline rapidly with an increasing number

> **A Fun Story**
>
> As a young engineer, working as cost engineer at that time, I was privileged to work with the consultant who was conducting a constructability study for our project (I was on the owner team). The study included visiting the region where the facilities were going to be built, identifying fabrication yards for modularization, labor and crafts sources, etc.
>
> At the end of site assessment, the consultant invited us for a drink and dinner afterword. At the hotel bar, our seating was low, on floor cushions. As the attendant was serving us the drinks, the whole tray with all of the drinks came down on me; showering me with the drinks, and shattered glass all around. (After numerous trips to the dry cleaner, I still could not clean that jacket.)
>
> The good news, as a young single man at that time, I got a date out of this incident—a reward since I claimed it was my fault and insisted that the hotel not fire the young attendant; nice guy.

* Technically this was a portfolio management office leading all projects in an area in which the owner had five petrochemical plants.

and impact of changes. The effect is not limited to the sum of the individual changes; there is also a hidden cumulative impact. In capital projects, it is also noted that the later in the project life cycle the change takes place, the more the significantly higher costs of the change and late changes could be the number one cause of failure. This is why capital projects do not like change (message to my Agile friends).

- **Organizing for success.** We will present our views on how to organize a project for success in Chapter 21, in Section IV.
- **PDRI (Project Definition Rating Index).** PDRI is a way to measure the level of development in the front end and to use the result as a part of the gates decisions. This could be an intermediate gate or the gate that is used to approve the funds for the project in order to proceed into the delivery phase (execution). The following is from *Project Management for Engineering and Construction* (Oberlender, 2014); the PRDI consists of 17 categories, 70 elements, and up to 1000 points. A score of below 200 is good. *"Based on an analysis of 40 projects, the research team found that projects scoring less than 200, out of 1000 total points, were significantly more successful than those that scored greater than 200."* p. 142
- **Others.** There are many other best practices, such as lessons learned, which we will cover in the closing stage of the tailored method, Chapter 19; planning for a startup would be part of the Operational Readiness Stage (Chapter 19); risk management, quality management, and team building are also topics we consider. Other best practices, such as alignment, is the focus of one of the stage gates in CAMMP™. In other words, the SUKAD CAMMP™ Model, which is the basis for the tailored approach for leading megaprojects, has incorporated many of these practices.

11.4 Independent Project Analysis

The following are not labeled as best practices or leading practices; however, we do view them as leading practices.

11.4.1 Front-End Loading (FEL)

The term *front-end loading* is what IPA uses to refer to the project work pre-final approval and sanctioning the project to go into the execution phase (delivery phase in CAMMP™). The front-end loading process (or front-end planning, as CII calls it) consists of more than one step (or phase), each leading to a stage gate. IPA calls these FEL 1, FEL 2, and FEL 3.* FEL 3 ends with the detailed scope definition, definitive estimate, and project final approval to move on to the execution work.

"I believe there is now a virtual consensus among professionals within the community of industries we serve that FEL is the single most important predictive indicator of project success." (Merrow, 2011).

* When we introduce the tailored approach, we will map these FEL phases to the CAMMP™ stages.

"I believe there is now a virtual consensus among professionals within the community of industries we serve that FEL is the single most important predictive indicator of project success" (Merrow, 2011). It is also interesting to listen to Merrow (2011) again, *"In more than 300 megaprojects, only a few had project directors that considered FEL not very important. All were inexperienced project directors, which is unusual for megaprojects, and all were disastrous failures."*

11.4.2 FEL Index

FEL consists of phases, as we discussed earlier, and the FEL Index is the scoring mechanism used to measure the level of development, in a similar fashion to CII's PRDI. It was initially published in 1981. However, even in the 1960s Exxon Research and Engineering had a similar work process.* That process is now the standard within the oil, chemicals, pharmaceutical, and mineral industries (Merrow, 2011).

In *Industrial Megaprojects*, Merrow clarified and showed via various images (starting on page 216) that FEL is a direct indicator of success and failure. If FEL work is best-practical, then on average the project will underrun by a few percent. If the FEL is "poor," then expect as high as 60% overrun (Merrow, 2011). The FEL-to-cost growth relationship is similar for mega and other capital projects, but the curve for megaprojects is twice as steep as the curve for projects that are less than $500 million. This point reinforces a statement made earlier in the book, which is that organizations are not giving megaprojects the attention they deserve.

In addition to cost, Merrow also shows the FEL-to-schedule relationship, which is similar: "best practical" achieves their target schedule; poor rating shows schedule slip by up to 30% (meaning 13 months' delay for the average project). If you recall, the North American Project from the case study was nine months behind. Although that project did not have a FEL assessment, we would guess it was poor, based on our observations.

IPA assesses cost and schedule in addition to production performance. Here again, Merrow (2011) shows a strong correlation between the FEL Index and production performance.

11.4.3 Integrated Teams

Another strong message from *Industrial Megaprojects*—and we do concur based on experience from more than one project—is that integrated teams† can have a significant positive impact on projects. The performance is better with integrated teams versus not integrated, since the integration affects the clarity of objectives. For example, *"Only 22 percent of the projects that lacked integrated teams were successful, versus 51 percent of projects with integrated teams"* (Merrow, 2011). It is evident from this number that integrated teams are not a magic wand—they do not guarantee success, but they do make a significant difference.

---

* I started my career in project management (after a few years in engineering and after my master's degree) with Exxon Chemical, where I spent eight years and learned the process well. We used it on all types of projects from small and multiple projects to major and megaprojects.

† In this context, *integrated teams* refers to project owners' teams that incorporate business, technical, and project management resources within the team. Another form of integration would be when the owner's project management team may integrate fully or partially with the contractor team. If you recall, the Asian Project had partial integration among owner and contractor staff.

11.4.4 Capital Projects Rules

In his book *Capital Projects* (not limited to megaprojects), Barshop shared 12 essential rules. These rules were addressed to the executives responsible for capital projects. They include a recommendation for using the stage gate process and keeping the stage gates strong, the need to ensure project sponsor involvement, develop clear objectives, invest in the owner teams and provide them with the support they need, use a robust risk management process, and (the last advice to the executive) "be coachable" (Barshop, 2016).

11.5 Other Leading Practices

What I will present next is based on lessons from the trenches—the real world. Some of what I share could be embedded in the CII or IPA practices, but we want to emphasize them.

11.5.1 Intermittent Project Reviews

Barshop (2016) emphasized the need to use a stage gate process and to keep the gates strong. In other words, there is a need for strict control of the stage gates. What does that mean, and how can we make it happen?

Let me explain further. First, remember that each stage gate (pre-final authorization) has two purposes: The first one is to ensure that the work was completed following the organizational policies, guidelines, and requirements. Validating achievement of this purpose opens the door for the second purpose, which is to give the go-ahead to continue into the next phase. *Hold that thought.*

Another thought to hold is this. When we discussed the constructability of CII, we mentioned that one practice that helps achieve good constructability is to get construction experts involved in engineering, if not earlier.

Building on these two thoughts, we can present our proposal. *Remember, the core objective here is to strengthen the stage gates,* and the ultimate aim is elevating the front-end work to the appropriate level—what IPA calls "best practical."

So, how can we do that?

Well, the idea is not to close a stage unless the team (or management) has a high degree of confidence that there are no gaps or hidden black holes. To do this, let us look at the stage gate again and consider the alternative question: are we ready to move ahead? Who can answer this question best? Experts from the next stage. In other words, have experts from a given stage involved in the preceding stage or even two stages earlier. That is, do not depend on management alone or the steering committee decision to clear you through the stage gates unless experts from the succeeding stage are willing to accept the work of the preceding stage.

Have experts from a given stage involved in the preceding stage.

To clarify further, consider this sequence of stages for now;

preliminary engineering → detailed engineering → construction → startup

Then late in the preliminary engineering stage, and in preparation for the stage gate review, have a team of experts (could be from the same organization but from outside the project) review the progress of the stage work. This team could include a project manager, project control, and an engineering expert (or two). The team can perform a review similar to an audit, which allows them to gain comfort that the work is progressing per the company guidelines, and it should be ready to move to detailed engineering. If there are gaps, they must be fixed before proceeding.

Then repeat the process during engineering, with a similar review team, but instead of an engineering expert have a construction expert.

Once more, during construction (late in the stage work) repeat the process but with operation and maintenance experts in the review team.

The above practice might seem exhausting, but it should not. In our view, it is a must on all large, significant projects and megaprojects.

11.5.2 Sustainability

I do not want to expand on sustainability much here. I have a full chapter on this topic in *Project Management beyond Waterfall and Agile* (Ajam, 2017), but why not go to the source, Green Project Management? GPM® Global publishes a guide on sustainability, and it is free to download. The way CAMMP™ incorporates sustainability is to treat it as an added layer on the tailored approach. In other words, use the GPM model and embed its steps in the various stages of CAMMP™, as you will see when we introduce the tailored approach.

11.6 Chapter Closing

In closing this chapter, we want to stress the need to build and sustain the organizational project management function, maturity, and culture. We also want to emphasize the value of the work of CII and IPA. Their work, or derivatives of it, must be included in the organizational project management systems of organizations leading megaprojects.

Chapter 12

Organizational Competence and Capacity

12.1 Introduction

Throughout this book, I have been talking about organizational competence, capacity, maturity, and culture. Further, decades of practice in the industry, the case studies, and the literature review all confirm the need for organizations to build their **organizational project management system (OPMS)**. Some organizations establish project management offices (PMOs), thinking that is a solution, but a large percent of PMOs fail—and fail within a short time—which is a clear indicator that PMOs are not the answer (see textbox). The answer, in our view, is to build and sustain the **organizational project management function (OPMF)—the internal organizational function**. In other words, to build the organizational project management competence and capacity, which would enable organizations to lead capital and megaprojects, they need to build and sustain the OPMF. Notice the terms, *build* and *sustain,* in contrast to build and ignore, or build and forget.

> To enable organizations to lead capital and megaprojects, they need to build and sustain the organizational project management function.

> **A Few Words About PMOs**
>
> To be fair, not all PMO are failures.
>
> However, even if a PMO is not failing, that does not mean much, because the PMO mandate could be limited to reporting, support, or nothing more than another organizational layer without adding significant value.
>
> We have seen too many variations on the concept of the PMO with the same results. Agile PMO, Next Generation PMO, Value Realization Office, Vision Realization Office, International PMO, and the various other names that are in the market today—the PMO failure rate continues to be high, and the project failure rate is not something to envy.
>
> Let me put it this way: I have not seen or read any extensive research that demonstrates organizations with PMOs deliver projects with more successes than organizations without a PMO.

12.2 What Is the Organizational Project Management Function?

I need to explain what I mean by the term *organizational project management function.*

Organizations have organizational functions, such as information technology, human resources, finance, marketing, engineering, and other formal functions, hence the common use of the term *functional organization.* The number and type of functions are factors of the organization's type of work and its business and operations. These organizational functions are typically structured within formal sections, divisions, or departments.* However, in many, or the majority, of organizations, project management is not yet a formal organizational function. That means there is no project management division, section, or department.

Even project-related positions might not have a formal job description or grade in the human resource system. However, in recent years, we have started to see positions that carry the word "project" in the role, but in the past, even the titles Project Manager or Project Coordinator were not official positions in companies. In other words, organizations, represented via their executives, do not see project management as an independent function, equal in importance to HR or IT. This despite the fact that nowadays, all of these functions work on projects. We have marketing projects, we have software projects, we have finance projects, and maybe even project management certifications, but no formality to the role or the function.

Why?

To be direct, project management is still not respected enough by executives.

Why?

To give these organizations and their executives the benefit of the doubt, project management is a young domain; although we might have reached a mid-life crisis point, some organizations are still discovering the value of project management.

And?

Noting that many executives could be coming from finance or MBA programs, it is time to reflect on Nieto-Rodriguez's 2014 study on leading MBA programs around the world, in which he found only two out of the top 100 leading programs had project management as a core module. In other words, it is likely that these executives did not study project management, and probably never worked in a project manager or director role. Some of them might think that project management is nothing more than common sense; and consequently, the project manager role could be filled with "accidental" project managers.†

I know the above is a current reality. However, to be fair, some organizations working on capital and megaprojects are likely to have project management departments or divisions, so what I said above would not apply to them. However, even for this class of projects, some of these organizations do not have formal organizational project management function or even systems—no OPMF and no OPMS. If they do, a business unit for project management could be weak or limited to a few members who could have big titles, but in reality, are a liaison between the business and the contractors—more of clients' representatives. I have met professionals with less than five years' experience as project managers for a megaproject. I have worked with "clients' reps" that do not know how to read a schedule, understand earned value management or even a project life cycle. More than once, these situations are even complained about

---

* The names of these organizational "units" may vary from one organization to another.

† http://blog.sukad.com/do-executives-understand-the-risks-of-the-accidental-project-managers/

in workshops when we discuss planning, scheduling, or earned value management, since they manage through reports and PowerPoints.

The bottom line—organizations have a choice:

- Continue to ride the cycle of doom, outsource, over-depend on contractors and consultants, go through the illusion of risk transfers through fixed-price and low-bid contracts, and work on preparing their blame messages. Blame project management, blame the project management consultants (PMCs), blame the contractors or the subcontractors until we have no one else to blame except Murphy.*
- The other choice is to climb out of the cycle of doom and build and sustain the OPMF. In our view, the OPMF is built on a robust OPMS, with a transparent organizational knowledge base (OKB) and with a significant focus on the people aspects—maturity, behaviors, and culture—as we shared in an earlier chapter.

12.3 Why Do Organizations Need the OPMF?

It is clear from the literature review and research that the failure to properly develop a project before reaching full funding and execution (delivery phase in CAMMP™) is costing industrial firms and their shareholders billions if not trillions of dollars in shareholder value. Although this situation has been the current reality for a long time, we do not see any real or focused effort to make a change. What we have seen, at least in West Asia, is an increase in dependence on contractors, management consultants, and generic low-value certifications. As we mentioned earlier, even the globally leading professional associations are almost absent from this scene.

What is not helping is that the executives of many of these organizations (even the industrial firms) are business/finance focused or driven. Merrow confirms what Nieto-Rodriguez found (shared above), which is, *"I believe that those charged with the education of business professionals . . . bear a good deal of the responsibility for the failures of business professionals in industrial firms"* (Merrow, 2011, p. 251).

Here is more fuel to this fire: Merrow (2011) also found that the projects that used project management consultants (PMCs) were considerably more likely to fail than those that did not. For example, PMC-led projects experienced average cost growth of 40% and 38%, which would make them uncompetitive† in comparison to their peers in the industry.

Here is the conundrum: if an organization does not have an OPMF or the competence and capacity to manage projects, they have no choice but to use PMCs or depend on the general contractor. Maybe so, but there are other options. More on this shortly.

The core message here is the question: continue on a cycle of doom or seek a recovery? Make the answer your strategic decision and imperative.

---

* Just a few minutes ago, I had a friend share with me that the project manager and team were sacked on a billion-dollar outsourced project. I guess they needed scapegoats.

† *Uncompetitive* is the term used in benchmark assessments, in which projects are compared to similar projects in the same industry by competitors. An uncompetitive project is a project that costs more or takes longer than industry average.

12.4 How to Transition Out of the Over-Dependence

Merrow (2011) recommends the following if a project owner must use a PMC:

- Never allow the PMC to take any portion of the execution.
- Integrate the PMC with the owner team and owner personnel, taking key leadership positions like sub-project managers.
- Look for the skills, not the brand, even from second-tier contractors.

Let me add to the above and offer a transition or a transformational model.

1. First, make the strategic decision and accept that project management is imperative for success. Without this decision, you will fall into the knowing–doing gap.
2. In a continuation of the above decision, treat project management as a function, equal if not more important than other organizational functions. This means establish a division (organizational unit) for project delivery and project management, if you do not have one. Then, assign this unit an executive; the head of this division should be an executive in the C-suite, a Chief Projects Officer (CPO), for example.
3. If you do not have time to build the organizational project management function before the next big project, then hire a PMC on the next project but embed within the PMC a few of your own people who have the interest and passion of project management. If you do not have them in the organization, hire them. Hire the right people with a winning attitude and the willingness to build the OPMF. Stay away from PMOs, unless the PMO will be the OPMF and will build and sustain the OPMS and OKB.
4. Depending on the size of the project and the team, you might be able to have only 10% or 20% of the overall team with your staff. It is essential not to take this for granted and expect the PMC to train your people; training their replacement is a conflict of interest to them. So, you must insist on the PMC giving your people access to all critical information, and make sure this is in the contract, with corrective actions clauses in case the PMC fails to deliver. Also, make sure you hold your people accountable, and that means they have to get in the trenches and leave the master–slave mentality at home or in the past. In other words, plan the assimilation and manage it carefully.
5. On the next project, increase the percent or your staff to 50% or more.
6. By the third project, your teams should be ready to take over. If you spot shortages, fill them through supplemental resources or from second-tier contractors, as Merrow suggests.

In parallel to working projects (one, two, or three projects), you should have other project management senior professionals working with the CPO to build the generic OPMF and OPMS. Building the OPMS will be covered later in this chapter. Start working to reverse the cycle of doom.

> **A Few More Words About PMOs**
>
> I might sound silly here, and that is OK, it is my annoyance speaking.
>
> I do believe in the psychology of words and their effects on us, consciously or subliminally. When I hear the word "office," I think of a room with a manager or a few staff sharing. I do not see this work as a logical, practical, or even ethical alternative to "function."
>
> Yes, I said the word *ethical*, because if we know of a solution that does not have a good reputation despite how many bells and whistles we put on it, it would be an unethical practice.
>
> So, the first problem is why an office and not a business unit?
>
> Second, if the above is only semantics, as some PMO salespeople or PMO managers often tell us, then why is only project management an office and we do not use this term for HR, marketing, sales, IT, or the happiness department?

12.5 A Historical Perspective (Side-Tracking a Bit)

In *Industrial Megaprojects,* Merrow shared a fascinating historical perspective, which I will touch on here. What I am about to say links back to the study on MBA programs that I mentioned earlier (Nieto-Rodriguez, 2014) and to Merrow's 2011 findings.

Merrow found that outsourcing of technical expertise was sold by consultants, business schools, and contractors as a modern and necessary approach to making industrial firms leaner and less subject to cyclical swings (Merrow, 2011). This belief led many industrial firms to start dismantling or diluting their engineering, technical, and organizational project management functions, starting the downward trend, what I have been calling a cycle of doom. *"Business executives who set in motion a process that will destroy huge amounts of shareholder wealth are rarely held accountable for their errors and omissions. In the usual case, not the extraordinary one, the executive that championed a bad project at the outset has been promoted out of any line of responsibility and accountability for the poor result long before the depth of the problem becomes known"* (Merrow, 2011).* These are strong words, but they are a reality and should be viewed as a wakeup call.

How to build an organizational project management function (OPMF).

The organizational project management system (OPMS) is one of the steps.

West Asia Case, Historical Perspective

In West Asia (Middle East), fixed-price contracting is the dominant type (used on almost all projects). Merrow (2011) explains the possible reason for this dominance. In comparison, the IPA database shows 53% of projects, globally, were done through fixed-price contracts.

Why the fixed price dominance?

"A good many of the megaprojects built in the 1970s by Western contractors on reimbursable contracts overran hugely without any ability of the host governments to control."

So clients were burned on cost plus. However, instead of building their project management functions, they shifted to fixed price. In addition to fixed price, they outsourced project management. The consequence of shifting to fixed price is possibly costing these government organizations today up to 40% (Merrow, 2011). The cost of these projects is higher than global averages. In other words, these countries are still paying the price, but maybe not in openly glaring embarassment.

However, my friends, colleagues, clients, especially executives and ministers, those fiascos were about five or six decades ago, why have you not built the project management function and capacity to handle these megaprojects, or even the smaller projects?

12.6 Organizational Project Management Systems

Building the OPMF means ***building the engines of project management, which include building a tailored, fit-for-purpose, OPMS.*** However, before we discuss how to build, don't organizations already have an OPMS? I know I touched on this question before but did not elaborate.

A quick answer is, some do and some, well, read on.

Although many companies today have an OPMS, some are still without a decent process, or *"struggle to improve their capital investments' performance"* (Aschman, 2018).

The previous statement came from an IPA study. The study was to assess the project systems for their consortium members. The study, titled *"Why capital projects systems succeed or fail,"*

* Keep in mind that these megaprojects would typically last more than 6 years and maybe up to 10 years. However, executives' reassignment cycle could be one to three years, supporting the statement from Merrow. I have even observed, at least in one organization, that the senior managers were rotating on an 11–12 months basis.

came up with the following findings (Aschman, 2018). Only about 36% of the IPA members had project systems that would be classified as excellent or in continuous improvement. Another 13% had a system that was in need of either transformation or renewal. The rest, about 51%, had a system that was either mediocre, declining, or poor. Considering that IPA members would be organizations that care about improvements, the status of their systems is questionable.

Why?

Because they either had highly decentralized systems or fragmented systems, *"Centralization does not ensure success, but decentralized is incapable of success"* (Aschman, 2018). The lack of trust in the system put them into a **predictability trap**, inflating their cost estimates and schedule and adding excess contingency. Basically, making the cushion as big as possible could have helped them achieve their targets (and appear successful), but padding estimates led to poor competitiveness (their projects cost more and took longer than the industry average).*

In closing this part, building the OPMS is a must, but if you use a paint-by-numbers approach, you could spend a lot of money and produce something mediocre that ends up collecting dust, in the trash bin, or forgotten altogether (until the next executive decides to repeat the same process).†

I know, I have been putting a lot of content here before explaining the SUKAD approach for building the OPMF. One more part and we will zoom in on our target. Remember, we need the right mindset, which is not possible without knowing the pitfalls—so one more pitfall—a dangerous one, executives!

12.7 Executive and Senior Management

12.7.1 Learn the Fundamentals

Executives and senior management must learn the fundamentals of project management. I repeat, executives, even board members, *must* learn the fundamentals of project management. I am not saying they need to know every process, input, output, tool, etc. But they do need fundamental concepts to learn to understand and appreciate the strategic value of project management. If they do not want to, then they need to learn to respect project management and trust the proven experts—if they had not already made them redundant.

12.7.2 The Responsibility

The responsibility is not limited to the executives, but is a call to action to business schools and engineering schools as well. Business schools can benefit from a dose of practical project management, and engineering schools can benefit from business acumen. Both schools can also learn from practitioners, assuming they do not limit their programs to pure academics (more on this in Chapter 37).

---

* If you recall, the IPA project success/failure criteria include cost growth, cost competitiveness, schedule delays, schedule competitiveness, and production.

† In one organization we know, within an 11-year period they had built and dismantled four project management offices. In one case the "fight" was who the head of the PMO should report to—the COO or Deputy CEO—and the decision was not based on where it was most effective but which person the proposed head of the PMO liked more.

> **Ego or Challenging the Team?**
>
> *"[An executive] set up a daring and ambitious program with [an] inexperienced contractor to deliver the projects in 70% of industry average time at 70% of industry average costs. The result was a program overrun of numerous billions of dollars, and a full four-year delay on the company's largest and most important project."* (Merrow, 2011)
>
> The above was done despite the fact that the project community warned "it cannot be done."

Also, the board of directors needs to consider hiring executives with some project background. In other words, I want those making decisions on billions of dollars of shareholders' money to understand what is right and wrong in leading megaprojects. Boards of directors must enforce the establishment of an OPMF (business unit) and have chief project officers that are highly qualified. Organizations cannot afford accidental CPOs, transferring the accidental project manager syndrome upward.

12.7.3 The Consequences

I will share some examples on the challenge of having the wrong people on top.

It is not uncommon to observe executives wanting their pet projects or pushing a project beyond the norm or good practice. Refer to the box for a story shared in *Industrial Megaprojects* (Merrow, 2011, p. 3).

Another challenge caused by executives who do not understand project management and leading practices of a capital project is the **cost reduction syndrome** (our label) and its cousin, the schedule reduction syndrome. The key is to understand that scope, schedule, and cost are highly integrated—along with quality. Therefore, **asking for a reduction in cost, or schedule, leads to one of two outcomes**:

1. The project management teams, those who aim to please, manipulate the estimate to fit the desired number that an executive wants. In other words, if the scope is not cut, and the estimate was accurate, how can one reduce cost? They do a mathematical game and play with numbers until they fit. The **real numbers** will eventually show up and expose the manipulations, maybe after the team members have updated their CVs, or perhaps even left the project. By the time the damage is done, the executive asking for the reduction would have been promoted out of sight, long before anyone noticed.
2. If cost and schedule reduction is the culture, then the project management team will be proactive and hide additional time and cost in their original estimates, anticipating this song and dance. Team members realize it is a survival game and play it. Here is how it goes:
    - Teams produce inflated cost and schedule.
    - The executives pressure them to lower (reduce, compress, crash).

> **Another Fun Story**
>
> Years ago, I was in Thailand with my wife on vacation. Coming from the US, most guidebooks advised that we should negotiate for everything because all prices are inflated. One time, we got tired of walking and wanted to hire a Tuk Tuk. The driver asked for a price, so we started to negotiate. After a while, my wife and I noticed that our negotiation was about $0.25; a quarter of a US dollar!
>
> The story here is the same in project management; the Tuk Tuk driver inflates the price, knowing that tourists will negotiate, but will finally settle on a price that the driver wants, which may be higher than a fair price or what they would be happy with, depending on how tough the negotiators are.
>
> So, executives, please note that project teams know where to hide money in the cost estimate and time in the schedule. You can squeeze but you cannot get all of it out—unless you build the right culture.

- The team will reduce the numbers (faking a sweat) and become the heroes (and labeled as team players).

However, unbeknownst to the executives, the "agreed to" numbers are likely still higher than what is really needed. Benchmarking could expose this situation, but it is possible the company does not subscribe to this practice. The final outcome? Possibly the cost and schedule actual performance would end up matching the revised numbers. However, these numbers are likely higher than what is truly needed, as we already said. **This is the predictability trap mentioned earlier.** See the textbox (Under Budget but Costly) for a story. This is what is known as Parkinson's law: *"work expands to fill the time available for its completion"* (Parkinson's Law, n.d.). This concept is equally applicable to cost.

Under Budget but Costly

Once, we heard an executive brag about his organization's cost performance. He was saying something like, "We typically underrun our projects by 18%." Obviously, he thinks that is an achievement. He does not realize the danger of such a statement, such as, on average, this company is locking 18% of the capital cost of every project instead of investing it elsewhere.

The story does not end here. Once, this company participated in a benchmarking exercise. The result was that their projects took significantly longer (more than 20%), and cost more (by more than 10%), than the industry average. (I cannot share the source without exposing the company).

The bottom line: this organization appeared or was misled into believing that it had great project performance, but the reality was they were locking significant funds away from other investments, and their capital projects were performing worse than industry average.

Here is the closure: this organization stopped its participation in benchmarking exercises.

12.8 Building and Sustaining the Organizational Project Management Function

12.8.1 Overview

Building and sustaining the OPMF will require building the OPMS (the process), selecting the right tools and technology, and, most importantly, to sustain it, you need the people with the right cultural mindset at all organizational levels. This is not easy and cannot be done overnight or in a year or two.

SUKAD has developed a model, The Seven Elements of Project Management Maturity™, to help organizations build and sustain their OPMS. The SUKAD approach is to help organizations achieve higher maturity gradually, as a three–five years' program, with the use of many small and focused projects. This would be an agile, incremental approach in which we analyze-design-build-test-rollout-improve cycles, for each increment or components. The overall system consists of seven elements, as shown in Figure 12.1 and explained below. Each of the elements could have many components, and each component could be considered a micro or small project. Most of this content and the image come from *Project Management beyond Waterfall and Agile* (Ajam, 2017). However, we will keep the focus on the higher level, the elements, and will not go into the components.

Notice that the name of the model contains *project management maturity,* because the whole idea of the OPMS is that it will help to increase the level of maturity of the project management practice in organizations. Increasing the level of maturity has been proven to improve the cost and schedule performance and predictability (Ibbs and Reginato, 2002).

The model has seven elements, which include three core or fundamental elements (the center of the image), three differentiating elements (the three inside circles around the core), and the strategic aspects, which is the outer ring.

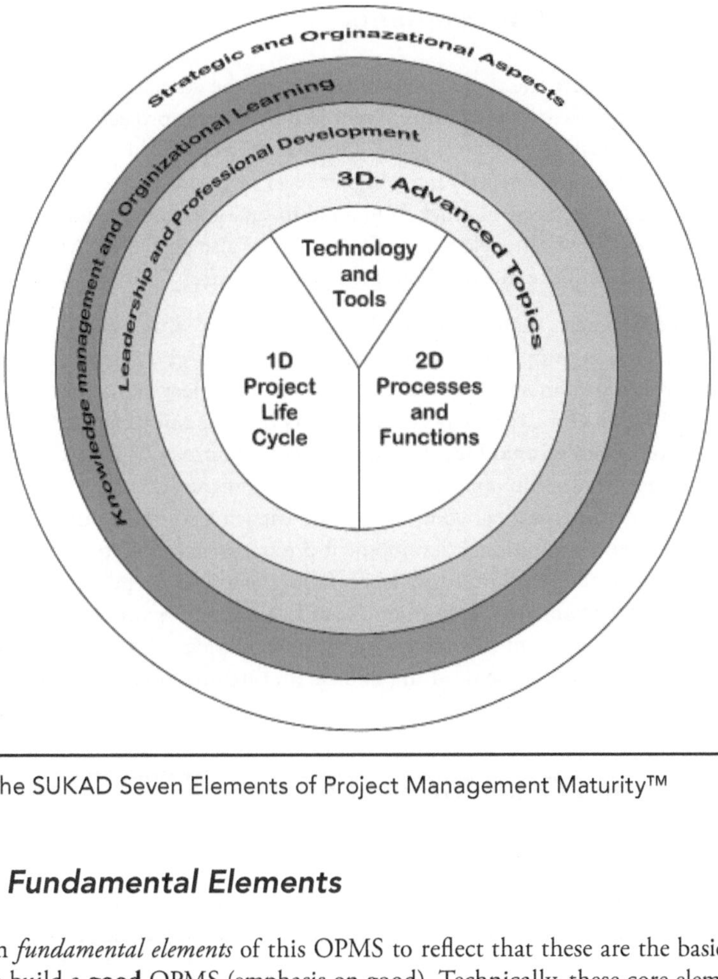

Figure 12.1 The SUKAD Seven Elements of Project Management Maturity™

12.8.2 The Fundamental Elements

We use the term *fundamental elements* of this OPMS to reflect that these are the basics, the essential core elements to build a **good** OPMS (emphasis on good). Technically, these core elements are part of CAMMP™.* We will expand on the CAMMP™-related elements in Section IV on the tailored approach, Chapter 19.

1. **1D, Project Life Cycle.** This element refers to the first dimension of the CAMMP™ Model, a project life cycle, stage-gate process.
2. **2D, Processes and Functions.** This is the second dimension, with the processes and functions as modified by CAMMP™. These would be the processes necessary to manage a stage of the project.
3. **Tools and Technology.** This is not unique to CAMMP™; instead, it covers any software or specific tools that might exist or be internally developed. This book does not address the tools and technology element.†

* CAMMP™ is a three-dimensional model that includes the core elements and the first ring surrounding the core.

† At the time of this writing, we are working with a group of colleagues and professionals from our network to establish a technology startup to develop an online platform for managing projects, which would be based on CAMMP™. We plan to have different versions, including one for large-complex projects. This is what we are calling the Uruk PPM Platform; Uruk will be a division of SUKAD Corp.

12.8.3 The Differentiating Elements

The differentiating elements are the three circles surrounding the core. The term *fundamental elements* emphasizes that these elements are the minimum required to manage projects and achieve good results. Remember, earlier we emphasized the term **good**. On the other hand, the term *differentiating elements* represents the elements that are essential for a higher level of performance and project management maturity. In other words, if we need to move from good to very good or a center of excellence, we cannot do it without these differentiating elements. The first one is also part of CAMMP™, but as advanced topics. The other two elements are mostly about people and culture.

4. **3D, Advanced Topics.** These topics are listed as the third dimension of CAMMP™ and deal with subjects such as competence, sustainability, best practices, and project success, which the author believes to be a must for advancing the field of project management. Although we call these advanced topics for a typical project, we think they are a must for capital and megaprojects.
5. **Leadership and Professional Development.** There are numerous resources and books on project management, leadership, and professional development. What we want to present here is that to build the organizational competence and capacity, we must build the leadership structure and formalize a professional development program, including project management career ladders, mentoring and coaching, and other leading practices for talent development.
6. **Knowledge Management and Organizational Learning.** Organizations cannot increase their level of project management maturity without organizational learning and managing knowledge. This is why we stress the need to establish an Organizational Knowledge Base (OKB) as the other side of the OPMS coin.

12.8.4 Strategic and Organizational Aspects

The **strategic element** is the outer circle.

7. **Strategic and Organizational Aspects.** This element is about the proper organizational aspects, such as project governance, portfolio management, and strategic planning.

12.8.5 Governance

Governance is part of The Seven Elements of Project Management Maturity™, the outer circle, but we want to address it separately here, because of its critical value for megaprojects. For megaprojects, governance starts with the board of directors (BOD) that has an appreciation and understanding of the strategic value of project management and that megaprojects should not be taken lightly or driven by inexperienced executives, project management-wise.

Maybe we should start with Merrow's direct words: *"To be blunt when it comes to the governance of large projects, most boards strike me as brain-dead. They are not asking the right questions, and they are not asking questions early enough in the process to deter bad decisions"* (Merrow, 2011, p. 13). This is one of the reasons we emphasized the need for the BOD to learn the fundamentals of project management. If they do not know, then we concur with what Merrow just said.*

* We have seen too many thirty-somethings on the boards of many companies—not only one. Assuming age is not an issue here (in terms of lack of experience) and the person is indeed competent, how can he or she balance 10 watermelons? I do wonder.

> **Three Flaws**
>
> *"There are three flaws in the dilution of the technical and project management functions in industrial firms.*
>
> 1. *Contractors cannot do owners' work (and vice versa); there is too much difference of focus. Basically, owner project management is what ensures obtaining the right input from R&D, from operations and maintenance to incorporate into the front-end loading.*
> 2. *Contractors cannot hold specialty resources; they cannot afford to, being people-driven (not asset and production driven) businesses.*
> 3. *Owners could not renew the expertise because as their experts age and leave or retire, no one is left behind."* (Merrow, 2011)

Earlier, I shared the findings by Merrow (2011) on the shift where the Organization for Economic Co-operation and Development (OECD) industrial firms downplayed, diluted, or eliminated their technical functions. The board of directors allowed that to happen as a result of lack of understanding or appreciation of those functions. We need to reverse this trend.

12.8.6 Executive Development

One of the differentiating element in SUKAD's Seven Elements of Project Management Maturity™ model is about leadership and professional development. Professional development should not be limited to staff but covers the whole organization, horizontally and vertically, all the way to the board of directors (obviously to different degrees and at different levels).

Technical professionals working on projects need to have a broad understanding of the project management process, especially as it relates to their roles. Cost engineers need in-depth knowledge of estimating and control, in addition to the general awareness of the whole process. Project managers need a great deal of development that could grow in terms of abilities and empowerment.

However, even senior managers, project sponsors, and executives need to understand project management. They need to know what it takes to build the culture of project management and be role models. They need to understand uncertainty and risk, and that fixed price might be good, but not all of the time. Executives and boards are the keys to enabling the culture to lift performance up or let their ignorance of project management create the culture that holds people and performance down.

Let me close by going back to the excellent reference: *"I am convinced that most business executives and project teams do not fully understand what they are signing up for when they start a megaproject. . . . It is very unlikely that they will have been specifically trained to deal with the peculiar problems posed by megaprojects, even if they are seasoned and well-trained business and project professionals"* (Merrow, 2011, p. 53).

12.9 Chapter Closing

I will keep these closing comments short, with a call for action.

- Do not build the PMO, build the OPMF; there is a big difference.*
- To build the OPMF, organizations need a robust, tailored, fit-for-purpose OPMS and its sidekick, the OKB.
- To sustain the OPMS and OKB, organizations must embed the right culture, starting with the board of directors.

* http://blog.sukad.com/rethink-the-project-management-function/

Chapter 13

Mega: Projects, Programs, Assets

13.1 Introduction

How do we view (or classify), manage, or lead megaprojects? Some practitioners think that a megaproject is a program or a big program. So are megaprojects projects or programs? Whether a project or program, how does *mega* relate to the term *asset*? We must understand how to classify these change initiatives. Notice, we called them "change initiatives" because, in the real world, every project or program is about delivering a change to the status quo. The creation of something from nothing is also a change. In this context, we are talking about organizational or physical change and not a project change, such as a change in the scope of a given project.

13.2 Project Versus Program

Based on our observations in the professional community, we often see confusion between what is a project and what is a program. One might view a sub-project as a project. There is a view that considers that a megaproject consists of many projects (Davies, Dodgson, and Gann, 2017), but that is the definition of a program. This drives back the question, does that mean that megaprojects are programs? Maybe, maybe not.

Part of the confusion, even by academics, is the view that projects with multiple contracts are programs, because the perspective is that each contract must be a project. I believe this might be the case in the paper by Davies, Dodgson, and Gann (2017), which is part of the *Oxford Handbook of Megaproject Management.* Sure, every contract could be a project—but for the contractor, not the project owner. In other words, just because we have many contracts does not make every contractual relation a project, and the collection would not be a program.

How to differentiate?

Let us say we have *X*, and *X* has much smaller *x*'s: *x*1, *x*2, *x*3, and *xn*. In this case, is *X* a project or a program? To answer this question, I offer my acid test.

For the product of *X* (the final output) to be operational, do we need to have all of the smaller *x*'s complete to allow going into beneficial service? Or, is there no need to have them all finished before beneficial service? If all of the smaller *x*'s have to be completed for Big *X* to go into operations, then Big *X* is a project, and the various small *x*'s are sub-projects. If each of the smaller *x*'s could go into useful operation on its own and produce value, then Big *X* is a program and the small *x*'s are projects.

Per the above, if all of the pieces are dependent on each other and we need to finish them all to have beneficial production, then we have a project with subprojects. In my view, this is the key differentiator.

The above might be a bit too academic for my taste, but it clarifies the definition and the use of these terms. Is that differentiation necessary in the real world? I would say, absolutely. On the surface, it may seem that managing a megaproject as a program with many projects is the same thing as managing a megaproject with sub-projects, but it is not. In the program environment, we can delay or speed up some of the projects without affecting the completion date and reaching beneficial operations. This is a reality because we can go live one project at a time, with the product of each project commissioned and started up, leading to production. However, in a project environment, each of the sub-projects must be completed or the whole project is delayed. Sure, we can test and pre-commission, and even commission the sub-projects (systems such as the power system, the sewer system, etc.), but we cannot start to produce the desired product until all are done! Therefore, in a project with sub-projects, we must have a highly integrated management and product delivery approach.

. . . in a project environment, each of the sub-projects must be completed or the whole project is delayed.

In conclusion, it is necessary to differentiate, because programs could give us more flexibility than projects. Which leads to a question, a point to ponder: can we use incremental delivery for megaprojects? More on this in Chapter 37.

The bottom line: If a megaproject is indeed a program, then manage it as a program and balance the work among the various projects. If it is a project, it will require a great deal of attention to deliver the entire product as needed.

13.3 The Asset Question

Regardless of the answer to the above and whether a megaproject is a project or a program, in both cases, the ultimate output—the product—is a production facility (or system, if not a facility) that would be considered an asset. Typically, financial people are concerned with this topic because an asset will go into the financial statement of the project owner. However, our interest here is not financing and "keeping the books," our interest is in delivering a productive asset.

Today, many practitioners and researchers look at the project from the perspective of project management or benefits realization, but are we measuring project success correctly? In Chapter 2, we briefly stated that the SUKAD success model is The Four Dimensions of Project Success™, consisting of a product's *technical success*, *project management success*, *project delivery success*, and *objectives success* (Ajam, 2017). We believe that most of the research and literature help us answer the first three dimensions with a degree of confidence, but is that enough?

Can we truly judge the success of a project from the first three dimensions only?

In some cases, maybe, but to have a clear understanding, one would need the fourth dimension. That fourth dimension is related to the success of the project objectives—in other words, the realization of the anticipated benefits. In a way, this dimension is about the success of the asset. One way to measure this final dimension is via the BCR (benefit–cost ratio), the NPV (net present value), and similar indicators. When organizations launch projects, they have the target BCR and NPV determined, which helps executives justify and approve the projects. However, do we have access to the right data or organizational information to calculate the actual BCR and NPV, post project completion? In other words, can we determine the success of the asset without having real production information and asset performance?

I will close with this statement: to truly assess the success of these change initiatives, we need all dimensions—most specifically, the fourth dimension. The fourth success dimension is a critical dimension to evaluate asset success. This point also relates to the question (in our last chapter) of whether we are under-reporting or over-reporting success or failure. If you recall, one of the academics' ongoing debates is how organizations approve projects, and do they underestimate the cost and over-report the benefits, where the end results could be overruns and underperformance (Flyvbjerg, Holm, and Buhl, 2002)?

13.4 Asset Management

If we agree that projects and programs are change initiatives, and capital and megaprojects deliver facilities that would be assets, then should we consider projects an integral part of asset management? If we answer yes, then how do we integrate?

Well, technically, a megaproject is likely approved to engineer and build a facility, which would be an asset to the organization and would show in their financial statements.

OK, but it's still not clear about the relevance to project management.

To answer this question, we must reflect back on the product life cycle. Think about this: is the product life cycle not equivalent to an asset life cycle? If you agree, then in building the facility, the product, the asset, we can limit our thinking to delivering the product within the cost and schedule parameters, ending up delivering a facility that does not work to the desired capacity or has too many "bugs" for the various things that could happen. Or we can build a facility for the lowest cost possible and end up paying or bankrupting a company to operate and maintain. So, we can put the blinders on and think project management is to deliver a product, end of story. Our success is delivering on time and on budget (or ahead and below, respectively). However, in mature and well-established organizations, this is not, or might not, be the best approach. The best approach is to manage the asset.

What does it mean, manage the asset?

It means that from the concept and selection of the solution, think strategically, think along with the full product life. In all of the other stages of the project life cycle, we must consider the life of the asset, the product life cycle. There are many reviews that state that the team can perform in the early stages of the project life cycle, such as value engineering, or value management, and sustainability practices. The design must take into consideration the cost to operate and maintain the facilities.

Before we close, reflect back on the cost reduction syndrome that we raised earlier. One of the consequences of that syndrome, if we cannot cut scope, is that we sacrifice quality. So a management decision to reduce the CAPEX (capital expenditures) could affect the quality of the product, which would likely impact the production quality or capacity. In other words, we lose on the revenues. Another situation is that quality challenges could increase the OPEX (operating expenses).

I think the point is clear: project management must be integrated into the holistic asset management approach.

13.5 Chapter Closing

Is it possible to use a program approach rather than a project approach? In other words, establish a program with multiple parallel, overlapping, or sequential projects? Maybe, maybe not. Does it matter? Yes, as we explained earlier. However, the most important message from this chapter is to repeat what I just wrote a minute ago—**project management must be integrated into the holistic asset management approach**.

> Project management must be integrated into the holistic asset management approach.

Chapter 14

Section III Closure

14.1 Introduction

In this section, we discussed many of the leading practices that would work well for capital projects and megaprojects. We also discussed the need for building organizational competence and capacity via building and sustaining the organizational project management function. Then we discussed the link of projects and programs to asset management. All of the above was about a given organization, public or private. Can we not elevate this concept to a national level? What I am about to propose applies to government projects—in particular, infrastructure projects.

My proposal is something like what the UK has created with the Infrastructure Project Authority, but what we propose is broader in scope. We suggest that governments (could be national for small countries, or statewide for large countries) establish a National Project Management Authority (NPMA). What would the mandate and value of the NPMA be? We have written a detailed white paper on this (Ajam, 2018), which we cannot practically fit here, so we will summarize it.

14.2 National Project Management Authority

The following is from the executive summary for the NPMA white paper (Ajam, 2018).

"… Is there a need for a national initiative for project management? Is there a value in standardizing processes and systems across all government agencies, ministries, and authorities? Does standardization mean imposing one-size-fits-all principles for all projects? Would generic training and certification programs lead to outstanding, best-in-class performance, or even enhanced organizational performance?"

The white paper provides answers to the above questions and outlines the SUKAD proposal for embracing project management as a national strategic initiative for all government projects. This pioneering initiative would enable and empower government organizations and their people to reach best-in-class performance in delivering successful projects consistently, contributing to national growth and advancement while reducing costs and increasing efficacy.

Today, project management applies to all aspects of life, and it touches on all sectors and all organizational levels, yet it is still not highly formalized and often is not based on leading practices that are tailored to government projects. It is important to stress that general project management knowledge, learning, and certifications do not deliver outstanding performance unless they are linked to an organizational project management systems (OPMS) or a national project management system (NPMS), which would require building a sustainable, comprehensive, tailored, and consistent approach for managing projects, programs, and portfolios. **Which government will take the lead and be a project management pioneer?** We will wait and see.

14.3 What Would the NPMA Do?

1. **Build it.** As we mentioned in the previous paragraph, the NPMA would build the national project management system (NPMS), with the tailored methods for the various types of projects. For example, what we propose in the book, or a modified version of what we propose, could be used as the tailored method for megaprojects. Our other book, *Project Management beyond Waterfall and Agile* (Ajam, 2017), can be used as a generic starting point for other projects.
2. **Roll it out.** In addition to building the system, it goes without saying that it would be rolled it out to the various government entities.
3. **Knowledge and learning.** The roll-out would be supported by a professional development program and maybe tailored certifications. The NPMA could also have an applied project management academy or partner with universities for such programs, with a focus on applied, practical, project-based learning. In line with this, a national knowledge base, inclusive of project files, performance information, lesson learned, and other relevant information, would be incorporated. We are in the era of big data, are we not? So this should not be too difficult to accomplish.
4. **Accountability.** The next part of the mandate would be to function as an authority, with the power to review, audit, and assess the various entities' use of the system and recommend corrective actions where necessary.
5. **Lead.** This could be an option: the NPMA would have an implementation arm capable of leading all strategic projects, such as megaprojects.
6. **Empower.** Many of the references stressed the challenges with infrastructure projects and covering topics such as optimism bias (Drummond, 2017) and underestimating costs of public works (Flyvbjerg, Holm, and Buhl, 2002). Even the study on Australian projects (Terrill and Danks, 2016) touches on the concept that politicians, who might not know anything about project management, could be worse than executives who lack that knowledge, because some politicians might be willing to manipulate the numbers for electoral purposes. Therefore, if governments want to avoid nightmarish projects, then they would need to empower the NPMA to overrule or stand up to politicians. Any decisions made by politicians would have to be based on the numbers, schedule, and estimated benefits provided by the NPMA. This mandate puts the responsibility on the NPMA to get their act together and earn trust, which they cannot do without the things that we have been talking about here: build the competence, implement best practices, consider the critical success factors and the tailored methods (covered in more detail in the next section).

14.4 Closing comments

In this section, we covered the critical concepts that we believe are necessary for the leadership of megaprojects. It is time for the next section, the tailored approach for megaprojects leadership, in which we also focus on the critical success factors (CSF) for building the tailored approach.

Section IV
The Tailored Approach for Megaproject Development

Chapter 15

Introduction to the Tailored Approach

[This is a transition chapter]

15.1 A Look Back

As we start a new section, let us take a minute to look back at what we have presented so far.

In Section I, we worked on setting the scene necessary for this book. We defined many of the common terms that we have been using throughout the book and listed the references that we have considered to support our work.

Section II presented the challenges and the state of the current practice of megaprojects management. In that section, we covered the challenges from our observations, our literature review, and case studies. The detailed case studies were specific to the petrochemical industry, but the other case studies covered more than one sector. It is interesting that the literature review included capital projects in general (*Capital Projects,* Barshop, 2016) and *Industrial Megaprojects* (Merrow 2011), which focused on megaprojects. Besides, the academic papers and reports that we have reviewed, especially from the *Oxford Handbook of Megaproject Management* (Oxford University, 2017), focused on infrastructure projects.

> Organizations must build and sustain the organizational project management function.

In Section III, we shifted the discussion to the critical concepts for megaproject leadership. These are the general concepts such as leading practices, organizational competence and capacity, and asset management. The core message in that section is that organizations must build and sustain the organizational project management function (OPMF) through the organizational project management system (OPMS), organizational knowledge base (OKB), and embedding the right behaviors and project management culture.

This will be a good point to reflect on the list of challenges (see Table 15.1, copied here for your convenience). Mark what we covered in the previous sections. Then keep it handy to mark what we will cover as we move ahead.

Table 15.1 The Documented Challenges, Grouped, and Where We Address Them in This Book

Group	Challenge	Where Do We Address This Topic?
Infrastructure	Politics in infrastructure	No dedicated chapters for these challenges since we cover them throughout the book. We believe that Section III and IV, especially the tailored approach with phases and stages, allow us to improve the organizational abilities to deal with complexity.
	Infrastructure unique challenges	
Complexity	The complexity of megaprojects	
	Complexity and number of stakeholders	
Procurement practices	Culture of fixed price	This topic will be part of a chapter in the tailored approach section.
	Culture of a low bid	
Project owner competence and capacity	Overdependence on contractors and management consultants	We will address most of the challenges related to the project owner throughout this book. We also dedicate Chapter 12 to this topic, Organizational Competence, and Capacity.
	Project owners' competence	
	Organizational competence and capacity	
	Absence of formal project management in the front end	
	Project management team size	
Risk and change management	Change management	These two groups of challenges are critical. We address them throughout the book, but we will also have dedicated parts in the chapters in Section IV (Tailored Approach). Some of these topics are also part of the Leading Practices Chapter.
	Lack of appreciation of uncertainty	
	The illusion of risk transfer	
Planning and control	Balancing the triple constraints	
	Scheduling	
	Cost reduction	
	Early mobilization of construction	
Vision and objectives	Clarity of objectives	This is the focus of the discovery and development phases in CAMMP™, and shaping; tailored approach section.
	The front end and scope definition	
	Optimism, extreme bias	Related to shaping, tailored approach section.
People aspects	Blame culture	We do address many of these challenges in the chapter on organizational competence, but also in the tailored approach section, especially the chapter on the people aspects.
	Lack of integration	
	Team leadership (dysfunctional team)	
	Turnover of project leadership	

(continues on next page)

Table 15.1 The Documented Challenges, Grouped, and Where We Address Them in This Book (cont.)

Group	Challenge	Where Do We Address This Topic?
Others	Permitting	These are related to shaping, which is the business theme in the tailored approach.
	Remote locations	
	The cart before the horse	
	Not learning the lessons	Will be addressed in with project closing; tailored approach section.
	Operational readiness challenges	CAMMP™ has a dedicated stage for operational readiness and another for initial operations; part of the tailored approach (also the simulation).

15.2 This Section

In this section, we present the tailored approach for leading megaprojects. We will do so in steps. Some of those steps apply to all projects, since they are related to The Customizable and Adaptable Methodology for Managing Projects™ (CAMMP™). However, we will maintain the focus on megaprojects. The steps include critical success factors (CSF) for a methodological approach, an overview of the CAMMP™ Model, and an introduction to CAMMP™ for Megaprojects.

However, this is not all. After these general chapters, we will cover the details of the tailored approach, stage by stage. We will close this section with a chapter on organizing the project for success (or the people aspects). With that closure, we will be ready to move into Part B and the simulation, using a hypothetical but realistic megaproject. You will notice that the tailored approach section and simulation section will have many similarities because, in both cases, we will follow the approach step-by-step, stage-by-stage. In this section, we will explain concepts, and in Part B, we will practice those concepts.

15.3 Organizational Project Management

In this chapter, I want to emphasize, repeat, and stress that methods, tools, processes, if they are collecting dust, serve no one. A reminder—"A fool with a tool is still a fool,"* or in this case, an OPMS that is not used, or is used by people who do not have the right level of expertise, competence, or care, will still produce failure. In other words, organizations must have what we have been advocating, repeatedly—they must embed the right project management mindset and culture.

Because projects are not done in a vacuum, and you can have the best project managers and a suitable method, but projects will still fail without the proper and mature OPMS and culture. Therefore, before we introduce the tailored approach, do not forget it is part of the OPMS and the SUKAD Seven Elements of Project Management Maturity™, which we shared in Chapter 2. For governments, I would also remind the reader of the proposed NPMA, presented in Chapter 14.

* The term is credited to Grady Booch per https://en.wikipedia.org/wiki/Grady_Booch.

15.4 Does CAMMP™ Apply to Megaprojects?

We will answer this question with significant details later; however, a quick answer here. We know, and we are confident, that CAMMP™ is an approach that applies to all industries, all types of projects, of varied sizes and complexities. However, we have stressed all along that CAMMP™ must be customized and adapted to industry, organization, and function, and finally adjusted based on the project size and complexity. Therefore, CAMMP™ is the methodological foundation for the tailored approach that covers the capital project industries, especially the large and complex and the megaprojects.

CAMMP™ is the methodological foundation for the tailored approach that covers megaprojects leadership.

15.5 Success One Step at a Time

In project management, I think it is highly possible to get lost in literature and models that focus on the project level, without paying enough attention to the stage level—especially the early stages of the project. We forget that a failed stage often means a failed project, because the consequences grow and multiply rapidly as we progress through the project life cycle. Therefore, any project management model, method, tailored approach, must give each stage the necessary attention. This is what we do in our model. This is why we need a project life cycle, stage-gate model.

15.6 Why a Stage-Gate Model?

Let me borrow from Paul Barshop for a minute here: *"[The stage-gate process] is the only approach that has ever been shown to work. . . . Using the stage-gate process effectively shaves about 25 percent off the project life cycle through better planning, reduced rework, and superior risk management"* (Barshop, 2016). One might ask whether, if the stage-gate process has been around since the 1960s, or 1980s at the latest, and it is more or less the standard for capital projects, why do projects still fail? This is the trillion-dollar question, is not it? As we said, a fool with a tool . . . we are not saying global organizations are fools, but some might be.* Otherwise, what can we call people and organizations that are not using what works? Well, some of them might not know it works, or they have seen incompetence rule, so they do not believe or trust that it works. The stage-gate process works! Therefore, we need to learn it and master it!

The stage-gate process works! Therefore, we need to learn it and master it!

Maybe I should say something about stage gate in innovative projects (research). Creative projects require a degree of experimentation and flexibility that might not be totally suitable for a "rigid" stage-gate

* I know some might have forgotten about Worldcom, Enron, Nokia, and a few others, but GE is almost falling apart, and that is still current era.

process (Lenfle and Loch, 2017). We would agree that rigidity for all projects is not appropriate. However, who is talking about rigidity? We still need a degree of discipline, but the discipline has to align with the context. Once I asked a colleague who works in the pharmaceutical industry how they practice disciplined project management in research and innovative environment. His responded that even scientists welcome a degree of discipline (management and control), but the organization allows specific "blue sky" projects to have greater flexibility, which enables experimentation and taking higher risks.

15.7 A Bit of Psychology

In closing this chapter, I'd like to fall back on a bit of psychology. The prevailing view of a project life cycle breaks it into two zones or phases, which some call *planning and execution* (Davies, Dodgson, and Gann, 2017), front-end planning and execution, or front-end loading and execution.*

I am one of those who believe in the power and psychology of the words used. What I am about to say might challenge the conventional terms.

1. First, if there is a front end, then where is the back end?
2. Second, is *planning* the proper term for a phase? We raise this question since planning is an ongoing process that is present in every phase of the project.
3. Third, is the latter part of the project only about implementation or execution? Where is the operational readiness and close out?†

Therefore, I could be silly here, but I am okay with that; these terms do not have the right feel to them and do not represent the true leadership of megaprojects. They give me the impression of limiting a complex project into what appears to be common sense terms but end up being misleading. Even if we restrict our thinking to technical, narrow-focused project management, the terms are not quite right. Think of what I just wrote with a bit of serious-humor mindset.

Regarding the front end, I am not too hung up on that term, and I have been using it, but CAMMP™ uses other terms, such as *Discovery Phase, Requirements Stage, Definition Stage,* and *Operational Readiness Stage.* However, I have a serious issue with the term *planning phase,* and I will touch on this point in the upcoming chapters. As you will see later, we not only prefer other structure and terms but encourage organizations to honestly go beyond the popular and common words and shift to terms that are reflective of the reality and complexity of project management. I am not saying the terms we are using are perfect; I am sure they do not fit every context.

---

* *Implementation* is another common term.

† I know, a few might say this is semantics or I am talking about the obvious. For example, it goes without saying that we will close projects, don't we? Do we formally close projects, capture the performance results, conduct success assessment, and document and share lessons learned? I know what the books and guides say, but are organizations doing this—and if yes, are they doing this properly?

Chapter 16

Critical Success Factors

16.1 Introduction

What would be the critical success factors or principles of a tailored approach for leading megaprojects?

In the previous section, we presented many recommendations and actions, and we can consider them all as critical success factors (CSFs), and you will see many of them repeated here in one way or another. One part of this chapter will discuss the general CAMMP™ CSFs, which apply to all types of projects. To do this, we summarize core concepts from *Project Management beyond Waterfall and Agile* (Ajam, 2017). Another part of this chapter will emphasize the CSFs in the context of megaprojects.

When we say CSFs in this context (what other references such as PRINCE2® might call *principles*), we are referring to the factors necessary for (1) building a methodological approach as the core element of an organizational project management system, and (2) developing the tailored approach for leading megaprojects to success.

16.2 CAMMP™ Critical Success Factors (Principles)

There are 11 CAMMP™ CSFs, and we will briefly discuss them here (Ajam, 2017).

16.2.1 Must Link Methods to Organizational System

Throughout this book, we have been making this statement. One can view this as the project management core philosophy statement. A vital principle for us in project management is the following: we view the management and leadership of a project not as a standalone action but as an integral part of organizational project management. In other words, as organizations manage projects, they must use the learning from other projects and continue to build the organizational knowledge base (OKB) and enhance the project management processes.

As organizations manage projects, they must use the learning from other projects, and continue to build the organizational knowledge base and enhance the project management processes.

16.2.2 Customize and Adapt to Industry and Project Type

As you recall, CAMMP™ is The Customizable and Adaptable Methodology for Managing Projects™. In other words, customizing and adapting the methodological approach are vital CSFs. If we link this CSF to the previous one, then the *customize and adapt* features are not limited to the methodology but expand to the whole organizational project management system (OPMS).

The customizable feature is to modify CAMMP™ to a specific sector or industry.

What do we mean by "to industry and project type?"

Industry, or maybe a whole sector, could have similar base methods—project life cycle models. For example, petroleum companies would have a high degree of similarity of their models (in terms of phases and stages). The same could be true for infrastructure or technology organizations. In addition to tailoring to an industry/sector, marketing projects might be different from IT, HR, or finance projects within one organization. Therefore, further tailoring is necessary.

The critical message here can be simplified by saying, one size does not fit all. Hence the need for tailoring, starting with these first two steps: customizing to industry and adapting to an organizational function.*

The adaptable feature is to modify a customized version of CAMMP™ to a specific type of project.

16.2.3 Tailor to a Given Project Classification

We view tailoring as a three-step process—customizing, adapting, and classifying. What do we mean by this third step? Let's zoom in on a capital project for a transportation authority, as an example. The transportation authority could have a small project, building a 5-km rural road, which could be considered a small project. Alternatively, the same authority could have a 500-km highway, or a metro system, which would be a megaproject. We cannot use the same approach for all types, hence the need to classify projects, and why megaproject classification is so essential.

For Project Management Institute (PMI®) fans, the sixth edition of the *PMBOK® Guide* (PMI, 2017) discusses tailoring. We also see various posts on social media on the subject of tailoring. However, in

* *Project Management beyond Waterfall and Agile* (Ajam, 2017) has two sections related to these concepts.

these contexts, tailoring is about the project management action to tailor the method. In other words, the project manager decides what processes to use or not use. This is not CAMMP™. If the project managers develop their own methods to manage projects, that is an indicator of a failed system or lack of an OPMS. Tailoring must be done at the organizational level, as a form of standardization. I know that some will argue, "a project is a unique endeavor" (PMI, 2017); sure, but that does not say, "and every project must have a unique method." In other words, these first three CSFs (and all of the others) are about standardization: developing tailored methods and OPMSs that are fit-for-purpose to the sector, industry, type of project, and its classification.

Tailoring must be done at the organizational level, as a form of standardization.

How do we classify projects?
We classify them based on size and complexity, a topic for the next chapter.

16.2.4 Follow a Stage-Gate Process

There is no need to elaborate much here—we have stressed throughout this book that in a project life cycle, the stage-gate process model is essential for all projects.

16.2.5 Integrate a Project Life Cycle with Processes

First, I must explain that the term *processes* in this context refers to the stage (project) management processes, such as those included in ISO 21500 and the *PMBOK® Guide.*[*] Notice, I used the term *stage* management processes, not *project* management processes. The reason for this is the confusion or the highly common misunderstanding by practitioners and certificate holders of these guides, thinking that the process groups are project phases. I have almost a whole chapter on this in my earlier book (Ajam, 2017), and I do not see a need to elaborate here.[†]

OK, what is the relevance here?

To manage projects, we must think at two levels—the *project* level, and the *stage* level. Keep in mind, project success is secured one stage at a time. Therefore, managing the project is done by going through the stages of the project life cycle (PLC), which would be tailored, as we discussed earlier, to fit the project industry, type, and classification. In other words, the PLC would be the product delivery process—the next CSF. However, success is one stage at a time, and here is where we need the processes, such as those advocated by PMI and ISO.

Remember to always think at two levels, at least—
the project level and the stage level.

[*] A few videos related to this point: https://www.youtube.com/watch?v=u3kAWCX6GEc; https://www.youtube.com/watch?v=03UJVrN3nt0; https://www.youtube.com/watch?v=-VIdLGQwARU.

[†] We also covered this via blog articles on http://blog.sukad.com.

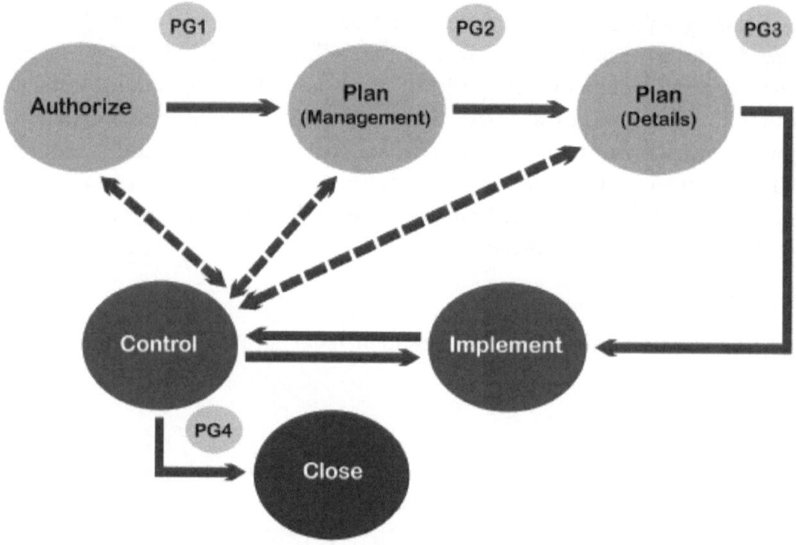

Figure 16.1 The CAMMP™ Six Processes and the Four Process Gates

Consequently, we consider that the processes are applicable to stage management. It would be good to reflect back on Chapter 2 on definitions and Figure 2.6 (duplicated here as Figure 16.1). To re-emphasize, to manage projects, we follow a project life cycle, stage-gate process. However, to manage each stage, we follow the processes of initiating, planning, etc. (PMI, 2017). Also, if you notice in Figure 16.1, we split planning into management planning and detailed planning.

16.2.6 Consider Product and Delivery Processes

This CSF is also related to something we discussed earlier in the book. Here is a refresher. Many practitioners of project management and even guides on the subject have a "project management purist mindset." This mindset might work to manage a stage, since the processes are generic and can apply in any industry and to **any project's stage**. However, unless you are a contractor or consultant, your business is not to manage projects, your business is to deliver outputs, products that lead to realizing the anticipated benefits from these facilities or assets.

> A purist view of project management does not work, especially for project owners' organizations.

Hence, we hold that a purist view of project management does not work, especially for project owners' organizations. Sure, the organizational project management function and personnel do not have to be engineers, quality experts, safety managers, and so on. However, they need to know what all of these things are and the logical sequence of events and stages for the industry they work in to deliver success; remember we want to lead megaprojects from concept to success (C2S).

The bottom line: CAMMP™ could have been CAMMPDP,* the "DP" for delivering products, but CAMMP™ will do; just think of the core purpose of project management, for project owners, is to deliver products successfully.

The core purpose of project management,
for project owners, is to deliver products, successfully!

16.2.7 Implement Project Success Model with Benefits Realization

In Chapter 2, we introduced the concept of project success, how to assess, and at what points. We will expand on SUKAD's Four Dimensions of Project Success™ in the next chapter.

16.2.8 Incorporate Industry Leading Practices

Project management guides such as ISO and PMI are about "good" or "common" practice—general practices that apply to various industries. However, for best practices or leading practices, we subscribe to the view of Construction Industry Institute (CII) and Independent Project Analysis (IPA) that best practices are those unique practices that make a significant and measurable difference to projects' performance. For capital projects and megaprojects, leading practices were covered in Chapter 11.

16.2.9 Capture Lessons Learned Throughout the Project

First, a sad story.

A few years ago, I was delivering a workshop to a client. We started talking about capturing lessons learned. Most of the participants were not aware of what is done in their organization and said they did not have a formal process for capturing lessons learned. One participant objected and said they had a process. I asked him, "How do you document them?" He said, "We just meet at the end and talk about them." He continued to say something like this: "We learned them, so why do we need to document them?" Sounds like he believes managing a megaproject is like managing a shed in the backyard.

Organizations cannot improve their organizational project management systems and processes without proper closure of projects, including lessons learned.

I have no idea how many times I have talked about maturity, behavior, culture in this book. Organizations cannot improve their organizational project management systems (OPMS) and processes without proper closure of projects, including lessons learned. If we do not learn the good, the bad, and the ugly, how can we improve?

* Customizable, Adaptable, Classifiable, Methodology for Managing Projects and Delivering Products.

What is essential is not to limit capturing lessons learned to the end, like our friend in the story above. I would not restrict them to the end of a stage either. Lessons learned, good and bad, must be an ongoing process. As soon as one encounters or identifies a lesson, it must be documented. Then, if the organization has the right level of respect for project management and maturity, the organization should have someone to own this process—a lessons learned coordinator. In that scenario, captured lessons from any project will be sent to this coordinator, to log and add to the database and circulate it to the relevant people in the organization.

16.2.10 Defined Roles and Responsibilities

Let me leave this topic to Chapter 21 on organizing the project at the end of this section.

16.2.11 Integrate Competence and Sustainability

I will also leave this to the discussion in the next chapter.

16.3 Megaprojects Critical Success Factors

It goes without saying that all of the above factors apply to megaprojects, maybe more so than any other type of project. However, in addition to the above, we will include factors that are highly needed for leading megaprojects, C2S. Some of these factors might be linked to the earlier CSF or to the leading practices discussed in the previous section. Nonetheless, we believe we need to stress them as independent topics.

16.3.1 Industrial Megaprojects Recommendations

We will list a few of the recommendations from the last chapter of *Industrial Megaprojects* (Merrow, 2011) and IPA. I will just list them, since all of these are addressed later.

- Address the business–technical divide.
- Formalize and institutionalize the shaping process.
- Develop the team staffing strategy.
- Remember, front-end loading (FEL) is still the world's best capital investment.*
- Restore professionalism (and sanity) to the owner–contractor relationship.

"The projects with the best FEL averaged more than four times the net present value (NPV) per dollar of investment of all the rest of the projects. The projects that did not achieve at least good FEL were usually NPV negative" (Merrow, 2011).

* Obviously, this could be a biased opinion since it is from IPA, and FEL/FEL Index are IPA products. I have not conducted or seen an independent study to prove or disprove this claim.

16.3.2 Barshop's Four Principle Practices

The following are from *Capital Projects* (Barshop, 2016):

- Clear Objectives
- Integrated teams
- Good project definition
- Robust project controls (and in real time)

16.3.3 Balancing the Triple Constraints

In recent years, we have read references, especially those for certification training, that the triple constraint concept is dead, and we must think of all "knowledge areas"* as the project constraints. We have a different perspective. We believe the triple constraints, or project constraints, concept is still vital, and the critical core of the science of project management. All other aspects and their impact would be manifested through these constraints.

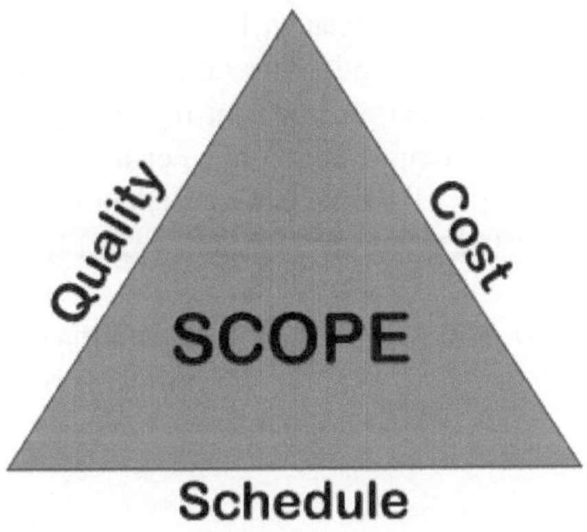

Figure 16.2 CAMMP™ Perspective on Triple Constraints

Notice in Figure 16.2 that scope is in the middle—it is mostly about the product scope, but it can include both product and project scope. Years ago, the executive director (sponsor) on a megaproject kept insisting that the organization must balance the constraints (including safety). The team resisted because, in the real world, one constraint will always be the controlling element. I am sure you have heard, "You want it fast, cheap, and good; choose two." In other words, it is impossible to balance the constraints. But is it? I do not have a certifiable answer, but maybe the sponsor on that megaproject is right, we should strike a balance.

What happens if we do not strike a balance?

Then we go with the motion or illusion of leading the management of the megaprojects, and we react to issues in favor of one of the constraints or another. I think this topic can be a huge research topic,

* *Knowledge areas* is the *PMBOK® Guide* term (PMI, 2017).

which is outside of our scope. In our view, it is essential to remember the following: define the scope and clarify objectives, decide on the quality strategy and standard, then develop the proper, predictable cost and schedule metrics. Easier said than done, but I have seen it done, and it requires the proper organizational maturity and culture.

What happens if we do not balance the triple constraints?

We start making tradeoffs. We may sacrifice quality for the sake of cost or time. Sacrificing quality has a humongous hidden effect, and it might not show until the start of commissioning or even production. Consequently, the impact could be massive and would significantly affect the rate of production, and in turn, reduce the benefits (profitability).

16.3.4 Shaping/Framing

Before I started the literature review, I had already defined the CAMMP™ Megaprojects model, since I had written an initial version of this book back in 2015. The literature reviews provided a lot of insights, as I am sure you have seen throughout this book's numerous references. However, the one concept that led me to modify the CAMMP™ MP model is the concept of *shaping* or *framing*. Many of the ideas of shaping, as addressed by Merrow or Barshop, are in the various stages of CAMMP™, especially in the discovery phase and development phase. However, I am now convinced that shaping should be an independent phase or stage, which is precisely what we have done. Shaping is a concept used in projects, especially in megaprojects, to properly define and understand the opportunity. It is a concept with broad actions, including site assessment, partnership agreement in case of joint ventures, study of the labor market or site conditions (especially in remote locations). For more information, refer to Chapter 19. For in depth understanding, Merrow (2011) would be an excellent reference. For megaprojects, this is a must; I have seen too many projects fail as a result of lack of proper shaping.

"The effective management of large projects requires a stable environment. The requirement to make major changes in the objectives, scope, precise location, or any other major element after the start of [the] detail definition phase can result in an unmanageable project. The turbulence in the project environment translates the changes and disruptions in execution and render projects failures" (Merrow, 2011). This is one of the core reasons to justify proper shaping.*

16.3.5 Control the Lock-in Effect

This may be more of a challenge, but we include it here to learn how to avoid falling into this trap.

One of the challenges of megaprojects is the *lock-in effect,* or not cancelling projects once they get moving; no one wants to take responsibility for the sunk costs. Some of the papers from the *Oxford Handbook of Megaproject Management* (Oxford, 2017), such as the Hirschman Hiding Hand (Hirschman, n.d.) or Drummond's Escalation of Commitment (Drummond, 2017), lead to the concept of lock-in. Simply, this is when an organization makes an initial decision and then finds it hard to retract it, even when the conditions change. The project owner might not recognize the changing conditions or may choose to ignore them. This is why the stage-gate process is essential and valuable, because when appropriately used, it helps validate the feasibility (or lack thereof) of continuing with the project. *"The history*

* I did not know what to call it at that time, except perhaps "site assessment," but the story I shared about my drinks shower and the resulting date was from a shaping process, since site assessment is part of that.

of megaprojects suggests that . . . decision makers may reinvest only to end up 'throwing good money after bad,' a phenomenon known as escalation of commitment" (Drummond, 2017, p. 195).

Let us reflect on this cycle of doom:

- Many project owners' organizations have lost their technical and project management capabilities (Merrow, 2011).
- This loss contributes to the lack of proper front-end planning or even poor planning (Merrow, 2011).
- The delusional optimism (or even deception) in the project objectives (Drummond, 2017) lead these organizations to sanction projects.
- Even organizations with sound project management systems might lose sight of the external environment and context. Sometimes, the "four sublimes" driving megaprojects take over (Flyvbjerg, 2017), even with the presence of the conflict of interest in some situations (see paragraph 16.3.6, below).
- All of these contribute to the lock-in effect and how escalation commitment starts (Drummond, 2017).

As project owners commit to going ahead, possibly on a faulty premise (whether intentionally or due to the lack of proper due diligence), it becomes harder and harder to stop. *"Regardless of what drives escalation, the simple truth is that once megaprojects gain traction, they are almost unstoppable"* (Drummond, 2017, p. 210).

Why is it hard to stop megaprojects that could be heading down the cycle of doom?

Drummond offers us a few contributing factors (2017, pp. 197–200). These include:

- Psychological factors, such as ego, optimism, among other influencers
- Social drivers, especially for projects in the public domain and political commitment
- Economic drivers where sometimes the cost to stop could be almost as much as the cost to complete the project

Then how do we manage the escalation of commitment?

Once again, Drummond offers many suggestions (2017, pp. 201–206), but what could be the most crucial factor is scrutiny before one even starts, which is the next topic.

16.3.6 Scrutiny of the Need for a Megaproject

We will stress this point in this section and also later, in Chapter 19. The point is, *are organizations (public or private) performing the proper due diligence that leads to a megaproject being the right answer to the business or community need?*

For example, Flyvbjerg (2017, p. 6) raises the question via the "four sublimes" that drive a megaproject. These are the *technological* (driven by engineers and technologists—"longest, tallest, fastest"), *political* ("building monuments to themselves"), *economic* (business and money drivers), and *aesthetic* drivers. Van Wee and Priemus question the decision making behind launching these projects (Van Wee and Priemus, 2017, pp. 18+). In their paper in *The Oxford Handbook of Megaproject Management* (Oxford, 2016), they raised a few questions about the ethics of the decisions and the potential conflict of interest that drives these decisions, especially in infrastructure projects. They questioned the quality of the cost estimate and benefits. *"Costs are more frequently underestimated than overestimated . . . overestimation of demand is more common than underestimation"* (p. 126). Van Wee and Priemus also raised another point, which is, *"the percentage increases of the cost estimates from the initial estimate to those when formal decision to build was made was much more than between the decision to build and project realization."* The same finding was shared in the study of the Australian project (Terrill and Danks, 2016).

Let me translate the last couple of statements with a few bullet points:

- The initial estimates, which organizations use to decide to continue with developing the project, are severely below what they should be (due to organizational competence or ethics). Even if it is a people competence issue, I believe the burden is on the organization and not the individuals.
- Organizations continue to develop the project and go through the front-end planning (loading), leading to the final approval. Although they could often discover that the cost is now much higher than initial estimates, they could still rationalize the decision to proceed into implementation. Again, the root cause could be competence and optimism bias, or maybe believing they have invested too much to stop the projects, which would be supported with the potentially inflated benefit and benefit–cost ration (BCR).
- Then Murphy visits during the implementation, or maybe Murphy has been living and hiding on the project all along. Cost and schedule overrun, exposing performance issues—which most likely reveal the faulty basis for the decision to continue—but the project continues into failure, and the contractors are blamed.

During my research for this book, the papers and books mentioned here stated the possibility of poor front-end planning, including poor benefit–cost analysis. The evidence shows that organizations often underestimate projects and might inflate the expected benefits, leading to a BCR (benefit-cost ratio) that is highly favorable only to discover the faulty logic or analysis later. I know that Love et al. (2019) expressed concern over the danger of deception. However, Barshop showed that part of the erosion of the NPV is due to a shortfall in sales (Barshop, 2016). The justification for mega-dam projects can also fit in this context (Ansar et al., 2017), where the business cases for these projects might ignore or de-emphasize other simpler or smaller alternatives; then one would wonder why large dam projects fail.

I conclude this section with a call to action for the tailored approach. The discovery phase of CAMMP™ is crucial, and it is where the success and failure of the megaproject are rooted. The concept for the megaproject must go through extensive scrutiny in terms of market analysis to verify that the demand exists at the level expected to prevent overestimating the benefits. Further, the organization must scrutinize the idea of economy of scale versus smaller projects; in other words, do not let the "four sublimes" drive the decision to go into a megaproject if other solutions could be adequately explored, such as through smaller projects or a program approach. A case in point here is the example of comparing the cost of BRT (Bus Rapid Transit) versus LRT (Light Rail Transit) (Love et al., 2017). Here are some numbers: (a) Australia CAPEX for BRT is $36.6 million per km compared to $63.5 million per km for LRT (73% higher); (b) the US Government Accountability Office found that LRT costs about 2.5 times that of BRT. Yet, we often see a preference for LRT.

16.4 Chapter Closing

In the previous section, we discussed leading practices and organizational competence. In this chapter, we focused on the critical success factors from CAMMP™, which would apply to all types of projects. We also covered other CSFs from the literature review. In other words, we have collected all of the ingredients for our feast. It is time to see how we incorporate, integrate, and blend all of these concepts in the tailored approach. This is the subject for the next few chapters.

Chapter 17

Integrated and Applied Project Management

17.1 CAMMP™ and the SUKAD Way™

We pronounce CAMMP™ "camp," and as you know by now, it is the abbreviation for The Customizable and Adaptable Methodology for Managing Projects™, which is **a methodological approach for managing a project**. This approach is one of the core solutions of **The SUKAD Way™ for Managing Projects**, which is the label we use for the project management innovation program that we have been running since 2007. CAMMP™ and other solutions of The SUKAD Way™ are central to this book's mission and the tailored approach. The standard CAMMP™ model that we use for this book is version 3, which was updated and published in December, 2017, as *Project Management beyond Waterfall and Agile* (Ajam, 2017).

CAMMP™ is pronounced "camp."

17.2 The SUKAD Way™

We will not go into a lot of detail here, only a brief summary is needed.

Since 2007, when we launched the SUKAD project management innovation program, under the label The SUKAD Way™, we have developed four leading solutions or approaches, as you will see here. CAMMP™ is the most developed* of the four. Please note two key facts:

* What we mean here is that CAMMP™ is the solution with many publications to support it. The rest of the solutions are integrated in these CAMMP™ books, such as this book and *Project Management beyond Waterfall and Agile*. All of these SUKAD Knowledge & Innovation "products" will be built into the Uruk PPM Platform that SUKAD Corp is developing as of this writing.

1. We propose that all four approaches be applied in an integrated way, which would enhance the organizational project's performance. Without this integration, you might achieve fair or even good results, but it would be a challenge to reach higher levels of maturity.
2. We did not reinvent the wheel; a good part of what goes into these solutions is available in the industry in one reference or another, although not in an identical form. We have built on or improved what exists, adding our perspective, with a focus on applied project management and integrating these concepts into a holistic platform.*

17.2.1 Organizational Project Management

Let us start with a higher (and most important) level, the organizational project management level.

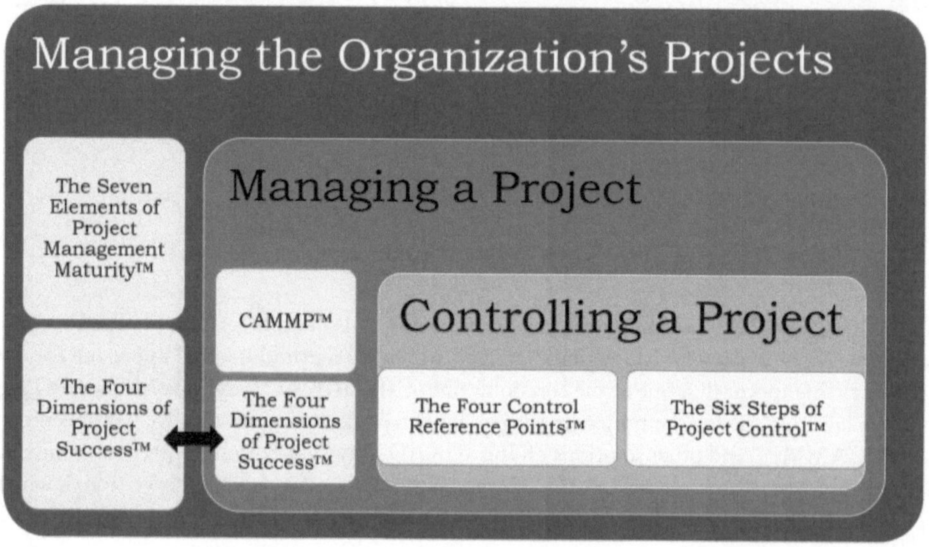

Figure 17.1 The SUKAD Way™ Integrated Project Management Solutions

For managing the overall organizational projects (and overall portfolio), we need to reflect on two essential parts, as you can see in Figure 17.1, The Seven Elements of Project Management Maturity™ and The Four Dimensions of Project Success™. We had discussed the seven elements in an earlier chapter, and we only touched on the project success model earlier, but we will expand on that later in this chapter.

Here we want to stress that The Four Dimensions of Project Success™ is shown twice in this image. The reason we do that is, although the model was designed for assessing the success of a given project, two of the dimensions cannot be measured during the project life cycle or before closing the project. A third dimension could be assessed (1) before closure on some projects, and (2) after closure for most projects. The fourth dimension cannot be evaluated until years after closing for capital projects and megaprojects. Therefore, the third and fourth dimensions are part of the organizational project management

* That will be the Uruk PPM Platform.

function, not part of the project manager scope. Maybe it would be assigned to a centralized project management leadership.

17.2.2 Managing a Project

The core part of managing a project effort is CAMMP™, which we will discuss in detail as we continue with this chapter. The second part (from the image) that we integrate into CAMMP™ is the project success model, as we have already explained. CAMMP™ is about managing a single project, using a tailored method derived from the CAMMP™ methodological approach.

Although the focus of CAMMP™ is on the overall management of a project, concept to success (C2S), using the project life cycle model, which includes control; we address controlling a project separately below.

17.2.3 Controlling a Project

To control a project, we consider two concepts. The first is about defining the baseline.

Where is the baseline of a project? Is it fixed at one point, such as the final authorization sanctioning a project to proceed into implementation, which is a common understanding? Or is it moving along the project life cycle?

To answer these questions, we propose the concept of The Four Control Reference Points™, which we will discuss in Chapter 20.

The first concept to address is about defining the baseline points along the project life cycle. The next thought is: at each of these points, how do we assess the project's performance and forecast the project? By forecast, we mean the updated expectations for completing the project in terms of cost, schedule, and other factors. We call this assessment and forecasting mechanism The Six Steps of Project Control™. This part is outside the scope of this book.

17.2.4 Other Resources

We have an appendix at the end of the book for additional resources that would help in applying The SUKAD Way™ and CAMMP™.

17.3 The CAMMP™ Three Dimensions

Let us expand on CAMMP™.

CAMMP™ is a three-dimensional methodological approach integrating the learning from various organizational guides and references, including from PMI® (Project Management Institute), IPMA® (International Project Management Association), ISO® (International Standards Organization), GPM® (Green Project Management Global), and others. Briefly, the first two dimensions integrate a project life cycle, with stage gate processes, with the processes to manage a stage. Remember, this is one of the critical success factors mentioned earlier. These two dimensions would enable organizations to achieve a good performance, but we need the third dimension, which is the advanced topics, to elevate the performance to a higher level. Let us briefly look at each of the dimensions; you will see more details in future chapters.

17.3.1 The First Dimension

The first dimension is the project life cycle (the horizontal dimension), which helps practitioners follow a project from idea to closure (and beyond); this is what we call concept-to-success (C2S) in the context of megaprojects. The project life cycle consists of phases, stages, stage deliverables, and stage gates. CAMMP™ offers a standard model that can be tailored. The intent of assigning "the first dimension" to the project life cycle is to emphasize that a project life cycle is the most basic form of a methodology and must be the starting point.

Please refer to Figure 2.7, Chapter 2 (page 22), which is the standard CAMMP™ Model. I know that I must have said the following more than once, but a *standard* model does not mean a *fixed* model. This is the model to use as the starting point for tailoring. The next chapter will detail the PLC for megaprojects.

17.3.2 The Second Dimension

The second dimension is the vertical dimension of the project life cycle. *It is about the application of the project management processes along the project life cycle in every stage or phase.* The CAMMP™ Model offers a modified version of that put forth by ISO and PMI. CAMMP™ offers six processes to manage each stage of a medium-to-large project, and these are a must for megaprojects. The six-stage management process was shown in Figure 2.6 in Chapter 2 (page 21). It is important to stress that each of the processes has many process steps (Ajam, 2017).*

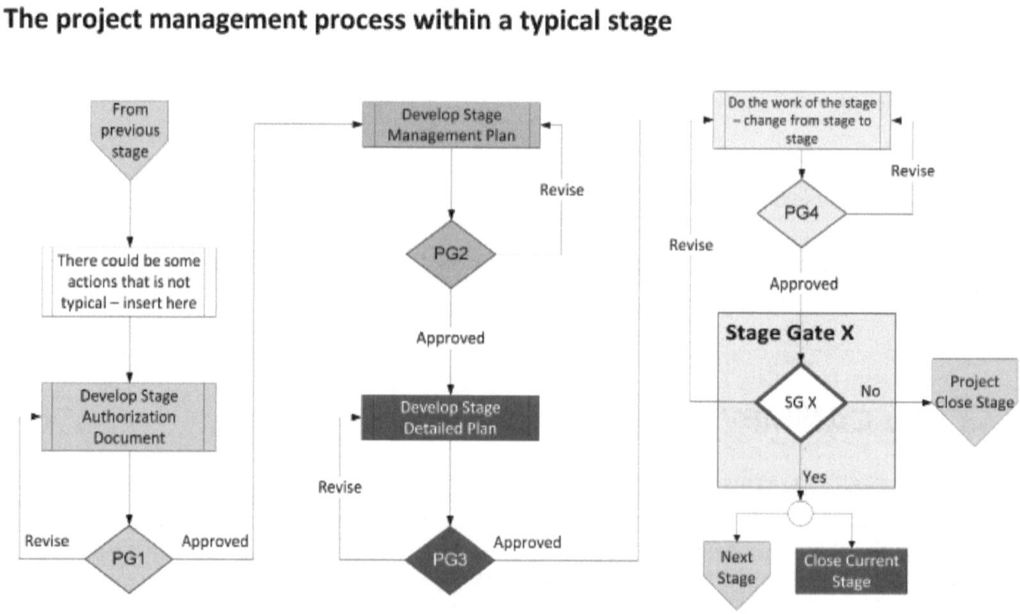

Figure 17.2 The Process Flow for Managing a Stage

* We will use these processes and process steps in the simulation part of this book. If you would like to get more into the original concept and details, we refer you to *Project Management beyond Waterfall and Agile* (Ajam, 2017).

Figure 17.2 presents another way of viewing the stage management processes. Briefly, for every stage we have an authorization process leading to a stage authorization document; a stage management process which provides the stage management plan; a detailed planning process to develop a stage detailed plan; the implementation process to do the stage-work; a controlling process from the start to the closing process. In this image, one can see the four process gates (within the stage), and at the end, the fourth process gate leads to the appropriate stage gate.

Two clarification points: First, each stage in a megaproject could be more extensive, cost more, and take longer than in a typical non-capital, non-megaproject. Second, this is what guides such as ISO 21500 and the *PMBOK® Guide* mean when they say the "process groups repeat" (PMI, 2017).

17.3.3 The Third Dimension

If one stops and goes back to the case studies presented in Chapter 8, we stated that one of the main differences and the factors that contributed to the success of one of the projects and the failure of the other was the organizational project management maturity. This maturity of practice is what we achieve through this third dimension. So what is the third dimension?

The third dimension consists of layers that organizations could implement along (and on top of) the project life cycle and the processes (the first two dimensions). These layers are what help organizations move from a fundamental methodological approach and transform it into a highly sophisticated and robust system that will enable and empower them to seek excellence. In other words, the primary approach (two dimensions) could be of value for organizations starting on their project management journey, and as they become comfortable with the approach, they can implement the third dimension, one layer at a time. What are these layers?

At this time, we have identified four layers.* These are:

- Project success, using the SUKAD The Four Dimensions of Project Success™.
- Competence, such as what we discussed in Chapter 12, but with a focus on using the work of IPMA, via their Organizational Competence Baseline and Individual Competence Baseline (no need to reinvent the wheel).
- Sustainability, and here we refer to the GPM guide for sustainability methods in delivering projects; again, no need to reinvent the wheel.
- Finally, the implementation of best practices, such as those from CII, IPA, and SUKAD, which we presented in Chapter 11.

17.3.4 Integrating the Three Dimensions

First, let us look at how to integrate the first two dimensions, presented in Figure 17.3.† Notice that the processes for managing a stage are in the middle, in the oval, and arrows show that they apply in every stage. Because some of the stages could be short and with limited resources (the Concept Stage, for example), while others could be huge (such as the Implementation Stage), one could expect the level of effort to manage a stage could vary greatly. Also, some of the efforts could be led by contractors, such as engineering or construction.

---

* In *Project Management beyond Waterfall and Agile*, we dedicated a chapter for each of these topics.

† This figure is from *Project Management beyond Waterfall and Agile* (Ajam, 2017).

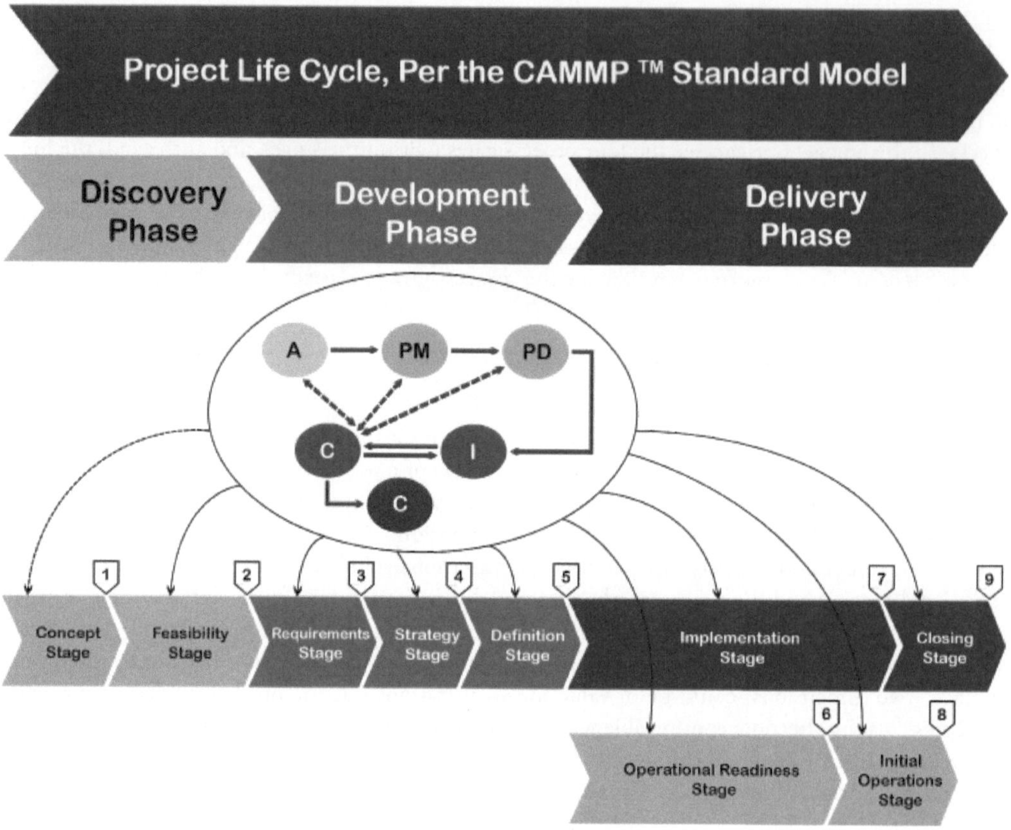

Figure 17.3 The CAMMP™ Model for a Typical Large-Complex Projects

17.4 Projects' Classification

In Chapter 16, on critical success factors, we introduced the concept of customizing and adapting a CAMMP™ standard model to fit industry and type of project. We also touched on further tailoring through a project classification approach. Again, this topic was covered extensively in *Project Management beyond Waterfall and Agile* (Ajam, 2017). We introduced a classification model with three major classifications. To determine these classifications, we used a scoring mechanism that includes a size category and a complexity category. Each of these categories consists of a few factors. Please keep in mind that this is a simplified approach, and we know that the subject of project complexity is a huge topic, with postgraduate programs focusing on complexity. In other words, the reader can choose to explore this topic to a high degree of in-depth knowledge, but that is outside our scope. An example of the fundamental scoring model is in Table 17.1.*

As you can see from the table, CAMMP™ initially considered three classifications. Please note, a project classification can be highly dependent on the context of the organizations adopting this approach. For

* This is modified from the version used in *Project Management beyond Waterfall and Agile,* adding megaprojects.

Table 17.1 The SUKAD Basic Scoring Model for Classifying Projects

Category	Factor	Small-Simple (SS)	Medium-Moderate (MM)	Large-Complex (LC)	Megaprojects (MP)
Size	Cost				
	Duration				
	People				
Complexity	The uniqueness of project work				
	Importance				
	Number of departments involved				
	New technology				
	Organizational impact				
Total Score		*[The sum of all points]*			

example, a small project in one organization could be a medium project in another, and a large project in yet another organization. The three original classes are (Ajam, 2017):

- **Small and simple projects.** These are typically small projects, requiring few resources, relatively short durations, and with limited complexity. For these projects, the recommendation is to use the CAMMP™ standard PLC without the processes—that is, use the first dimension only, which is more than enough.
- **Medium and moderate projects.** These would be more significant projects with moderate complexity. For these projects, the recommendation is to use the PLC (first dimension) with the processes of the second dimension but applied at the phase level, not the stage level, per Figure 17.4 (Ajam, 2017).
- **Large and complex projects.** As the name indicates, these would be large and complex projects—remember, in the context of an organization. For these projects, we recommend using the integration of the first and second dimension, with the processes of the second dimension applied at the stage level, per Figure 17.3, presented earlier in this chapter.

Since the publishing of *Project Management beyond Waterfall and Agile*, we decided to expand on the above, per the following additions:

- **Micro projects.** This class was extracted from the small-simple project class and is highly similar, but some of the process steps are taken out, because they might not be needed.*
- **Megaprojects.** This class was extracted from the large-complex class because, as we have already seen in this book, megaprojects are extremely large and extremely complex. Similar to large-complex projects, the processes of the second dimension will repeat for every stage. similar to large-complex, for this megaproject class, we also have three phases, but we have modified the stages significantly. More on this in the next chapter.

* Outside the scope of this book, but in case you are interested, check out some of our e-books.

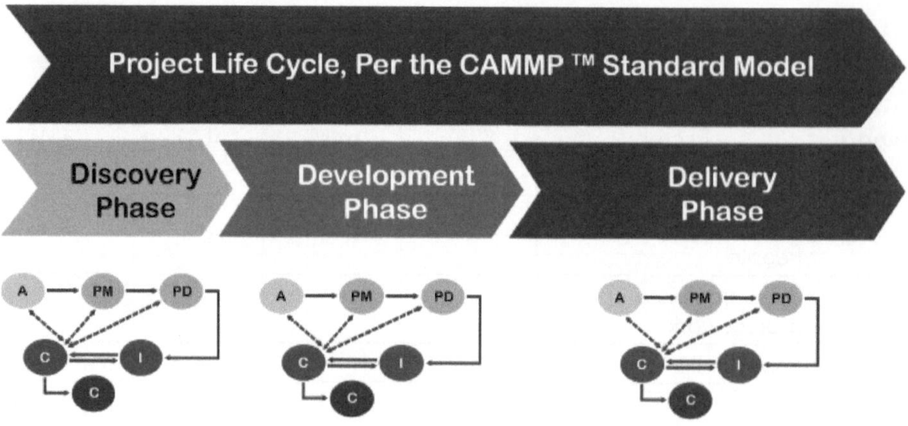

Figure 17.4 The CAMMP™ Standard Model for Medium to Moderate Complexity Projects

17.5 Project Success

17.5.1 Non-SUKAD Concepts

Reflecting back on all of the references in the literature review, outside IPA (Independent Project Analysis) books, I could not recall finding a clear definition of project success. What was implied is that overrun means failure. Defining how to determine success and failure is highly variable among industries and organizations. We have seen the simplistic view, that customer satisfaction is enough to determine success, which might be OK for a service provider but not a project owner. Others, like Oberlender (2014, p. 319), offer us the keys to successful projects,* but no criteria, unless we missed it. Others confuse or consider project management performance as the only dimension to measure project success, which would be the project management perspective. However, project owners are not in the business of building and managing projects for the sake of managing; their core business is not project management. They launch a project to realize the benefits. Hence the need for a different perspective.

IPA does offer a project success or failure model that includes five factors, which we covered in Chapter 6. A reminder: the model consisted of cost growth, cost competitiveness, schedule slip, schedule competitiveness, and production in the first year after startup and commissioning. To compare the above to the SUKAD model, the first four factors would be part of one of the four dimensions (project management success), and the last factor is a part of the third dimension (project delivery success).

Let us move to SUKAD's The Four Dimensions of Project Success™.

17.5.2 Perspectives

Organizations must view projects in line with a strategic approach, which means that organizations should not limit their views to the *project* life cycle but must go beyond that, touching on the *product*

* These are: good field construction representatives from the owner, good detailed construction schedule prepared by contractor with involvement from superintendent (not imposed by owner/designer), good project control system, and good communication (Oberlender, 2014).

life cycle. In other words, a project is a business venture, rather than an output that the project team produces before demobilizing. Ideally, project owner organizations must use an asset management process covering the whole **product** life cycle. However, for now, another way to think of this approach is that **CAMMP™ includes product delivery**—the acquisition part of the product life cycle. If you recall, Figure 17.1 at the start of this chapter showed The Four Dimensions of Project Success™ as an integral part of Organizational Project Management and Managing a Project.

In line with the above, it is essential to emphasize that, when it comes to success, CAMMP™ views projects from the project owner's perspective. It also takes the view of the organization, rather than the project manager or the organizational project management function.

17.5.3 The Four Dimensions of Project Success™

Here we will summarize a chapter from *Project Management beyond Waterfall and Agile* (Ajam, 2017). However, what we present here also includes additional content relevant to megaprojects.

Product Delivery Success: Technical

This first dimension of success is about whether the team delivered the product (output) of the project successfully; this is about technical success (scope and quality). In other words, the key questions are: did the team complete the work, deliver the product, and meet the technical standards and specifications?

Figure 17.5 presents this view, highlighting the following key points:

- The client (internal or external) works with the project management team to determine the success of the product delivery.
- The team can measure this dimension with a focus on quality, scope, and technical specifications.
- The measurement is in comparison to the project detailed plan (PDP); as you will see later, this is the output of the Definition Stage and the point of final authorization.

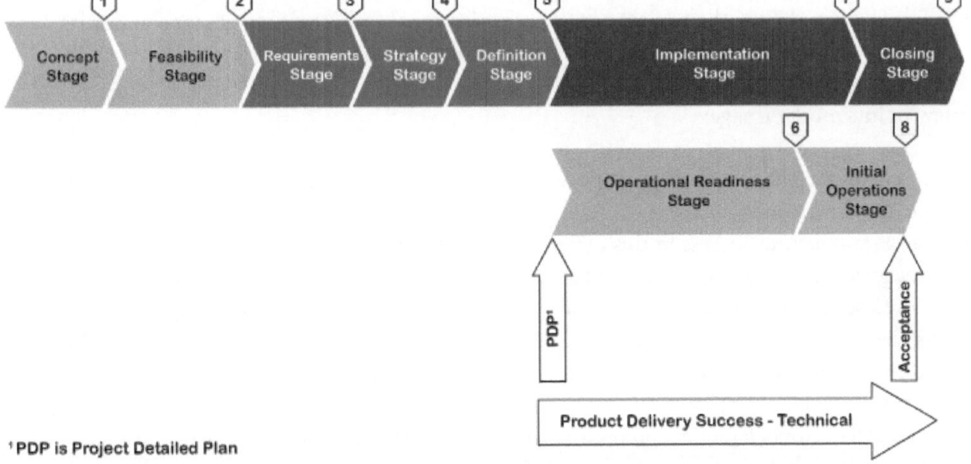

Figure 17.5 Project Success, First Dimension, Product Delivery Success

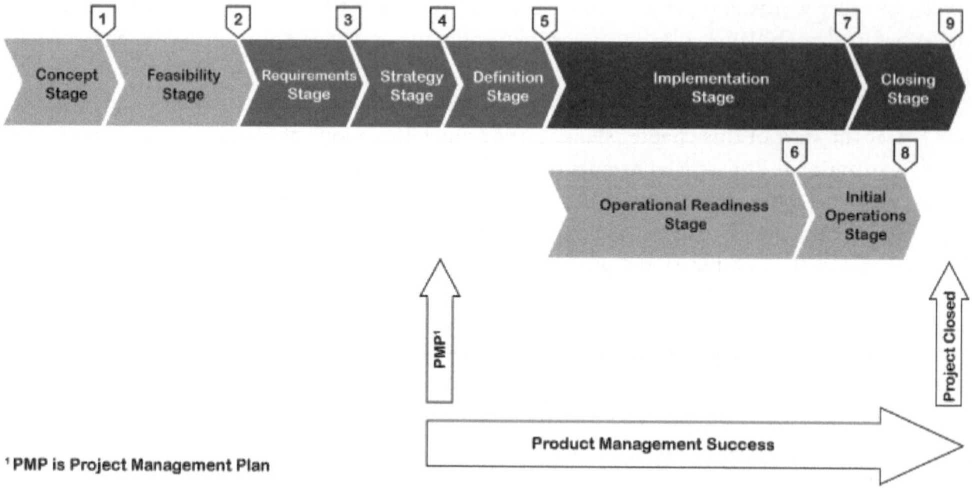

Figure 17.6 Project Success, Second Dimension, Project Management Success

- Because the measurement is per the PDP, this is mostly a technical assessment of the product compared to the specifications.
- The team performs this assessment at final acceptance (Stage Gate 8 per the standard model).

Project Management Success

Figure 17.6 shows the second dimension—project management success. This dimension assesses whether the project management team delivered the project successfully per the Project Management Plan (PMP). The focus here is primarily on time and cost, in addition to any other compliance requirements. Therefore, some of the questions would be:

- Did the team deliver the project per the organizational project management standards?
- Did they deliver it per the established performance metrics and parameters?

A few points on the image:

- The client (internal or external) works with the project management team and project management office to determine success.
- The team measures success in this dimension in comparison to the PMP, which is the output of the Strategy Stage, at Stage Gate 4. It is important to emphasize that Merrow (2011) proposes that this should be the last point for the business to decide whether to go ahead with the project or not. In other words, his view is that the Definition Stage should be to finalize the technical details and develop the plan for the delivery phase. We agree. The idea here is that the business should have completed the shaping process at this point and should be in a position to make the final decision. Further, although the project could still be stopped (or not approved) at Stage Gate 5 (end of Definition Stage), organizations should not wait until that point, because about 3–5% of the project costs would be spent during this stage.
- The team performs the success measurement during the Close Stage, with Stage Gate 9.

There are three crucial things to note before continuing.

- **Note 1:** Notice that in discussing the second dimension, there was no mention of the project's being ahead of schedule or below budget. Instead, the terms used are "per the established metrics and parameters."

Why is this?

The answer would be a lengthy discussion, but let's just say, when management insists on below-budget and ahead-of-schedule performance, they are indirectly evoking the survival instinct of the project management team, which leads to their inflating the cost and schedule estimates. Inflating these estimates becomes a self-fulfilling prophecy, and the final cost is likely to end up higher than necessary but within the bloated budget. However, staying within the parameters, which could be ±5% or 10%, would allow for more realistic estimates. You may want to consider revisiting Chapter 12 for the discussion on the executive and senior management need to understand the fundamentals of project management.

- **Note 2:** The team can perform success measurement on Dimensions 1 and 2 before closing the project at SG9, and they would be able to include the results as a part of the project close-out report. However, Dimensions 3 and 4 would be a different story, as discussed below.
- **Note 3:** Four of the IPA factors to assess failure and success are related to this dimension. In terms of the criteria, you choose your own. For example, IPA classifies a project a success or failure. Failure is missing one of the factors: for instance, in the cost area, an overrun of more than 25%. I have seen some organizations use the label *overrun* once actual cost exceeds budget with no tolerance factor (this is related to Note 1 on the "below budget, ahead of schedule" mindset).

Project Delivery Success

Figure 17.7 represents the third dimension.

This dimension is the most difficult to explain. The purpose of the third dimension is to assess whether the team *delivered the project successfully* per the various plans, specifically:

- Delivering product success per the project detailed plan (PDP) (the first dimension).

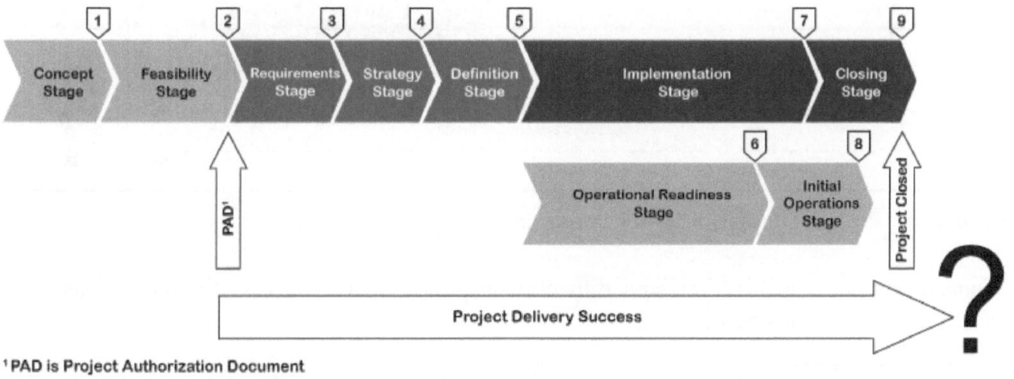

Figure 17.7 Project Success, Third Dimension, Project Delivery Success

- Delivering the project per the PMP (the second dimension).
- Achieving other factors specified in the Project Authorization Document (PAD).

The first two bullets above refer to the first two project success dimensions, which are required input to this third dimension. Failing to deliver the product per the PDP and the PMP is a project delivery failure. To repeat: if one of the dimensions is not met, CAMMP™ considers project delivery is not a success. Now, assuming the first two dimensions are in line with the criteria, the third bullet refers to other factors, which would be project or organization specific.

For example: In the case of an industrial facility, the success criteria can include customer satisfaction, achieving production capacity by the end of the first year, no issues or gaps in the operating and maintenance procedures, etc. The IPA fifth factor in the success criteria is related to this dimension.

When can the team assess the project delivery success?

It is possible in some situations to evaluate this dimension at project close, but it is not likely. For most projects, the organization might have to wait for months or even years to be sure about this dimension, which is represented by the "?" symbol in Figure 17.7. The organization can measure the third dimension in comparison to the PAD.

Some practitioners might say this is not the project manager's responsibility.

Right, and this why we said this dimension could be part of the OPM rather than within the project.

Objectives Success

The fourth and last dimension is the business* or objectives success, as illustrated in Figure 17.8.

This dimension is designed to assess whether the project outcome resulted in delivering the expected capabilities, which in turn led to realizing the benefits that the organization outlined at the start of the project. In other words, this dimension has a direct link to the concept of *benefits realization and asset performance.*

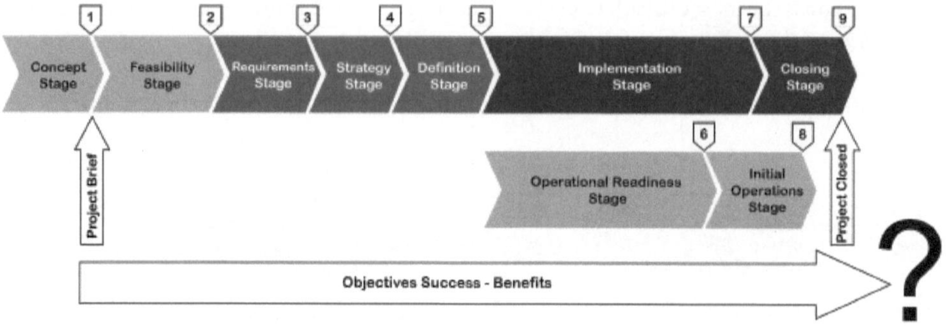

Figure 17.8 Project Success, Fourth Dimension, Objectives Success

Some may argue that this is the same dimension as project delivery success. On some projects, this may be the case, but not always.

* The use of the term *business* here is in the general context and is not specific to for-profit projects. If the project is for a not-for-profit, non-governmental organization (NGO), or a government, then the driver for the project is a social or community need, not financial returns.

> **Are We Assessing Success Properly?**
>
> In reference to the various resources in our literature review, unless we missed it, we are not sure if any of those books or papers tackled this fourth dimension. The IPA criteria is mostly about the second and third dimension. Are most other references that are discussing benefit–cost ration (BCR) and net present value (NPV) obtaining the actual production performance to be able to calculate properly?
>
> We will revisit this question at the end of the book.

Why?

The fourth dimension is concerned with the expected benefits and the justification behind the project. The third dimension, by contrast, is in relation to the PAD. There is a possibility that the PAD was wrong, and the sponsor did not develop the PAD properly to reflect the essence of the project. Therefore, the project team could deliver a good project according to the PAD, but from the organization's perspective, they delivered the wrong project. If one refers to the examples that were used in the third dimension and reflects on them per the fourth dimension, we notice the differences.

In the case of an industrial plant, the success criteria could be achieving the return on investment, the anticipated NPV or BCR. In an information system project, the criteria could be reducing staff or increasing productivity. In a marketing project, project success could be increasing leads by X%.

17.5.4 When to Define the Criteria for Success

Regardless of which model you use for measuring success, when does the team define the criteria for project success? Here are our recommendations:

- On the project level, the team must determine the success criteria as early as the Feasibility Study, but no later than project requirements, Project Management Plan, and conclusion of shaping the project.
- In terms of the stage level, it should be part of the authorized process.

If the team does not define the success criteria for the various dimensions, the project will fall into the trap of "completion = success," which is not valid from an organizational perspective and will not help the organization achieve a higher level of project management maturity.

17.5.5 A Project Success Score Card

We have presented The Four Dimensions of Project Success™ and the need to establish the criteria for success (for all dimensions) early on. We also identified where on the project life cycle the team (or organization) could perform the success assessment.

One of the reasons for the four dimensions is to enable the organization to zoom in on the pain points on the project—most importantly, to identify consistent challenges related to one of the dimensions. For example, an organization might do a good job in the project management dimension but consistently fail to achieve the objectives. That is an indication that the failure is not in cost and schedule performance but in setting clear objectives and adequately carrying out the early work and decision making. So, weaknesses in this area would trigger special attention and maybe a task force to investigate and remediate the situation. Similarly, if the problems are in the delivery phase (implementation), then perhaps the core problems lie with the procurement and executing processes. In other words, this can be viewed as an application of the cause and effect diagram (fishbone, Ishikawa), where the system is the whole project and the components would be a technical component, project management component, and so on.

Take a moment to reflect on where we have come. We needed the four dimensions on the project level to evaluate success from various perspectives. This approach allows us to zoom in on the gaps, especially if they are consistent across projects—a trend. Then we need the criteria for the various dimensions.

Now it is time to reflect that most of the literature looks at the question of project success in an either/or scenario: success or failure. However, in Chapter 2, we introduced a four-grade approach for assessing project performance. Therefore, we now put these pieces of the puzzle into Table 17.2, which presents a possible project success scoring dashboard. The columns are for the dimensions and the rows for the assessed grade or score. The actual scoring could be Yes/No or a scale from 1 to 5 or 1 to 10; it is your choice, and this should be clearly defined in the OPMS.

Table 17.2 A Possible Dashboard for Project Success

	D1 Product	D2 Project Management	D3 Project Delivery	D4 Objectives
Successful	Yes			Yes
Partial Success		Yes	Yes	
Partial Failure				
Failed				

For example, using a Yes/No approach to explain the dashboard, let consider a few scenarios.

- A Yes for Product Technical Success Dimension in the successful row means that for this dimension, the team delivered per the set specifications and criteria; we are not showing these criteria here since they vary from one domain to another.
- Staying positive, a Yes for D2, Project Management success, but in the partial success row, meaning that the project was completed slightly outside the set parameters. For example, if the cost parameter is that ±10% of the set budget (or schedule) is a success, the actual performance of up to ±15% might be considered partial success, and maybe a ±20% would be a partial failure, and anything more than ±25% would be a failure.
- I want to stay positive, so we can start to break the cycle of doom. Let us say that despite the partial success of D2 and D3, the project achieved its objectives, D4.

The key here is that the criteria and thresholds should be established in the organizational project management system, not on a project-by-project basis. Also, as we said in Chapter 2, an organization can choose a three-grade model, a five-grade model, or fall back to the two-grade model.

17.6 Closing the Chapter

In this chapter, we continue with the process of explaining the foundation for the tailored approach for leading megaprojects. In the previous chapter, we covered the critical success factors, and in this chapter, we discussed The SUKAD Way™ and how to use it as the starting point for developing a capital projects model first, then upgrade it to the megaprojects model, via CAMMP™ MP. We also presented the project classification model and how to define and assess project success.

It is time for CAMMP™ MP, The Customizable and Adaptable Methodology for Managing Projects™ for megaprojects leadership, concept to success!

Chapter 18

CAMMP™ for Megaprojects

18.1 Introduction

We have been using the term CAMMP™ MP (CAMMP™ for Megaprojects) to represent this particular project class. CAMMP™ MP is the SUKAD proposed tailored approach for leading megaprojects, from concept to success. It is based on CAMMP™ for large-complex projects that we mentioned earlier but will have additional components. In this chapter, we start with generic tailoring of CAMMP™ to fit a typical capital project. Then we will shift the focus from managing a project to integrating the project work (facilities development) and project management into a broader, more strategic perspective—an asset management perspective. We will do this more extensive coverage via what we call the *four streams of the tailored approach*. Then, we will transform those streams to introduce the proposed project life cycle for megaprojects. Before closing this chapter, we will cover two other essential topics: the first is about the *phase/stage management process*, and the second is dedicated to *gates relevant concepts*. The remaining chapters in this section will continue to expand and elaborate the SUKAD tailored approach for leading megaprojects!

CAMMP™ MP is the SUKAD proposed tailored approach for leading megaprojects, from concept to success!

18.2 A Capital Project Model

18.2.1 An Industry Perspective

The Construction Industry Institute (CII), Independent Project Analysis (IPA), and other sources present to us a capital project model (project life cycle) in two primary parts. Simplistically, these are the *front-end planning* (or loading) part and the *execution* part. These two parts are for the development of the facilities, also known as the *acquisition phase* of the product (asset) life cycle. However, some of these models often ignore the early work, the Concept Stage, or what some might refer to as the *ideation stage.*

The PMI and ISO guides even skip the whole first phase—pre-authorization—which would include the Concept and Feasibility Stages (the Discovery Phase in CAMMP™). They view this work as pre-project and outside the scope of project management and not part of the responsibilities of the project manager. In addition to these two primary parts, some references mention a third zone, which is *operations.* During the operation period, which could be 25 years or more, there could be many other projects, each with its own project life cycle, as we presented in Chapter 2.

If we expand on this simplistic, two-part view, IPA and CII have similar models that split the front end into three major stages. IPA calls these stages FEL-1 (early work), FEL-2 (scope development), and FEL-3 (definition) (Merrow, 2011). For the front end, Barshop presents a model that consists of Assess, Select, and Define Stages (Barshop, 2016). We will link these to the CAMMP™ MP Model later.

18.2.2 A SUKAD Perspective

The following graphic is another image from *Project Management beyond Waterfall and Agile,* and it presents possible tailoring of CAMMP™ to fit a typical capital project. If we compare this model loosely to the IPA Model, we can conclude that:

- FEL-1 would be somewhat equivalent to the CAMMP™ Feasibility Stage.
- FEL-2 is in alignment with the CAMMP™ Requirements Stage (Design Basis Stage in the image).
- FEL-3, the Definition Stage, ties to preliminary engineering in the image.

IPA also introduces the concept of shaping or framing the opportunity. CAMMP™ did not have this as a specific stage, because most of its elements are part of the Concept, Feasibility, and Requirements Stages, but for megaprojects, we believe it should be separated, as you will see later. We will not expand on the descriptions of all of these stages or the model at this time.

Figure 18.1 is a possible project life cycle for a traditional capital project. The front end will be from the start to Stage Gate (SG) 5. The execution part will be after SG5. Notice that there are a lot of similarities with the CAMMP™ standard model, except we use different terms for some of the stages, and we have split the Implementation Stage into detailed Engineering and Construction Stages. Please note, this is NOT a set-in-stone model, it is only a possibility for capital projects.

18.3 The Four Streams of the Tailored Approach

Here, we will start to convert the capital project model that we just discussed. We will expose this model for further tailoring as we transform it from a typical capital project model to a megaproject model. The next part will cover this expanded project life cycle, but for now, we must address the asset perspective, which is to consider that every capital and megaproject will produce an asset.

Our slogan in this book has been, "A tailored approach for leading megaprojects, concept to success." By success, we made it clear that it is not limited to project management success but encompasses The Four Dimensions of Project Success™, including the objectives' success; in other words, the *asset* success. We can deliver a project per the inflated cost and schedule estimates, which would be a project management success but possibly a competitive failure. Also, it could result in production failure, or lack of demand, and the asset or business would fail to deliver an added shareholders' value. Therefore,

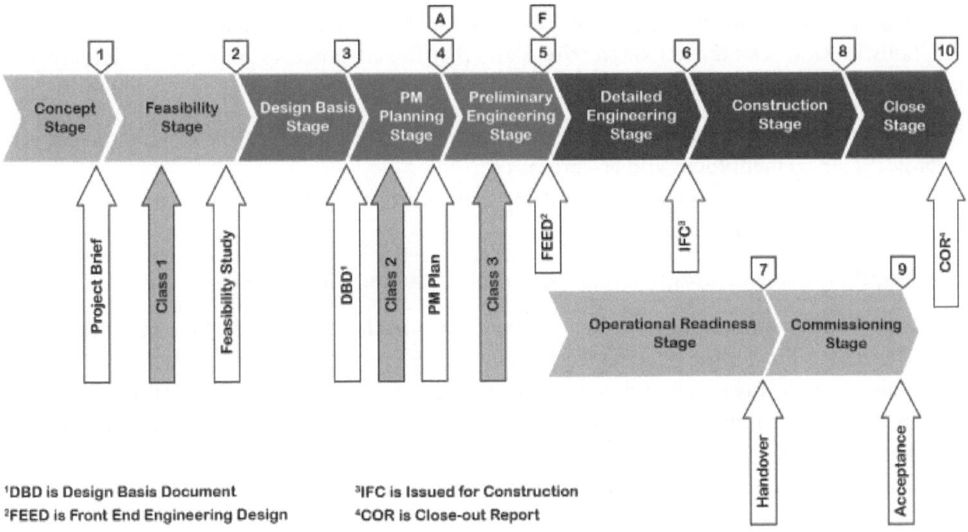

Figure 18.1 A Possible Project Life Cycle for a Typical Capital Project

we must think holistically, which is an asset perspective. Strategically, we view the asset perspective via four streams, as we present in Figure 18.2.

A tailored approach for leading megaprojects, concept to success

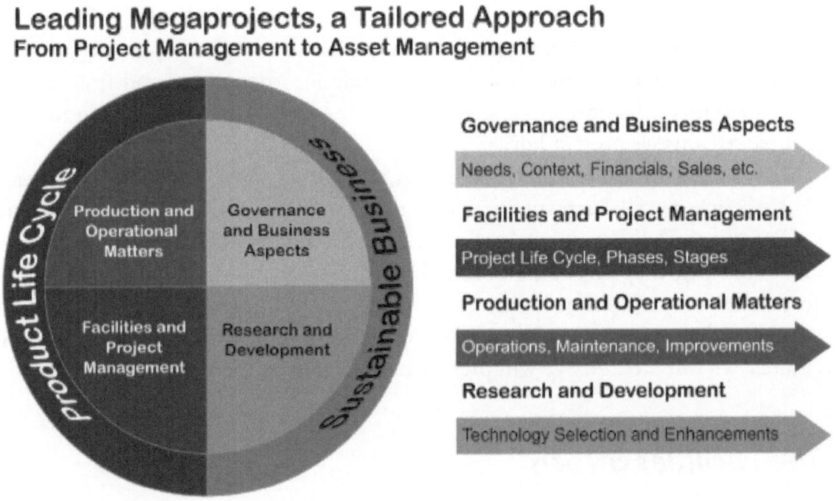

Figure 18.2 The Four Streams of the Tailored Approach Per the Asset Perspective

The image presents more or less the same information via two styles, and it is worth pausing to study it. The "holistic" view is on the left via the circle, and the "linear" process view is on the right. In general, what these views represent is the following:

- The facilities stream is about the acquisition phase of the asset. This is the stream that gets the project management focus and attention.
- The production stream is everything related to operational readiness, operation, and maintenance of the asset over its product life.

Basically, these first two streams would be part of the product life cycle.

- Then we add two streams under the *sustainable business* label, which include the business aspects of the venture, including governance and partners' relations, among other factors.
- We also include a research and development stream that contributes to sustaining the business. This is about technology and process research for the industry and the licensed technology for the asset facilities. This topic is over my head, and I do not have expertise in this area. Therefore, I would rather leave it out of this book rather than ruin it.

With the above, we highlighted the key concepts, and it is time to continue to elaborate.

18.4 The Proposed Project Life Cycle

18.4.1 Overview

As we have said from the beginning, leading megaprojects requires the integration of many fields of study and subject matter expertise, which is beyond the reach of a single reference or this book. We also made it clear that we aim to produce a tailored approach for leading megaprojects, concept to success. We do so by providing an umbrella view—a broader perspective to traditional project management, in which we elevate the focus toward an asset management approach. The conclusion of all of the above is CAMMP™ MP, which focuses on project leadership and product (asset) delivery. CAMMP™ MP is a more comprehensive methodological approach than a typical capital project or a large complex project delivery model.

In Figure 18.3, we present the project life cycle foundation that we expand from the four streams concept that we presented; the only missing part here is the technical stream or research and development. These three streams we present horizontally.

In CAMMP™ MP, we stick to the three phases, which we show vertically. In other words, this model consists of three parallel streams that link to the three CAMMP™ phases: Discovery Phase, Development Phase, and Delivery Phase. In this part, we will briefly explain the streams, but, once again, the details are for the next chapter. An important point to stress at this time—although the three streams are shown in parallel, they should be highly coordinated, if not integrated. The integration can be recognized by showing the stage gates on more than one stream, indicating that the decision would need input from more than one stream. We will present more details in the next chapter.

18.4.2 The Facilities Stream

This stream, which we represent in the top row of the figure (below the names of the phases), is the closest to the traditional project life cycle, which traditional project management is best equipped to handle. Here are a few comments about this stream:

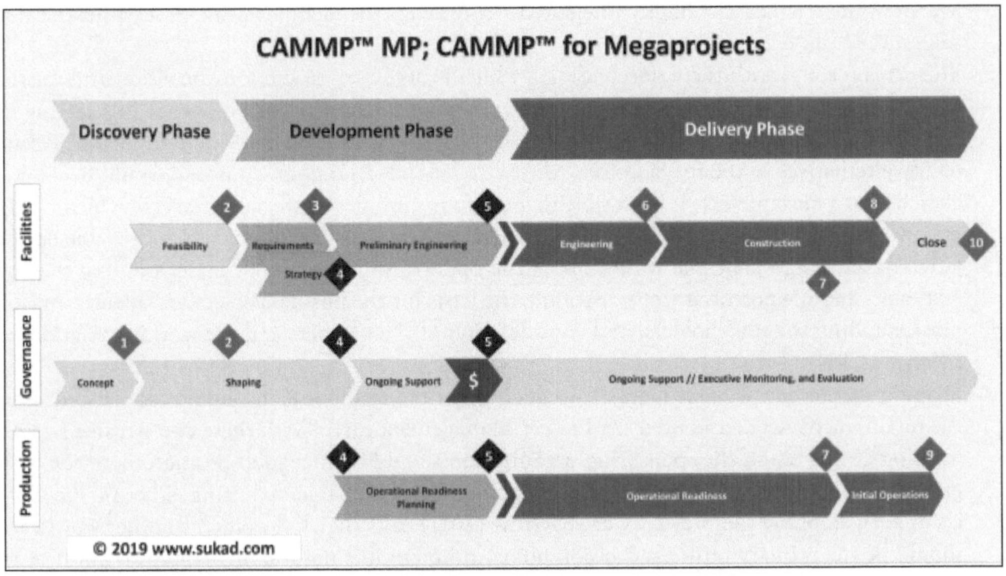

Figure 18.3 CAMMP™ for Megaprojects, the Tailored Approach for Leading Megaprojects

- It spans the three phases, like all the other streams. However, notice that instead of having the Concept Stage here, we include it in the next stream.
- The stages are Feasibility, Requirements, Strategy, Preliminary Engineering (Definition), Engineering, Construction, and Close.
- The Close Stage requires input from all streams.
- There is also an unmarked narrow part between Stage Gate 5 and the start of detailed engineering; this reflects a pause or limited effort in parallel to the business effort of securing funds. In other words, there could be a lag or a transition period between preliminary engineering and detailed engineering.
- We are not showing procurement here, because procurement of engineered items could be with engineering, and purchases of bulks, consumables, and subcontracts could be part of the construction phase.
- I think it is clear that a project manager with experience in facilities development would lead this work.* For megaprojects, the need is for a senior project manager or even a higher level, a project director.

18.4.3 The Governance (and Business Aspects) Stream

We chose the term *governance* for this stream for two reasons: (1) to expand on the concept of framing (Barshop, 2016) or shaping (Merrow, 2011) of the opportunity, and (2) to emphasize the governance aspects of megaprojects change initiatives.

- The governance stream is mostly driven by the business director and business divisions.
- The project starts in this stream with the initial concept identified by the business.

* Chapter 21 expands on project organization and the people aspects.

- We view Governance as a highly integrated stream with the facilities work, at least for the early stages, the front end.
- This stream starts with the Concept Stage, which is necessary to support the vision and mission of the organization and the business objectives, through the proper due diligence in selecting the right solution. The proper due diligence and scrutiny of the decision could determine the presence of any alternatives to the megaproject. Maybe an extended program approach would be a better choice than a megaproject. It is possible that the four sublimes of megaprojects* (Flyvbjerg, 2017) are pushing toward a megaproject for which a basic, lower-cost solution could be a viable option.
- Governance would continue with shaping the opportunity in terms of understanding the context, assessing the potential value, assessing the basis for the business objectives, identifying and understanding the stakeholders, and consideration of the partners and the working relationship (Merrow, 2011).
- Shaping should end at Stage Gate 4, at which point the facilities team should have also finished the requirements work and even the Project Management Plan. With these two streams laterally meeting at this point, the sponsoring organization should be in a good position to decide if the project is still viable and continue, or the viability is questionable, requiring either further studies or even stopping the project, what Merrow (2011) calls the "kill point." In other words, this should be the primary business decision point, although it is not the final approval point or the point of full funding. One of the reasons for making this crucial decision at this point is that the cost expended so far is about 1% or less, whereas the cost of the next stage, the Definition Stage, would be about 3% to 5%, and in some cases higher, especially if there is a need for committing the procurement of long-lead equipment.
- After full authorization at Stage Gate 5, the governance stream role will shift to an ongoing support role, which includes getting the necessary business and technical support from the various organizational functions, providing oversight—maybe through a steering committee, and handling the decision making for any changes in the objectives, although changing objectives at this point could have a significant adverse effect on the project's performance.

18.4.4 The Production Stream

We show this stream formalizing and starting at Stage Gate 4. We could have extended this stream to start with the Concept Stage, since it is critical to have input from the operational functions. However, we considered that these inputs are taking place as an integral part of the business and shaping process.

- As the project reaches the business decision point, Stage Gate 4, the integrated team would be ready to go into the Definition Stage, which is mostly about planning for the delivery phase.
- The delivery phase includes Operational Readiness and Initial Operations Stages. Therefore, there is a need to activate the Production Stream, starting with Operational Readiness Planning, in parallel to the Definition Stage.
- Since this work is necessary, and due to the uniqueness of the operational actions, we recommend assigning a leader for this work, such as an operational readiness director. More details about these stages are included in the next chapter.

In closing this section, we can see the need for a business director, a project management director, and an operational readiness director. All of these would report to a project executive, although in some cases, the business director could shift into the overall project executive role.

---

* Technological, political, economic, and aesthetic drivers.

18.5 The Phase/Stage Management Process

The previous part introduced the project life cycle for megaprojects, but at the highest level—the main streams. The next chapter will dive down to the phases and stages level. It is also essential to discuss the phase or stage management process and the organizational project management functions, which we will do in Chapter 20.

18.6 Gates Relevant Concepts

18.6.1 Types of Gates

Did you realize that not all gates are created as equals?

We have already discussed *stage gates* and *process gates,* briefly. Therefore, we have at least two types of gates. There are four process gates that the organization can use to help in managing a given stage or phase.* The other type is the stage gates along the project life cycle. In the CAMMP™ standard model, there are nine gates; there are ten in the proposed CAMMP™ MP. However, if you carefully examine these stage gates, you will notice the following:

- All stage gates, except the last one, have two functions: approve the work done before the gate, and authorize the continuation of the work.
- All stage gates in the front end, up to Stage Gate 5, could be used to terminate the project.† In other words, after verifying that the prior work is done correctly, the function of the stage gate should be to answer this question, "Should this project continue?"
- On the other hand, most of the stage gates after SG5 would also be about verifying prior work, like all of the other stage gates, but the second function would change from, "Should this project continue?" to, "Are we ready to continue?" For example, Stage Gate 6,‡ at the end of engineering, would be to verify that engineering has progressed enough, and the team would be ready to move to the construction stage.

18.6.2 Different Types of Stage Gates?

If you refer to Figure 18.3, shared earlier, you will notice that SG4 and SG5 have darker shading than the rest of the stage gates. These are highly specialized stage gates, with significant integration between the facilities stream and the governance stream.

Why is that?

- SG5 represents the final authorization and full formal funding of the project. For many

Rough Costs Along the Project Life Cycle

Barshop (2011) states that average costs for the early stages of the project, per the IPA database, are:

- Assess Stage (CAMMP™ Feasibility Stage) cost is about 0.5% of the total project cost.
- Select Stage (Requirement Stage) cost is 1.5%.
- Define (definition) Stage cost is 3.0%, excluding long lead equipment. For megaprojects this could be higher: 5% to 10%.

The above are only capital costs and exclude any other working capital or shaping costs.

* Refer to Figure 2.6 on page 21.

† That does not mean the project could not be stopped after SG5, but the chance is significantly reduced.

‡ Figure 18.3.

organizations, this is the ultimate control point and what they consider the official baseline. Most of the literature review refers to this point when discussing cost and schedule performance.

- SG4 can be as crucial as SG5, if not more so. We mentioned earlier that Merrow (2011) views this as the critical gate and the point at which the **business decision** should be made. In other words, Merrow advocates that the Definition Stage should be the stage focusing on execution planning rather than the business decision. It is also common that, at this point (SG4), the team would seek advanced funding to cover the cost of the Definition Stage and possibly the commitment for long-lead items.
- SG2 is also vital, in that it functions as the initial authorization, but there is no formal funding at this point.

18.6.3 A Common Pitfall

Many business professionals assume that the stage gate process is limited to the project management and engineering functions, and the process is structured to meet an engineering purpose. However, the stage gates are primarily to serve a business purpose, which is to allow the executive management, at specific points in the development process, *"to make decisions to stop, recycle, or proceed"* (Merrow, 2011). In other words, the challenge is to get the businesspeople interested enough in participating in decision making, **effectively**. I stress *effectively,* which implies in-depth review and analysis, challenging the decision-making process to build **real confidence** in the viability of the megaproject—*real* confidence instead of *the illusion of* confidence. We believe that the CAMMP™ MP model emphasizes and mandates a process to avoid this pitfall.

18.6.4 Another Pitfall

Merrow argues that the gate monitoring passage from FEL-1 to FEL-2 is the least well-managed of all the gates. *"In too many companies, what would constitute an acceptable package for this gate is not well established. This weakness in the FEL-1 stage foreshadows the weakness we see in the shaping process for megaprojects"* (Merrow, 2011). CAMMP™ and CAMMP™ MP both stress the need to have a strong gate here, which is Stage Gate 2. This gate concludes the Feasibility Study work and requires a formal Project Authorization Document as the last step in the discovery phase.

18.6.5 Engineering Stage Gate

If you recall, the standard CAMMP™ model has nine stage gates, whereas the CAMMP™ MP has 10 stages. The added stage gate is the result of splitting the Implementation Stage. Therefore, there is another gate between engineering and construction. In Figure 18.1, we show this stage gate at the end of engineering. This would be appropriate in situations where the procurement approach follows the concept of separating engineering from construction, such as the use of DBB (Design–Bid–Build). In DB (Design–Build) or EPC (Engineering–Procurement–Construction), where the design (engineering) and build (construction) overlap significantly, due to one DB or EPC contract, this stage gate (SG6 in the model) would be earlier in the engineering phase, but at a critical point at which the team needs to decide readiness for effective field mobilization.

"Owners have largely abandoned a decision gate in their work process just before committing to field execution of the project" (Merrow, 2011), and the CAMMP™ MP model advocates a reversal of this practice.

Merrow holds that the absence of this stage gate could have an impact on construction, because if the engineering is not progressed enough, the project owner must have the courage to slow down the project. The case study in Chapter 7 shows that the NAP project was a victim of not having the engineering gate and mobilizing to construction too soon. This approach also supports the mixed contracting* approach, since the construction contractor is separate from engineering and procurement.

18.6.6 Stage Gates Criteria

The detailed criteria for the various stage gates are outside the scope of this book, because it may vary from one industry to another within the capital projects industries—industrials, buildings, and infrastructure. What is important to stress, which also helps in avoiding the pitfall discussed, is the need for the valid and active interchange of information between the proposed streams that we mentioned earlier. If you need more in-depth knowledge to define the criteria in your environment, we suggest the work of CII and IPA. However, we must stress at this point the following: most of the stages could be approved by a steering committee or are best left to the executive committee. However, the financial gates that we mentioned earlier, SG4 and SG5, are likely limited to the board of directors.

18.7 Chapter Summary

With this chapter, we started to get into the core of the tailored approach for leading megaprojects. We started with a basic capital project model, tailored from the standard CAMMP™ model. Then we introduced the four streams of the tailored approach, which we use as a stepping stone of transforming traditional project management into CAMMP™ Megaprojects, which views projects as integral parts of assets, and consequently, managing and leading megaprojects would integrate into an asset management perspective.

We also introduced the project life cycle for CAMMP™ MP with three of the four streams; the research and development stream is outside the scope of this book. We briefly discussed the stages of each of the three main streams: facilities, governance, and production. We also emphasized the need to work these streams in a highly coordinated way, or, it would be more appropriate to say, using an integrated approach.

In the next chapter, we will discuss the phases and stages in more detail than we could do in this chapter. Then, Chapter 20 will shift focus to address the core organizational project management functions in the context of CAMMP™ MP and the stage management process. These include topics such as cost, estimating, control, and change management, among other functions. The last chapter addresses the project organization and the people aspects.

* More on the mixed contracting approach in Chapter 20.

Chapter 19

CAMMP™ MP: Concept to Success

19.1 Introduction

We chose our catchphrase, *Concept to Success,* for the title of this chapter, because it is time to go through the project life cycle, step by step, or stage by stage. We will follow the stages from the concept and business need (or trigger) all the way to success. If you recall, the ultimate success cannot be measured until after the facility is in operation and the asset is producing shareholders value.

The structure of this chapter is straightforward. On the following pages, we re-share a previous image and add another image.* This next image shows the sequence of the stages with the streams. The streams are the horizontal swimlanes, whereas we display the CAMMP™ phases vertically. This is a very high-level perspective, limited to the stages of the project life cycle. However, it is a map for this chapter. For better clarity, we include in the graphic the names of the main stages' deliverables and the stage gates. A flowchart with more detail, adding the stage management processes and other vital steps, is available for download from the SUKAD website.† An abbreviated version of the flowchart is shown in Figure 19.2.

In this sequence, we unfortunately have to stop at the end of the project life cycle, with closing the project at Stage Gate 10. However, keep in mind that the team can only assess two of the success dimensions before closing the project life cycle. The other two dimensions are the responsibility of the organizational management and would happen after going into full operation, which is also not shown here. Let us present the images:

Each of the following chapter's parts will cover one of the three phases. Within each part, I will include an overview of the phase, then briefly explain each stage. Keep in mind these critical points as you review these parts:

* We will also repost the project life cycle so the reader can see these two images sides by side.

† Refer to Appendix A for details.

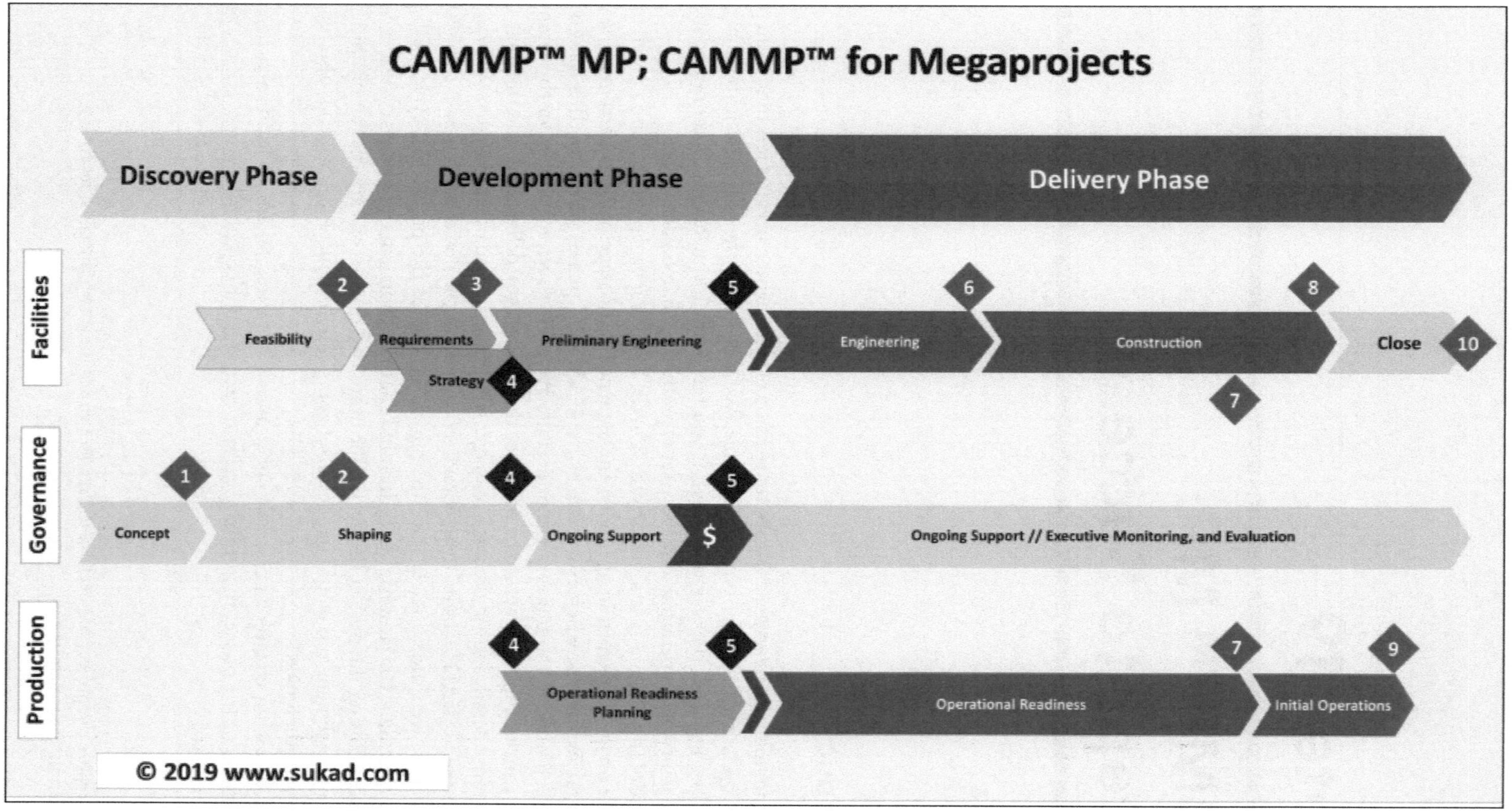

Figure 19.1 CAMMP™ for Megaprojects, Project Life Cycle

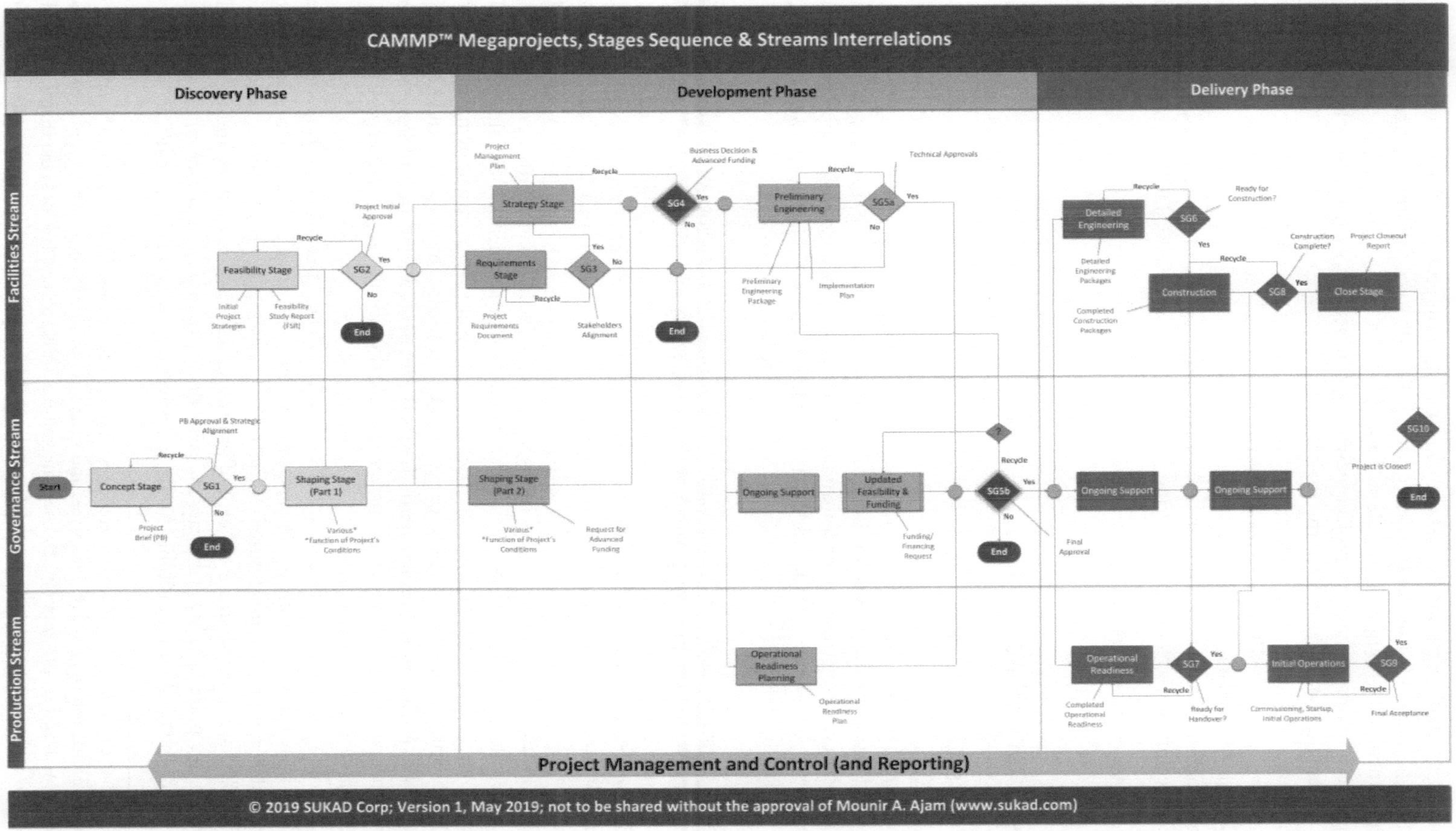

Figure 19.2 CAMMP™ for Megaprojects, Streams, Stages Sequence with Stage Deliverables and Stage Gates

- The management of each stage follows the stage management processes that we will discuss in the next chapter. In other words, for each stage, we should authorize, plan (management and details), implement, control throughout, and close the stage.
- Discussions of the organizational project management functions, such as cost, schedule, risk, and other topics, are also the subjects of the next chapter.
- For the details of the content of each stage deliverable, we must regretfully refer you to an external source. It would add a 100 page to this book if we were to include here. Therefore, there is no need to repeat here if available elsewhere.*
- The last point is to reiterate what we have said all along: the technical processes, such as market research, technology selection, engineering, and construction, are outside the scope of this book.

This is the longest chapter in this book, so get ready.

19.2 Discovery Phase

19.2.1 Phase Overview

We chose the term *Discovery Phase* because this is indeed about discovery—discovering the needs that trigger the consideration for the project, exploring the options that could help the organization meet the desired objectives, developing the justification and the business case, and closing with proper feasibility and initial authorization. Notice that the CAMMP™ standard model differentiates between the business case and the Feasibility Study; this differentiation is crucial for megaprojects.

Why?

We see the business case as the need, justification, and selection of the solution, which we summarize via the Project Brief. If the organization performs the right due diligence at this point, it only means that the project is *justified,* but we are not sure yet if it is *feasible.* Approving the project based on needs and concept could trigger the lock-in mindset, which could possibly contribute to the megaproject's potential failure. We think one of the challenges of megaprojects is the satisfaction with "justified," which leads organizations to rush into the Development Phase in preparation for the implementation work (execution), or what CAMMP™ calls the Delivery Phase. In response, we advocate the need for a project Feasibility Study, as an independent study from the business case and needs assessment. Therefore, we will have a business case as an essential part of the Concept Stage and a Feasibility Study with its own dedicated stage.

The Feasibility Study could be in two major parts: one part focuses on revisiting and expanding on the business case and justifications (market analysis, selection of desired solution, business, and economic risks), and the other part, often missing, is to answer this question: *does the organization have the resources, capabilities, and proper realistic strategies to deliver success?*

To answer the above question, we must first reflect on why we chose the term *concept to success.* We did so for the following reasons: (1) to stress that success is the end objective, and (2) to emphasize that the end is not limited to completion and closure but to a formal assessment of performance, which we want to be a success. We have touched on the concept of project success more than once, and we are likely to visit it again. The key point here is to advocate the criticality of taking time in each phase and stage of the project. Hence the need for a Feasibility Study to answer the question posted earlier.

* For the details of every stage, your best reference is *Project Management beyond Waterfall and Agile* (Ajam, 2017). For additional information, the templates and instructions, in Appendix A we provide the link to our website, where we host this information. By the time you read this content, it is possible that the Uruk PPM Platform will be live and the sequence, templates, and all of the steps will be built into this online solution.

Once again, the Discovery Phase splits the business case and its development from the feasibility of the project, the organization's ability to deliver the final product successfully, and the capability to produce per the expected demand and market needs. The Discovery Phase closure is triggered by passing through Stage Gate (SG)2, leading to the formal and initial Project Authorization Document. If you recall the discussion on gates in the earlier chapter, this is a crucial gate, although it often is not given the needed attention.

One last statement here: the standard model provides us with two stages, concept and feasibility. That original model did not emphasize the separate streams that we clearly identify for CAMMP™ MP. Therefore, for megaprojects, we propose the following:

- We shifted the Concept Stage to the Governance Stream.
- Upon concept completion, symbolized via an approved Project Brief at SG1, the Feasibility Study work could start. In addition to the feasibility, we will also begin the shaping process.
- We are including the Shaping Stage in the Discovery Phase, although a significant part would expand into the development phase. This fact is what is leading us to split shaping into two parts (in the flowchart). Part 1 is in parallel and tightly integrated with the Feasibility Stage, and Part 2 is aligned with the Requirements and Strategy Stages.

19.2.2 Concept Stage

The Concept Stage in the standard CAMMP™ Model includes:

- Documenting the idea for the project, project description, or statement of work.
- Outlining the conceptual business case (justification).
- Determining alignment to the organizational strategic objectives.
- If all are acceptable and a priority, management could approve the idea and authorize a Feasibility Study.

The output of this stage is a document we call the Project Brief. The Project Brief would go through the first stage gate. This may sound as simple as 1, 2, 3, but it is not, at least for the megaproject. Let us elaborate.

The Project Brief

The idea for a project could come from anyone within the organization. In theory, there are two sources for ideas. The first source is top down, meaning all projects must be initiated from the strategic planning process because the strategic plans must have strategic initiatives, which executives would convert to portfolios or strategic programs and, in turn, projects. This approach is the ideal scenario for organizations that genuinely follow a strategy formulation and implementation process. In this scenario, every project aligns with the organization's strategic direction, which is a requirement for the Project Brief.

The second source is bottom up. Ideas could come from an employee, a strategic business unit, a department, or executive management. It could be a new concept or a modification of existing one. Organizations should have a process for idea generation, review, and disposition, which could be part of the business planning process or strategic planning. Some organizations could even have an innovation management process, in which ideas could come from anywhere, including from customers.

Most megaprojects will likely come via the first route—the strategic business plan.

Regardless of what idea management system exists, the idea for a new project must go through an idea management system or assessment process. Therefore, documenting and justifying the idea would require significant effort for megaprojects, and here the team must be careful to ensure the right level of effort that would be suitable for the complexity and massive size of megaprojects. Once the team completes the assessment, noting that there could be more than one solution to achieve the strategic and business objectives, it would be time for the first stage gate, at which executive management should scrutinize the work to avoid the optimism or skepticism biases. Optimism bias could approve a project that has a questionable benefits–cost scenario, while skepticism bias could stop a viable project.

A key point to remember here—it should be understood that at this point no one has all of the answers, and there could still be a few options to consider, which the team could carry over to the next stage. Another critical point that could be specific to CAMMP™ is this: the business case is the justification that is assessed and included in the Project Brief. This business case focuses on needs and justification, but that has not been validated commercially yet—this is also for the Feasibility Stage.

> There is nothing as wasteful as doing with great efficiency
> that which does not have to be done at all —Anonymous

Stage Gate 1 (SG1)

It is time to bring strategic project management into the discussion.

This first stage gate, SG1, is about approving the Project Brief, but it is primarily about strategic alignment. **Every project must be in line with the organizational strategic direction and objectives.** If organizations do not consider this strategic view as a priority, then project management would not be highly effective in contributing to long-term organizational success. Remember, projects are value-improving change initiatives in organizations, and management should consider them carefully. At this stage gate, the decision makers could (1) end* the project, (2) recycle for updates, modifications, or corrections, or (3) approve it to proceed to the Feasibility Stage and the Shaping Stage.

Who approves the Project Brief? Another way of asking the question is: Who would be at the stage gate?

This is a function of organizational size and culture. In some organizations, empowered line managers or senior executives might be able to approve up to certain pre-authorized limits or other predefined criteria. For megaprojects, this could be executive management, the board of directors, a government ministry, or any other high-level decision maker.

What Is Next?

With approval at SG1, the organization will appoint a project sponsor to assume leadership for the project. The project sponsor is senior management's representative and the person responsible for the project overall, but not the day-to-day project management (Englund, 2006). For megaprojects, we refer to this position as the *project executive.* Alternatively, the organization may choose to have a business director at this point acting as the project executive. However, we recommend segregating these two roles. The business director will be responsible for all Governance Stream actions, whereas the project executive would be the one accountable for the whole venture.†

* In the flowchart "end" represents ending the work, either permanently (project canceled) or temporarily (putting the project on hold).

† We will elaborate on the project organization in Chapter 21. For now, think of the following: each of the three streams will have a director (business director, project director, and operational readiness director). All three will report to the project executive.

In addition to the business director, we advocate that the organization have project management involved at this point. This involvement is best represented by the appointment of a project director, the person accountable for the Facilities Stream. The business director and project director will work closely throughout the project. The aim is to have a healthy marriage (as healthy as it can be, until the end "do us part"). The only sad part here is the divorce at the end of the project, but hopefully, it would be amicable.

19.2.3 Feasibility Stage

Overview and Sequence

The main purpose of this stage is to validate the idea and to authorize the project. Note the intentional shift from *idea* to *project*, which is to emphasize that only after this stage is there a project authorized to proceed. Validating the idea is accomplished via a comprehensive Feasibility Study that will consider various factors.

Let me elaborate on the primary reason we separate the business case from the Feasibility Study. Even though many projects could have a strong business case (need), at times that is not enough to approve a project, because the organization might not be able to deliver it successfully owing to a variety of factors, which could be resources, expertise, or something else. Therefore, the CAMMP™ viewpoint is that organizations should pursue a project if it is justified (there is a business case) and the organization's management believes that the project can be delivered successfully—in other words, the organization would realize the expected benefits. That is the purpose and definition of the Feasibility Study and why it is a critical stage. Organizations cannot afford to skip performing a proper Feasibility Study—it would be a waste of effort and resources if the organization only discovers that a project is not feasible during project development or delivery phases.

There are three main stage deliverables in this stage: the Feasibility Study Report (FSR), the Initial Project Strategies,* and the Project Authorization Document (PAD). There is a stage gate at the end of the study, which will separate the first two deliverables from the third. The sequence of events for this stage is simple. Please note that the same general sequence will apply for the upcoming stages.†

- Executive management will authorize the project director to start working. This authorization is formalized per SG1 documentation or through a separate stage authorization document.
- The project director will identify and mobilize the necessary project management personnel to help in planning the stage by developing a Stage Management Plan. Refer to the "Project Management Team" textbox.
- Once the stage management plan is complete, there is likely a need for a Stage Detailed Plan. The PMT and other lead specialists with expertise in feasibility studies will develop this plan. The project director, project management personnel, and technical leads will be the core project team (PT).

> **Project Management Team**
>
> The project management team (PMT) consists of the personnel that will support the project director in managing the stage and project work. The team members may include planners; cost, schedule, and procurement specialists; and other personnel as required.
>
> For megaprojects, the PMT will start with a few people in the first stage of the project and grow as the project proceeds into the other stages, peaking during the Implementation Stage.

* This could be part of or independent of the Feasibility Study. In the standard CAMMP™ model, we do not have this deliverable.

† A useful reference is the part of stage management processes in the next chapter.

- Once the team finalizes the Stage Detailed Plan, the PMT will identify and mobilize the feasibility team—the technical and functional personnel who will conduct research and develop the Feasibility Study. This action will lead to acquiring the remainder of the team members to complete the project team for the stage. *Stage team* would be a more appropriate term, but we will use the term *project team* (PT) to represent the stage teams and overall project team.

> **Technical and Functional Personnel**
>
> The technical and functional personnel are likely to change from one stage to another. These are the personnel that will do the technical or product-related processes and work.
>
> For example, for the Feasibility Study, the project may need experts in market research, financial modeling, economy, and sustainability. The number and disciplines of these resources would be a function of the project domain, type, and size.

- The PT will perform the Feasibility Study and prepare the final report with their recommendations. The work will be carried out by the technical and functional personnel but managed by the project director and supported by the PMT members.
- The PT will submit the Feasibility Study Report (FSR) to management* for review and decision making at Stage Gate 2.
- If executives approve the project at Stage Gate 2, the project executive will issue the Project Authorization Document (PAD), the authorization for the whole project. Obviously, if no approval is given, the project would be stopped and closed.

One more comment: The main difference for megaprojects in comparison to what we just presented is that the input for SG2 would include the FSR and various documents from the Shaping Stage, Part 1. We will also have the initial strategies report.

The Feasibility Study Report

The FSR will include the content per the outline presented in Figure 19.3. This is similar to other project classes, such as large-complex projects. For a description of every item, once again we refer you to Appendix A, where we provide the links to the location of the details. Here, we will address some of these factors from a megaproject perspective.

- One possible difference between large-complex projects and megaprojects is that, due to the addition of the Shaping Stage for megaprojects, some of these items could shift to the Shaping Stage or could be performed jointly.
- Internal cross evaluations are designed to identify whether there is a similar project going on in a different division of the organization. However, because these are megaprojects with high visibility, it would be unlikely that something like this could happen by more than one division.
- For megaprojects, the most likely scenario is that the project owner organization will not have all of the professionals necessary to handle the project. It is almost a certainty that engineering and construction will be left to specialized contractors. The vital question would be, does the project owner have the necessary resources in terms of availability (or lack) of competent resources and adequate numbers to handle the critical work that should be the owner's responsibility? Project management competence and capacity will be a significant factor to assess as well.

* We use the term *management* in this book to represent executive management, which may include the board of directors, depending on the context. In other words, these are the decision makers.

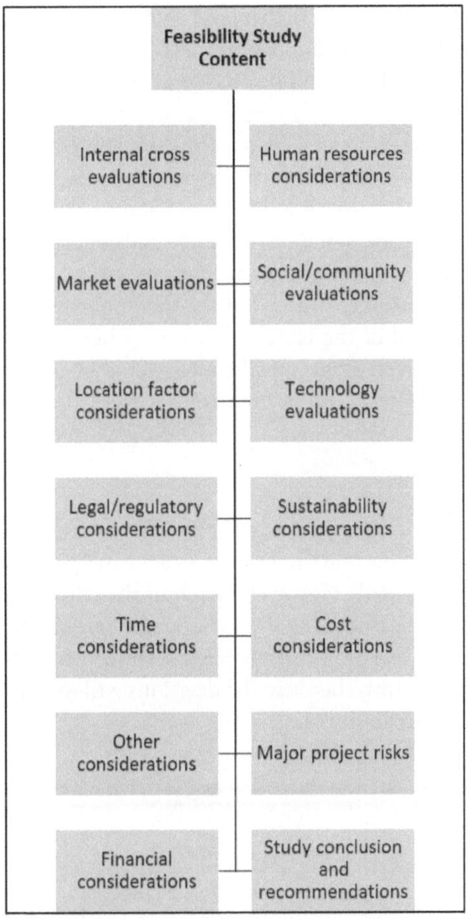

Figure 19.3 Feasibility Study Report

- Market research is often vital for some projects, such as commercial projects releasing new products or services into the market. *More on this in the Shaping Stage.*
- Social/community evaluations: *More on this in the Shaping Stage.*
- Location factor considerations: *More on this in the Shaping Stage.*
- Technology evaluations: *More on this in the Shaping Stage.*
- Legal/regulatory considerations: *More on this in the Shaping Stage.*
- Sustainability considerations: Today, sustainability considerations are becoming the trend and the right thing to do. The best reference for more details on sustainability on projects would be GPM® (Green Project Management Global) and their guide (GPM, 2012) and method.
- Time (schedule) considerations: Clearly, the challenges associated with scheduling should be evident from the chapters in Section II on the current state of practice. Although during the Feasibility Study, there are not enough details to allow the PMT to produce a detailed schedule, the organization still needs a rough schedule, because it will impact the result of the Feasibility Study. Management only needs an executive summary, but one that is dependable and includes a reasonably predictable completion date. As a UC Berkeley study shows (Ibbs and Reginato, 2002),

predictability is one of the benefits of a higher level of organizational project management maturity. In other words, an organization with a competent organizational project management function (organizational and individuals) should be able to determine the overall project duration within reasonable control limits.*

- Cost considerations: Cost is also an area of concern, especially because the majority of megaprojects experience overrun, as we have seen from the literature review. Similar to schedule, the organizational project management maturity is a direct contributor to cost predictability.
- Other considerations: What are some of the other considerations that would be unique to the type of project considered?
- Major project risks: The Feasibility Study and the shaping work address the overall project risks, such as market, legal, and all of the factors that we list here. The key message is that the focus would be on the overall project. Stage-specific risks would be part of every stage detailed plan.
- Financial considerations: *More on this in the Shaping Stage.*

Initial Strategy Section

The CAMMP™ standard model includes a Strategy Stage. The primary purpose of the Strategy Stage is to consider strategies for various important factors and produce a Project Management Plan. This same stage is also included in CAMMP™ MP. However, owing to the criticality of megaprojects, it would be necessary to perform an initial assessment for the various strategies at this time, then revise or update them in the Strategy Stage. This initial project strategies assessment could be part of the Feasibility Stage, the Shaping Stage, or performed jointly, because the decisions will impact the facilities and governance streams. We include this in the Feasibility Stage.

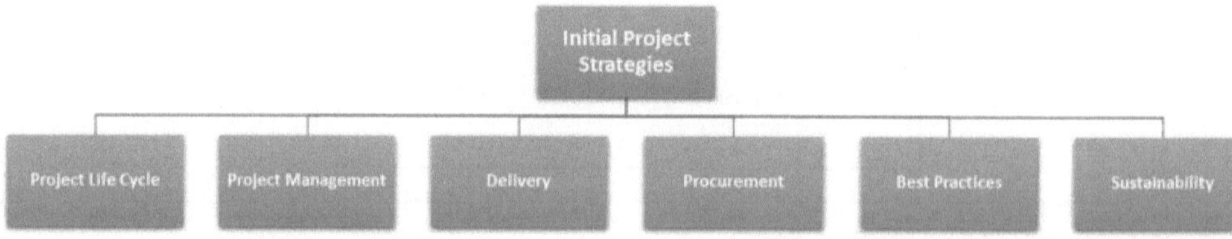

Figure 19.4 The Project Strategies, Initial Assessment

The strategy section (and initial project strategies assessment) includes the items shown in Figure 19.4. Depending on the nature of the project, the team could add or delete from this list. We will now briefly discuss each item on this chart.

- **Project Life Cycle.** If an organization has (and it should) a standard project life cycle for these projects, which could be similar to what we present in this book, then there is nothing to do here. Therefore, the team could adopt the tailored method. However, we still recommend that the team take a look at their standard megaproject life cycle and decide if any modifications are necessary.
- **Project management strategy.** At this early stage, maybe the only question to answer is, will the project owner have the competence and capacity to handle this project, or will the project management

* The reasonable control limits for schedule developed at the Feasibility Stage could be –10% to +25% (or a narrower range). The numbers could vary greatly depending on project type, industry, and organizational maturity.

have to be partially or fully outsourced to a consultancy firm? A follow-up question would be, if outsourced, when would the starting point be—as soon as possible or only for the delivery phase?

- **Delivery strategy.** What are the general project conditions, site location, remoteness, congested areas, etc.? Does location (or other factors) justify pre-assembly or modularization? How about labor availability? One of the objectives of the shaping strategy is to conduct site assessments, and these could be some of the topics to address.
- **Procurement strategy.** What is the high-level procurement approach? What could be done internally, and what would require outsourcing? Can the project owner organization handle the preliminary engineering work (licensed technology), or will even that be outsourced? Will the Implementation Stage be treated as one stage with one contractor, or split, separating engineering from construction? What is the initial consideration of the commercial nature of the contracts? Will the host country require a certain percentage of expenditures in local hires or local components?
- **Best Practices Strategy.** CII (Construction Industry Institute) offers a set of seventeen best practices. IPA (Independent Project Analysis) has what they call *value-improving practices.* Will you apply some of these practices on your project? Which ones, and when along the project life cycle? Would the project consider benchmarking support or assessments?
- **Sustainability Strategy.** It might be a challenge to address sustainability on megaprojects, but it is a must, in our view. I do not think we have stressed this point before: sustainability is about the 3Ps—people, planet, and prosperity. How would the project address these points?

If you have noticed, all of this work is still in the Feasibility Stage.

Why?

Because we honestly believe that megaproject success and failure are pre-determined here. In other words, if you do not put the train on the right track, could you be facing a derailment? Could you be triggering a cycle of doom? We do not want that, do you?

The megaproject success and failure is pre-determined in the discovery phase.

Stage Gate Two (SG2)

The Feasibility Study is an essential requirement within the project life cycle and leads to Stage Gate 2 (SG2). This control point provides management with the opportunity to validate that the team has performed a proper Feasibility Study. If management is satisfied with the level of effort and comfortable that the study is unbiased, then management would consider the result of the study and recommendations. **The vital question here: is the project feasible?**

Satisfying the feasibility question leads to the second objective of SG2, which is to decide if the project is a priority and should proceed to the next stage. In other words, this is the time for the initial go/no-go decision. If there is a go decision, the next step is for the project executive to develop and issue the PAD, officially launching the project. If the decision is not to proceed, the team will perform project closure, documenting the justification for stopping the project and any lessons learned. It is also possible that an organization could decide to hold a feasible project for future development.

As we have been saying, in addition to the Feasibility Study report, for megaprojects, we have the initial project strategies as input to SG2. Further, we also have the output from the first part of the shaping work that would help contribute to the decision-making process. We will discuss shaping soon.

Finally, SG2 will lead to the initial approval, which is represented via the PAD.

Project Authorization Document (PAD)

The PAD authorizes the whole project; it is what PMI and ISO label the *project charter.*

What does authorization mean?

- It announces to the organization that management has authorized a new project and is allowing the use of company resources for the project.
- It identifies the project executive, project director, possibly the business director, and their authorities.
- It serves as a critical reference and control document for future work. CAMMP™ uses it as the base for the third project success dimension, project delivery success. It is also the second control reference point.*

> **Project Prioritization**
>
> If a project is feasible, management will have enough information to make a decision on whether to authorize the project. However, before they make the final decision, they would consider the project to be part of a pool of other projects, a portfolio of projects and programs. In this case, management should approve the high-priority projects using whatever criteria they believe appropriate for their organization's strategic objectives.
>
> This concept is project prioritization, which is part of portfolio management.

Does this mean the project manager can proceed with the project all the way to completion?

Usually, the proper answer would be *NO.* What the PAD tells the organization and the project director is that executive management has a clear intent to take the project to completion, yet they are only *effectively* authorizing the next stage. The decision of whether to continue with the project will be a subject of the subsequent stage gates.

For the content of the PAD, please refer to the external resources mentioned in Appendix A.

19.2.4 Shaping Stage, Part 1

Overview

As we have stated earlier, we added this stage, separated from feasibility and requirements, because for megaprojects, this can be tremendous effort. We have adopted the term *shaping* from *Industrial Megaprojects* (Merrow, 2011), where Merrow talks about "shaping the opportunity" or "shaping the project." In *Capital Projects,* Barshop (2016) discusses a similar concept that he calls *framing.* This effort should be driven by the project executive or business director (Governance Stream), in close collaboration with the project director (Facilities Stream). The Shaping Stage straddles CAMMP™'s Feasibility Stage and Requirements Stage. We could have considered it as a separate phase, but we prefer to keep it the way we have shown it, in the project life cycle model and the stage sequences flowchart.

Transferred from Feasibility

Part of this work is integrated or related to the Feasibility Stage, such as the items you see next. These are included in the Feasibility Stage in the CAMMP™ standard model but separated in CAMMP™ for Megaprojects.

- **Market research.** Market research is often vital for certain projects, such as commercial projects releasing new products or services into the market. For megaprojects, misreading the market could be the leading factor in setting the project on a path of success or failure. Barshop (2016) clearly

* In the next chapter we will discuss the control reference points.

states that, on average, capital projects experience about 10% reduction in their net present value (NPV) due to lower than expected sales. Barshop also states that three out of 10 projects in the IPA database had issues related to market analysis.

- **Social/community evaluations.** This assessment can be critical on projects in remote locations or projects that would have a significant impact on a community, which is to be expected in the mega world. It is also possible that for remote locations, the project might have to provide a lot of infrastructure facilities to accommodate the local communities or host countries. The Shaping Stage should include extensive assessments of the location of the project, which would cover these topics.
- **Location factor considerations.** The location factor can be a significant item. Merrow (2011) has shown differences in the performance of megaprojects based on site conditions, such as remote or non-remote. Many of the factors related to this matter are things such as the need for building infrastructure facilities independent of the production facilities. Some of these could be needed for access, but others could be required in response to political requirements, or even from NGOs and community groups.
- **Technology evaluations.** The technology evaluation depends on the domain or type of project. These might be factors related to the technical or technological aspects that the team must consider during this assessment. This factor is vital when the project requires the use of new, possibly unproven, technology, which would add to the challenge of the project. Another related item here is that some of the technology that would be used on the project would be licensed, and the licensor may have significant terms and conditions. The research done by IPA also highlights this point as one of the factors contributing to the project's performance.
- **Legal/regulatory considerations.** Megaprojects will definitely have to comply with legal and regulatory requirements, permitting, safety, and environment laws and regulation, among numerous other factors. In Chapter 6, we had listed that permitting is one of the challenges identified by Merrow (2011), and it can vary from one country to another. Further, this area can be affected by political uncertainty or corruption.
- **Financial considerations.** Financial considerations depend on all of the other factors. Financial considerations are about project financing or funding and economic returns. Due to the size and cost of megaprojects, this topic leads to two significant challenges. The first is financing from banks, which can bring with it a host of restrictive terms and conditions. The other challenge is the need for partnering—joint venture—which has many uncertainties that the partners should address carefully. These include preference for types of contracts, financing, development and delivery strategy, integration of teams, which organizational project management system to use, and so on.

If you were familiar with CAMMP™ before this book, you will notice that, although all of these topics are in the CAMMP™ standard model, they are highly sensitive and have significant impact in the mega world, justifying the Shaping Stage.

Shaping from Literature Review

Although we have adopted this term from Merrow (2011), other references have used this term.

"Analyzing many projects, we noted that shaping occurs through progressive problem solving over multiple episodes" (Miller, Lessard, and Sakhrani, 2017). The authors identified six episodes* for the shaping process; the first three align to shaping as presented by Merrow (2011); the other three are linked to project

* *Episode* could be a loose equivalent to *stage.*

definition, implementation, and operation.* During the shaping process, many issues, challenges, and actions are happening, which may resemble rolling a huge rock uphill.†

Merrow (2011) suggests that organizations must formalize and institutionalize the shaping process, which helps with the proper discipline, which is often absent on the business side of the project owners' organizations. He also suggests that the board of directors must establish a standard work process and hold the businesspeople accountable.

"Opportunity shaping is a business-led process by which sponsors‡ evaluate the key attributes of a potential project, develop and gather information that is needed for key decisions, and then allocate the value of the project to the various stakeholders to make the project environmentally stable enough for successful execution while holding enough of the projects' value for themselves to make the venture worthwhile" (Merrow, 2011, p. 54).

Shaping requires significant scrutiny of the decisions and decision-making process. Organizations must understand the implications in terms of risk, resources, and reputation, among other factors. Merrow presents us with a shaping model. We will list the steps, but for more in-depth learning, we advise you to seek the source, especially if you want to use it to build your organizational shaping standard. The model steps are:

1. Understanding the context
2. Assessing the potential
3. Assessing comparative advantage—the basis for business objective
4. Identifying and understanding the stakeholders
5. Partnering

Reinforcing these points,

"Projects fail for reasons having nothing to do with shaping, but not very often. Shaping errors and omission are the most common root cause of many of our failures" (Merrow, 2011). Further, Merrow groups the errors and omissions into five groups, with the fourth one being by far the most common.

1. Failure to achieve full stakeholder alignment
2. Ceding so much value to other stakeholders that the project has no value
3. Failure to develop coherent objectives
4. Impracticable cost, schedule, and quality trade-offs
5. Overly conservative targets

19.3 Development Phase

19.3.1 Phase Overview

The Development Phase is necessary for developing the concept of the authorized project to ready it for the Delivery Phase. In this phase, the level of effort is more than it was for the previous phase, but still low in comparison to the next phase. The CAMMP™ standard model splits this phase into three stages—Requirements, Strategy, and Definition. For CAMMP™ MP, we continue to use the names of the first two stages, but we will use the Preliminary Engineering Stage instead of the Definition Stage.

Also, for megaprojects, the Shaping Stage continues in parallel to the Requirements and Strategy Stages. The core idea is that shaping should end with FEL-2 (front-end loading second stage), which

* We will not cover the details here because it is outside the direct scope of our work.
† The concept is represented via Figure 10.1 in Miller, Lessard, and Sakhrani, 2017.
‡ Merrow (2011) uses the term *sponsor* as a reference to what we have been calling *project owner*; it is not intended to be the sponsor or senior executive.

Merrow calls the *scope development stage.* This stage is equivalent to the CAMMP™'s Requirements Stage. However, since (1) CAMMP™ has a Strategy Stage, (2) in this stage we have an updated cost and schedule estimates, and (3) the recommendation is that the business decision should be made at the end of the shaping process, the FEL-2 decision would be best aligned to the CAMMP™ SG4, which is what we show in the flowchart presented earlier.

After the business decision at SG4, it is time for the preliminary engineering stage, which is equivalent to FEL-3 per IPA and also a similar stage per CII. In other words, SUKAD, IPA, and CII agree that this last stage in the front end is the project definition. This stage is about execution planning; in CAMMP™, it is detailed planning, which is necessary for the Delivery Phase. The detailed planning will lead to SG5, which is the point of the final decision to approve the project and authorize full funds. Other names for this stage are *Front-End Engineering Design* (FEED), *Architectural Design,* among others.

Needless to say, this is a critical phase and requires significant effort. The good news is that the cost of the Development Phase is low, maybe to 3% to 5% of the final expected capital expenditures (CAPEX). In some cases, organizations may approve up to 10% (see textbox). However, in terms of duration, this is also shorter than the Delivery Phase.* The cost is low primarily because a lot of work could be internal and still on paper (or PCs), with a lower staffing level in comparison to engineering. Also, there are no major procurements and no construction yet.

Project Prioritization
Long Lead Items

Long lead item is a term we use to refer to engineered items (equipment) that take a long time to order, design, fabricate, test, and deliver to site. If the organization waits until construction, or even the engineering stage, to order, delivery might take a long time, which would extend the project schedule. Therefore, project owners might decide to order these items early, during the preliminary engineering stage and even before final budget approval. This is a fast-tracking approach.

Due to the above, project owners typically request increasing the advanced funding from about 3% to 5%, up to 10%. To mitigate the risks associated with this action, procurement would include a cancellation clause in the order for the long lead item.

This concept is project prioritization, which is part of portfolio management.

19.3.2 Shaping Stage, Part 2

The work in shaping the opportunity will continue from the Discovery Phase. Some of the critical actions that could happen here include onsite assessments—access, labor, contracting, political stability, etc. Also, for projects in host countries, there would be many challenges to resolve, including goodwill and stakeholders' value, political and legal matters, and many other factors. Another challenge to cross here would be the partnership agreement and alignment, because it is highly likely that the project will require joint ventures between two or even more companies. The project could also be a form of public–private partnership. **All of these must be resolved before reaching Stage Gate 4.**

In all transparency, we are not elaborating a great deal on shaping the opportunity, because this is adapted from Merrow (2011) and we do not want to get into an infringement situation, plus why not go to the source?

* The IPA database for megaprojects, as of 2011, shows that the typical duration for FEL-2 and FEL-3 (Development Phase in CAMMP™) would be about 23 months for the average megaproject (which is about $3.2 billion). The execution stage, which is equivalent to the CAMMP™ Implementation Stage, would be about 43 months. There is no information on the Discovery Phase duration, which is likely longer than one year.

19.3.3 Requirements Stage

Stage Overview

The focus of the Requirements Stage is to define the project requirements—all the needs of all the relevant stakeholders. The stage concludes with the submittal of the project requirements document (PRD) for review and decisions at SG3, which CAMMP™ labels the *stakeholders alignment* gate. Other names we have encountered for this stage are *design basis, design basis memorandum, design basis scoping paper, scope development, basic requirements, business analysis, master planning,* and there could be other names.

We avoid using the word *scope* in the name, because it is our experience that the term is confusing. For example, are we talking about product scope, project's high-level scope, or detailed scope? Project scope or contract scope? and the list can go on. Therefore, the purpose of this stage is to define all of the requirements, including clarifying the product scope, initial project scope, technical requirements, site requirements, contracting and project management requirements, etc. In other words, in this stage, the facilities team works with the business team to explore the expectations of all of the relevant stakeholders, internal and external, to define all of the requirements.

Finally, the project team will use the output of this stage to finalize the Strategy Stage and project management planning work. Also, the output will be used to go into the Definition Stage for detailed planning for the rest of the project work.

Project Requirements Document

Figure 19.5 lists the various components of the Project Requirements Document (PRD). This list is not exclusive; in some domains, there might be more or fewer topics to consider. Organizational preferences also have an impact on the PRD structure and content. We will not go into the details of every item in this list. Also, keep in mind that a good deal of the work here is domain dependent—for example, industrial facilities are different from buildings or infrastructure facilities. What we list are the general concepts, with some elaboration as needed.

- **Description of the product.** The first step is to describe the output of the project—the product. At this point, any additional information that the team develops or gathers will be included in the description. The team is not yet ready to produce the detailed scope of work; that is part of the

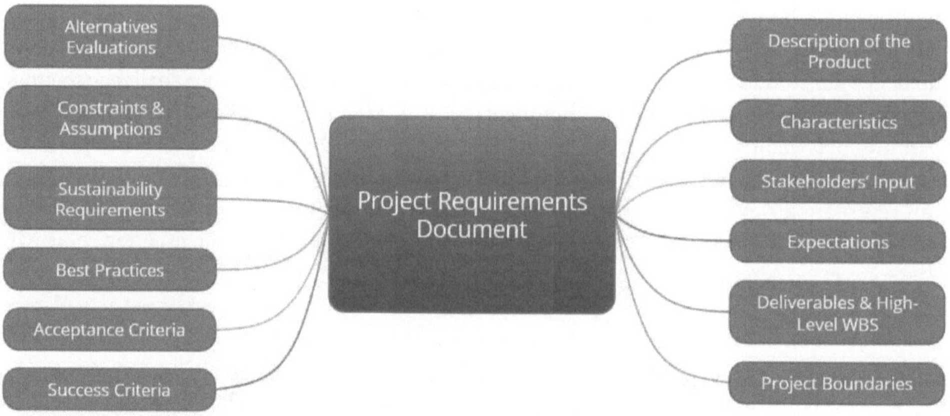

Figure 19.5 Project Requirements Document

Project Detailed Plan (PDP)—the preliminary engineering package in megaprojects. Therefore, the need is to provide enough detail to clarify the requirements and use the information later for the preliminary engineering and detailed planning packages.

- **Expectations.** The project team needs to be able to distinguish between *expectations* and *requirements*. Expectations are implicit, which, if not explored and appropriately identified, will likely lead to a lack of clarity, misunderstandings, delivering less than an optimal product, and ultimately the possible dissatisfaction of the various stakeholders. Therefore, the challenge for the project team is to work with the business, operations, and all relevant stakeholders—internal or external—to **explore for the expectations**. Many tools and techniques can be used to explore for the deep-rooted expectations, but these are outside the scope of this work. Next, the project team must do their best to **transform those implicit expectations into specific requirements**.
- **Deliverables/high-level breakdown structure.** From the above characteristics and requirements identification, one can develop a list of the key deliverables for the project. A couple of tools would be highly effective here: a product breakdown structure (PBS), showing the various components of the proposed facilities, then developing a work breakdown structure (WBS) to help the team define the work required to deliver the product and all of its components.
- **Project boundaries.** Sometimes, one of the best ways to define something is to identify what it is not. In project management, that brings us to the closest thing to *project boundaries,* also referred to by some as *project framing*—in other words, what the project scope includes and what it does not. For megaprojects, this can be critical, because there will be many parties and stakeholders involved in the various stages and at any given point in time.
- **Alternatives evaluations.** The evaluation of alternatives to the project, scope, schedule, location, and procurement approach takes place at various stages of the project, and in each stage, the alternatives can be different. Some of the options could be due to the nature of the project. Regardless of what opportunities the team considers, it would be best to address these alternatives during the development of the project requirements document (PRD).

  For a typical project, the team and management **should not** leave significant alternatives open beyond the PRD. However, for megaprojects, we would say, "the team **must not** have significant alternatives beyond the PRD." Megaprojects are not flexible to justify changing routes or significant scope; I think more than one reference uses the term *fragile* for megaprojects. If you recall, one of the major contributing factors to the failure of the North American Project (in Chapter 7) is the addition and deletion of major product scope items upon starting the implementation work. I think it is clear that if there are unresolved issues and alternatives beyond this point, the project will be at a higher level of uncertainty, which is a threat to project objectives.
- **Best practices.** Because some best practices could create alternatives, many of these practices must be implemented at this point, before finalizing the requirements. A few best practices, in particular, could be significant, such as the change management philosophy (i.e., zero-change policy), constructability, partnering, and integrated teams.
- **Acceptance and success criteria.** I think it is clear that completion, acceptance, and success are different things. Just because one completes work, and it could be acceptable, that does not mean success. We have discussed project success already. The team will also need to define the acceptance criteria.

Value Improving and Best Practices

Value Improving Practices (VIP) and Best Practices (BP) are similar concepts. They both deal with practices that are not commonly used by the average organization, and if organizations use them properly, they can expect significant performance improvement in the areas of safety, quality, cost, schedule, performance, etc.

VIP/BP should be limited in number, because they are often differentiating factors between typical performance and best-in-class performance.

Stage Gate 3 (SG3)

With the PRD complete, this is an opportunity for executive management, the client, end users, operations and maintenance, project management, and other appropriate stakeholders to sit down again and review the work progress and the PRD. The main objective is for the stakeholders to reach an agreement on the project and to align on the expectations, requirements, and major deliverables. For some types of projects, end users' representatives might be working with a project management team as an integrated team. In those situations, the alignment would take place on an ongoing basis; therefore, the organization would be able to expedite decisions at SG3. The decision, in that case, would just be a formality.

Importance of This Stage Gate

Unfortunately, it has been proven time and again that communication gaps exist; the lessons are often not learned. If organizations do not ensure that all sides are talking about the same thing, the risk is that the project management team could run in a direction that might not be the one required by executive management. Therefore, it is necessary to define the requirements and key deliverables and agree to them before the team moves forward and only discovers the mistakes later. This clarification is a crucial point for significant projects and megaprojects, because a considerable amount of work and substantial cost would be required in the next stage. Therefore, it is vital to be sure of the scope, objectives, and other factors. It is crucial to have the alignment of all relevant stakeholders—this is the core purpose of SG3.

You will notice that for megaprojects, per the Figure 19.2, SG3 and SG4 are close together. The idea is that SG3 is focused on all requirements and technical aspects with the relevant stakeholders, whereas SG4 concentrates on the business aspects and business decision; more on this in the next stage.

One last statement: Changes after this point can start to become costly. Therefore, if possible, this would be the point for activating the CII change management best practice and even considering freezing the scope at this point. Freezing the scope would mean that no changes will be allowed after this point unless the change is critical to respond to high-impact threats.

19.3.4 Strategy Stage

Stage Management Process

This part is only a reminder that for every stage of the megaproject life cycle, we use the stage management processes to manage the stage work. These are *Authorize, Plan Management, Plan Details, Implement, Control,* and *Close.* We have not discussed these processes in the stages' explanations in this chapter, hence the need for a reminder. The simulation in Part B will show how to apply these concepts.

Stage Overview

If you refer to the common literature or even the IPA model, you will not often see an independent stage similar to this stage. However, some of its work could be incorporated into other stages, such as FEL-2, per our earlier discussion. However, we believe that a dedicated stage for what we will discuss here is necessary, hence the inclusion of the *Strategy Stage.* The Strategy Stage is required to emphasize the need to develop the Project Management Plan (PMP) as an essential stage deliverable, which incorporates the delivery strategy, especially for projects with some level of complexity. In this context, *delivery* represents the upcoming work, the *Delivery Phase.*

As before, the output of this stage will lead to another gate, SG4. Before the stage gate, it will be necessary for the project management team (PMT) to have updated cost and schedule information to help management decide whether to continue with the project. For megaprojects, notice we still have the Shaping Stage happening in parallel in the Governance Stream. Therefore, the business team will use the updated cost and schedule information to update the feasibility. In other words, SG4 requires input from both streams—Facilities and Governance.* It is also good to remember that, since this point will coincide with closing the Shaping Stage, it is the recommended point for the business decision to go or no-go. If you work in an organization that depends on the equivalent of SG5 (end of definition) for this decision, your practice might not give this gate, SG4, enough attention. We recommend a change in the mindset and consider this gate to be the critical business decision point.

We recommend a change in the mindset and consider SG4 to be the critical business decision point.

Project Management Plan

Per the CAMMP™ approach, the Project Management Plan (PMP) has four distinct sections, and each section has many subtopics. This plan is quite comprehensive and could be a substantial document for a large and complex project. We will continue with the approach that we have been following, which is not to discuss all of the details here because most of the content could be significantly detailed in other references. We remind the reader that our objective is to provide the "big umbrella," broad enough to cover the full project life cycle with brief explanations of all-important topics. However, for in-depth knowledge, especially for those empowered to build their organizational project management systems, you must use the other references that we mention here or their equivalent.

Section 1. The Strategy Section

In the Discovery Phase, we touched on this stage, and we explained that for megaprojects it would be necessary to explore these various strategies, at least an initial high-level evaluation, which we call the Initial Project Strategies. That is important because the decisions on some, or all, of the strategy topics could have an impact on the requirements and project management planning. Now would be the time to revisit the initial strategies and confirm them or update them if new information would justify a change. In order not to repeat the same questions, please refer to the initial strategy section (page ##).

Section 2. The How-To Section

The how-to section† includes the core components of the PMP. The team must focus on six core questions to complete this section:

1. **How do we define the project?**
2. How do we manage the project?
3. How do we control the project?

* Technically, there could also be input from the Production Stream. However, it is likely that the organization would not have mobilized this team yet (as we show in the flowchart). Therefore, operation's input would happen through the business and technical functions.

† In the context of the Project Management Plan, the "how to" comprises who, what, where, when, and how.

Figure 19.6 The Project Management Plan with Sections and Subsections

4. How do we manage change on the project?
5. **How do we get ready for operations?**
6. **How do we close the project?**

Obviously, the project team could add other questions if necessary, depending on the project type and circumstances. If you reflect on these questions, you will notice the following:

- Three of these questions (1, 5, and 6, bolded for emphasis) have a link to a project stage, and the rest of the questions are not limited to a stage. For example, *define* is related to the Definition Stage, *operations* to the Operational Readiness Stage, *close* is linked to the Close Stage. On the other hand, *manage, control,* and *change* apply to all stages, although they are most active during the Delivery Phase owing to the high cost of this stage work.
- Therefore, these questions are about the "think before you act" approach, in which the project management team works to transform the strategies already identified into a plan of action. This section outlines how the team (the extended team*) will need to handle all of the work from this point to closure.
- Also notice that these questions are about *how to do* something, they are not about *doing.* In other words, how to define the project is only answering the how.† The answer might be, *"will use internal resources, and these are x, y, and z,"* or *"will hire a licensor to develop the onsite process facilities design package."* On the other hand, when asked to "define," this would be the development of the design package.

Concerning these questions, we have already touched on how to define. For the others, we will follow a similar approach. Here, I must reflect back on the need for the organizational project management function (OPMF) and the organizational project management system (OPMS) discussed in earlier chapters. If

* More on the extended team later.

† Remember, this includes the who, what, where, when, and how.

the organization has productively invested in building and is sustaining its OPMF—raised the maturity and embraced the proper megaproject culture—most of these questions will have their foundation in the OPMS. The project director and PMT will use what exists and apply what is appropriate for the context.

Let us provide brief answers to these questions:

1. **How do we define the project?** If the organization has not eliminated its technical expertise, they could define the project (develop the front-end design package; do preliminary engineering) with their internal resources and technical functions. However, we believe that most organizations do not have the competence or capacity to handle this work anymore, and will have to outsource, most likely using technology licensors.
2. **How do we manage the project?** We have already made the case and will do so again: owners who manage their projects directly do better; this is the clear message from IPA via Merrow (2011) and Barshop (2016). Therefore, if you have already built the competence and capacity inside your organization, great—use your people. If not, follow the advice we shared in Chapter 12 on how to gradually build the OPMF now, not waiting for a New Year's resolution that will evaporate on January 2.
3. **How do we control the project?** Project control is a subject in the next chapter.
4. **How do we manage change on the project?** Also, a subject in the next chapter.
5. **How do we get ready for operations**? If you do not want to fall into the trap of forgetting operational readiness, remember the section in Chapter 8, Do We Need to Operate This Facility? See textbox for more examples. So how to get ready for the startup, commissioning, and operating the facilities developed by a megaproject? Typically, operations personnel perform the work related to operational readiness, with the support of the project management team.* The primary consideration at this stage is to decide how to perform this work. What CAMMP™ MP proposes is starting the planning for the operational readiness activities in parallel to the preliminary engineering stage. Further, CAMMP™ MP also emphasizes the need for the Production Stream; refer to Figures 19.1 and 19.2. With these additions, CAMMP™ provides the mechanism for a disciplined approach.
6. **How do we close the project?** Unfortunately, some organizations do not put effort into a formal project close—they view it as one of the least important project activities. There is no glory after the work is complete. However, from a strategic perspective, the project work is not complete until the team formally closes the project. Building and sustaining the proper OPMS and increasing maturity requires formal project closure. Therefore, the team must adequately plan for this before everyone jumps ship.

From a strategic perspective, the project work is not complete until the team formally closes the project.

Section 3. People Aspects Section

We will cover this topic in Chapter 21.

* In some situations, clients might have let go of the necessary functions that can perform this work. Therefore, they end up depending on turnkey projects in which the contractor will perform these activities and hand over an operating facility.

Section 4. Class 2 Estimate Section

The term *Class 2 Estimate* will be more transparent as we address estimating in the next chapter. For now, let's think of this as the second, full project estimate along the project life cycle. The first—the Class 1 Estimate*—is completed during the Feasibility Stage. Because cost estimating and control and schedule development are subjects of the next chapter, let us postpone the discussion.

> **Megaproject Case Studies**
>
> A nuclear power plant is under development. A news item reports that the project (facilities) will be complete on a certain date, but commissioning and start-up will be delayed many months because the organization needs more time to get ready for operations. Remember, this is a megaproject costing billions of dollars, and this delay will have negative consequences in terms of additional cost and delayed revenues and services, not to mention reputation.
>
> In a similar case study, a major museum project has been delayed. A report came out saying the facility will be delivered this year but "we do not know when the museum will open."
>
> These scenarios are often common challenges, because organizations focus on completing a facility and do not pay enough attention to what must happen next. Another challenge is that some practitioners do not consider this work as project management work.

Stage Gate 4 (SG4)

As a refresher, SG1 was about determining if the project is in alignment with the organization's strategic plan and objectives. If so, then the organization considers the project for further evaluation and performs a Feasibility Study. SG2 was about verifying that the project is viable (feasible) and whether management decides to proceed and authorize the project. SG3's purpose was to ensure alignment among the various stakeholders on the project requirements. At SG4, the main concern is for executive management to review the PMP and to determine if this plan will lead to the effective delivery of the project in line with the objectives that management has outlined earlier.

Furthermore, management will review the updated Feasibility Study and other output from the shaping process to determine if the project is still viable and the organization still expects to realize the benefits. Finally, management may need to approve a certain amount as an "advance fund." Advance funding is typically a financial and corporate governance requirement for capital-intensive projects. The team will need advance funds for expenditures in the next stage, primarily if the project team outsources this work. Therefore, they must request advanced funding, which is part of the Governance Stream.

Further, another reminder: if your organization agrees with what we presented earlier (which was recommended by Merrow [2011], and we agree with it) and would treat this stage gate as the point of a final business decision, then added scrutiny is essential. The idea is that the board of directors and executive management must be confident that this project will not be stopped beyond this point. This is a vital decision point. That does not mean the project could not be canceled later. SG5 is another opportunity, and the cost would still be relatively low at that point. However, beyond that point, if the organization becomes aware and certain that the project is failing and draining shareholder value, then it should be stopped regardless of the sunk amount. There are many examples of megaprojects stopped during implementation; we won't mention any to avoid embarrassment.

* AACE International and other organizations use a different classification system. For example, the AACEI estimate classification system offers five classes of project estimate, starting with the Class V, which is the conceptual or order of magnitude estimate.

19.3.5 Preliminary Engineering Stage

Stage Overview

The CAMMP™ standard model has a Definition Stage, which is the last stage in the Development Phase. However, other terms for the stage could be the Preliminary Engineering Stage, Architectural Design Stage, or Front-End Engineering Design (FEED) Stage (FEL-3), among other names. For CAMMP™ MP, we will use the name *Preliminary Engineering Stage*, which is more appropriate for the capital megaproject context. So, what is this stage all about?

It is about the answers to the first essential "how-to" question from the previous stage: the question of how to define the project, which occurs within the development of the project detailed plan (PDP), or in this context, the preliminary engineering package. Also, via this stage, the team is focusing on "implementation planning"—that is, planning for the Delivery Phase. In addition to the preliminary engineering work, the business team will revisiting the feasibility and business case and prepare the final funding package. Finally, all will come together at SG5.

Notice in the flowchart that we show SG5 in two parts—SG5a and SG5b. These could be separated or combined, depending on the organizational process. The intention is that SG5a will focus on the technical work in the preliminary engineering package and implementation planning (Facilities Stream), and if all is OK, the business team will proceed to SG5b, with the final funding request (Governance Stream).

Notice that we show three major deliverables: the preliminary engineering package and the implementation plan, in the Facilities Stream; both happen in parallel in the Preliminary Engineering Stage, and the request for final funding is part of the Governance Stream. In closing, just think about this: the output of this stage is an elaboration on the Project Requirements Document (PRD) and Project Management Plan (PMP). In other words, it combines the work of the prior two stages and expands on them significantly.

Project Detailed Plan

We use the term *Project Detailed Plan* (PDP) here, which is per the CAMMP™ standard model. However, because there are two major deliverables in this stage (remember, the third deliverable is in the Governance Stream), the output of this stage would be two sections: (1) the preliminary engineering package (whether outsourced or done internally), and (2) the implementation plan done by the project owner's team or their project management consultant, if project management is also outsourced. For small to medium projects, all of this work could be in one package. However, it can be three sections split between two streams. The split is partially due to the type of work and the fact that each package could be the responsibility of a different party. Refer to Figure 19.7 for a general breakdown.

Preliminary Engineering

The preliminary engineering scope is significant engineering work—technology and licensor work for industrial facilities; architectural design for buildings projects; and other specialized disciplines for infrastructure projects (civil engineering and its various sub-disciplines). In other words, remember what we have been saying all along—**the leadership of megaprojects requires the integration of many disciplines,** and it is not possible to address them here. If you refer to Figure 19.7, you will notice that in the Preliminary Engineering section, we have three parts: the technical scope of work (based on the type of project), the relevant quality standards and specifications for the various disciplines, and engineering risk management or the risk assessment that would be appropriate for this package. We could add other parts, but all of the other required work would likely be in one of these three parts.

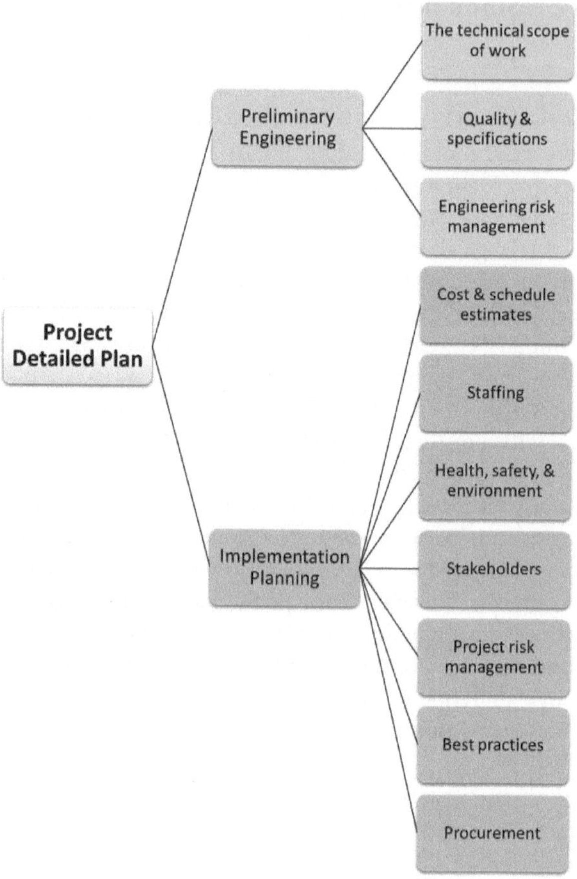

Figure 19.7 Project Detailed Plan

Implementation Planning

In terms of implementation planning, we have several items. It is likely that these work items would be handled by the project management team, whether by internal owner staff or outsourced. You will notice that the headers (topics) are almost identical to the PRD and PMP, which is to be expected because this work would expand on the earlier stages. As the image shows, this work includes updated cost and schedule estimates*; staffing; health, safety, and environmental planning; among the other topics. Notice we have another risk management assessment here, which is for all of the work in this section of the PDP. Some best practices are applicable at this time. In terms of procurement, because the strategy was defined earlier, this could be limited to final planning and carrying out the procurement actions.

19.3.6 Operational Readiness Planning Stage

The CAMMP™ standard model does not have this stage, because this work might not be significant enough to have an independent stage. The Strategy Stage includes how to get ready for operations, which

* These would be per the final estimate, which would be used for final approval.

could be enough for those small or medium projects. However, for megaprojects, the organization should not wait until funding to start planning or working the operational readiness actions. We could have combined this work with the PDP above; however, it would be best to separate it because the leader for this work would not be the project director. Instead, we would have an operational readiness director (or operations director)—someone with significant experience in the operations of the type of facilities the project is building.

The vital point here is to consider operational readiness planning to identify all of the things that will need to go into the implementation contracts and financial package.

19.3.7 Request for Final Funding

Stage Gate 5 (SG5)

The title for SG5 is the Final Approval, reflecting that in most industries, once the team produces the PDP, inclusive of what we have discussed above, management would have enough information to make this final decision. The reason for the use of the word "final" here is that the chance of canceling the project after this point is extremely low.

As we said in the stage overview, in the flowchart, we show SG5a and SG5b; we could have combined them, but we separated them because different people could be involved in each. We could have one stage gate review, or maybe three separate reviews before reaching the financial gate. This situation would typically be addressed by the OPMS.

Financial Planning and Funding Request

In parallel to the preliminary engineering work, the governance (business) team could be working on an update of the Feasibility Study based on the updated cost estimate, schedule, procurement, risk management, and the other items from the Preliminary Engineering Stage. Further, megaprojects take so long—possibly about three years would have elapsed since the original idea and about two years from starting the development phase—that there could be other changes in technology, business, market, competition, etc. Consequently, the business team should be updating the business basis, just in case something drastic happened since SG4 and the business decision to go.

Once executive management is convinced that this project is still viable, the business team would finalize the funding request and pursue the final approval and funding or financing. Project financing for megaprojects could have books written about it, so it is outside our scope here.

19.4 Delivery Phase

19.4.1 Phase Overview

The Stages

The CAMMP™ standard model has four stages in this phase—Implementation, Operational Readiness, Initial Operations, and Close. For megaprojects, we show five stages, because we split the Implementation Stage into two stages.

Why the split of Implementation Stage?

1. For capital projects, and in particular megaprojects, the Implementation Stage would be split into two major stages: Detailed Engineering (engineering) and Construction. Various references

may use the term D/B or design/build* for these two stages. Other references use EPC (engineering–procurement–construction).† If you noticed, EPC implies three stages, but in practice, there are only two. The reason for two, not three, stages is because the common practice is that engineered items would be procured by engineering staff, so is part of the engineering stage; whereas bulks, consumables, and other construction-related items would be the responsibility of the construction contractor(s). However, it is possible in some situations to split procurement as an independent stage.

2. We propose the split even though the work could be done by one EPC or D/B contractor, which is a common practice. The reason for this is to add a stage gate in engineering, which we call *Ready for Construction Gate.* We use this stage gate to enforce the discipline that is often missing and not to start construction until ready. The North American Project in Chapter 7 suffered from mobilizing the construction work too early.
3. One of the reasons for lack of discipline about engineering is that construction transition is common, especially if there is one EPC or D/B contract on a fixed-price basis. Therefore, the client project team must include clear guidelines for this stage gate in these types of contracts. Otherwise, the owner is leaving a big window open for Murphy and the construction contractor to come in, eat your dinner and your Baklawa.

The above work and the Close Stage are part of the Facilities Stream. The Close Stage is here because most of the closeout work is related to this stream. However, significant input would be required from the other two streams.

The other stages are related to operations and are in the Production Stream. They are separated because they would be led by an operations director and operations personnel. For the Governance Stream, there is no major work to list, and the business team would be supporting the project team. That does not mean there are no business activities ongoing, just that they would typically be related to post-project completion.

The Stage Gates

There are five stage gates spread around the stage (in the whole Delivery Phase). However, the stage gates now are unlikely to be used for project cancellation. They are mostly about readiness to move to the next step. For example, SG6 is about readiness for construction, SG7 is about readiness for handover, and SG8 is about preparedness to end the implementation work and go to closure. SG9 could happen a year later to verify the final acceptance, which would trigger completing the closeout activities. Finally, Stage Gate 10 is to confirm closure.

19.4.2 Engineering Stage

For the engineering stage, there is not much to say, because the vast majority of the work is technical engineering work. Therefore, any good engineering reference will be useful, and Oberlender (2014) would also be a good reference. It is important to say that with this stage, the work effort will be significant, with hundreds of engineers and other resources working on the project. The stage management process will follow the CAMMP™ six processes that we have discussed earlier, except in this stage, most of the planning and implementation actions will be done by contractors' personnel.

---

* Term is common in buildings and infrastructure projects.

† Term is common for industrial facilities.

The management of engineering by the owner or a designated project management consultancy will depend greatly on the type of contact. For example, the EPC contract would be different than if we had split the work into separate contracts and various parties. The commercial nature of the contract will also impact the owner's management; for example, a fixed price would require different focus than cost-plus, although the overall level of project management effort could be similar.

Finally, it is essential to state that, owing to the high level of expenditures, project control would become prominent and critical. We view project control as the navigators on the project, yet, as we have seen in the North American Project, the owners did not see the need for project control until the damage was done. Their view was, "Why do we need project control on a fixed price contract?"

In the Figures 19.1 and 19.2, we show SG6 at the end. This will likely be the case if we are using two separate contractors, one for engineering and one for construction. In that scenario, SG6 could be at the end or at the 90% design point.* However, in the real world, the organization can likely fast track the work and release engineering in packages—for example, site preparation first, civil works next, etc. In those scenarios, SG6 could be a series of smaller gates, one for each package. What is important is to enforce the discipline that no package should be released unless it is substantially complete.† Also, for situations in which we have one D/B or EPC contract, that stage gate could be at the 60% design point, or something like that. Early approval of construction work has significant risks, and the team must be able to manage those risks.

19.4.3 Construction Stage

I can repeat almost the same thing here as what we said for engineering, except that instead of hundreds of engineers and project management staff, now we are talking about thousands. Construction also has the added complexity of safety, health, and environmental matters because the exposure to incidents, injuries, and even death increase dramatically—hence the need for safety best practice.

Toward the end of construction, it is also likely to release the packages or systems in a scheduled sequence, and not all of the work at one time. Therefore, SG7, Ready for Handover, could also be subdivided into smaller gates, with one for each corresponding system. Depending on the nature of the contract, it is common to have the construction company handle the pre-commissioning or testing before the handover at SG7. At SG7 (whether one review for all work or partial review) the operation team would take over the completed facilities to start the initial operations. However, there could still be some snag-list items that do not impact the commissioning and startup work. The construction team would need to close them before reaching SG8, which would grant the contractor the completion certificates and demobilize them.

However, going through SG7 also requires completion of the work in the Operational Readiness Stage, which is next.

19.4.4 Operational Readiness Stage

Every project that will result in a product and that requires operations will need an Operational Readiness Stage. Just to be clear, a *product* could be a facility, software, or an organizational system, whereas the term *operations* refers to the operational activities of the new system, including maintenance and upgrades.

---

* This is a likely scenario and not a standard, hence the need to follow your own OPMS.

† When a project is delayed and contractor payments are based on milestones, such as IFC-Issued for construction, it is common for engineering to take short-cuts and issue packages as IFC but "with holds." The holds can be extensive, which makes these IFC releases somewhat useless. Therefore, the quality standard should be clear to help the team avoid falling in this trap.

The focus of this stage is on all the work that the teams have to complete before the handover and provisional acceptance. These activities include developing procedures, training personnel, testing, and pre-commissioning. This stage is about anything that allows the starting of initial or full operations of the new product. In some situations, initial operations could be a pilot period after the handover. Whether or not there is a pilot, and regardless of whether it is at this stage or as a separate stage, such a project would likely include an operations period—after the handover, and until the client, operations, or end users grant final acceptance.

There are many deliverables in this stage, which are a function of the project type, but the critical deliverable required from the stage gate would be provisional acceptance and handover. As discussed earlier, SG7 is the crucial stage gate here, in which management, sponsor, implementation team, and operational team review the work completed and decide if they can exercise the transfer of custody via a formal handover from implementation to the operational team. The team accomplishes the custody transfer via a provisional acceptance certificate or a similar document.

The flowchart and project life cycle do not show that this stage could be subdivided into two sub-stages: pre-handover and handover.

- The pre-handover activities could vary considerably from one project to another and one domain to another. In the process industry (oil and gas, petrochemical) and industrial projects, pre-handover activities can require a considerable effort. For an industrial project, some of the deliverables would be operations procedures, safety procedures, pre-commissioning activities, commissioning procedures, staff training, maintenance procedures, the inclusion of the new facilities into the organization's change management system, and any other work that might be necessary. Once the operational readiness team accomplishes all of these activities and the work packages from construction are complete, operational and project teams would be ready to initiate the handover activities and transition the custody to operations.
- The handover activities are those that the teams perform around the time of handover. These activities include verifying work packages and product completion and verifying the scope and product completion through inspections, testing, and reviews. To perform the handover reviews, the project management team, operations team, and service provider team will work together. The objective of these reviews and inspections is to ensure that all items are complete per the project's requirements and are ready for sign-off—provisional acceptance—by the client. At this time, there might be some remaining issues that are not complete; these are "punch-list items," "snag-list items," or "exceptions list." If it is necessary to perform these items to meet the handover requirements, the implementation team must finish the work and verify completion before the client can accept custody. On the other hand, if these items do not affect initial operations, then operations personnel can proceed through the handover process and take custody, on the condition that the implementation team will complete these remaining items post-handover and before final acceptance.

19.4.5 Initial Operations Stage

Every project that will result in a product and requires operations will need an Operational Readiness Stage. It may also require an Initial Operations Stage. The term *initial operations* represents (1) a pilot period, (2) soft launch, (3) operations at minimum capacity, or (4) commissioning and start-up. The term is a function of the organizational preference and type of project.

The focus of this stage is to transition from work-in-progress mode to final status and operational mode. This transition period could last hours or days for simple projects and many months for industrial facilities and megaprojects. The term *initial operations* is used to stress that this is not the permanent status, meaning no continuous or full-scale operation has been achieved yet. What do initial operations include? It all depends on the industry and the projects.

Commonly, the initial operation starts after the client accepts the output at the handover point, SG7. As mentioned earlier, at SG7, there is provisional or interim acceptance. It is also the point at which the project team transfers the custody of the facility to the client's operations team, which performs various commissioning, or start-up, activities in final preparation for the operational activities.

Once the operations team is ready and satisfied that all individual components of the products are okay and functioning as expected, they start formal operation of the facility, which is defined here as the initial operation. Initial operation and normal operation are essentially the same in terms of the work, or what is happening physically; the only difference is that the initial operation takes place before final acceptance and possibly at reduced capacity. Operational personnel would closely monitor the initial operation, looking for potential issues. This period is necessary to build a level of comfort that all is working well and as expected.

The work of IPA as documented by Merrow (2011) clearly states that many of the potential problems that happen here have their root causes in the front end owing to lack of clarity of objectives; in accelerating the front-end development, which would leave gaps that would show in engineering or construction; and ultimately in initial operations. The challenge for project management is that work and the problems that could happen are often outside the responsibility for the project team. This situation creates a vacuum of knowledge, because not a single reference we reviewed actually documents the potential cost of fixing the problems. Barshop (2016) did document that about 5% of the reduction in the Net Present Value (on average) is due to production shortfalls, but this is for all capital projects, not megaprojects. IPA success and failure criteria also measure production shortfall stemming from significant technical problems going into the second year of production after startup (Merrow, 2011).

19.4.6 Close Stage

Stage Overview

The Close Stage is the last stage of the standard project lifecycle of the Customizable and Adaptable Methodology for Managing Projects™. The focus of this stage is on properly closing the project and includes all of the related activities—gathering the final lessons learned, updating employee files, documenting project performance metrics, updating organizational records, etc. During the Close Stage, the organization should also include project success evaluation and benefits realization, where possible.

Although closure can start as soon as possible after completing the construction work, final closure cannot happen until after final acceptance at SG9. The major deliverable is the project close-out report. Although there are other items at this stage, they could be either part of the close-out report or independent deliverables. It is important to remember that this is the project closure. However, the CAMMP™ stage management process includes a close process for every stage. In other words, if the team had the discipline to close every stage, this final closure would be mostly about gathering all of the project records and conducting any final closure actions.

Importance of Proper Project Closure

It is time to re-stress the significance of this stage and the close-out report. As mentioned earlier, some organizations neglect this stage, partially because they re-assign the project director and team almost as soon as the project is handed over. In these organizations, project completion coincides with the end of the implementation work or acceptance of the product. On the other hand, CAMMP™ views project closure as an essential component of project delivery and a necessary enabling factor for organizations to achieve a higher level of organizational project management maturity. No organization can achieve a higher level of maturity without learning from the past and pursuing continual improvements. Continual improvement requires proper project closure, inclusive of lessons learned and update of the organizational records.

Keep in mind that significant effort and work takes place between the idea and project completion, which is crucial for organizational learning. **If project teams do not document the learning in a ready-to-access manner, a wealth of information is totally lost**. Capturing the learning should not happen at the end only—it should happen throughout the project. The CAMMP™ approach advocates proper closure by including the Close Stage as part of the Delivery Phase. Another factor is the introduction of SG10. These actions clearly state that the project is delivered to the owner organization only after it is closed and passes through this final gate. Without going through this gate, the project is not complete.

The Close-Out Report

What are the major components of the close-out report? It depends on organizational policies. In the CAMMP™ Model, the final report consists of the following topics. I will expand on them because of their critical value to the organizational maturity and to sustaining the organizational project management system. Also, most of what we are covering here applies at the stage level.

Closing Resources and Procurement

One of the critical things to do in closing a stage or the project is closing the resources (people and physical) along with procurement (contracts and purchases). In terms of the people aspects, resources need to be released, with a record of the employees' performance, new skills acquired, commendations and suggestions for improvement, along with other individual performance assessment.

Concerning the physical resources, it is possible to have surplus materials, some of which could be handed over to the maintenance department as spares and parts, others will not be of use to the organization and could be sold as scrap surplus. Further, the team needs to close any open purchase orders and contracts if they were not yet closed in the previous stages. The PMT can also develop and document the performance of their vendors and contractors for future reference.

Performance Reconciliation

The term *performance reconciliation* refers to the final project performance reconciliation report, which is an essential document for organizational learning. This report documents what happened throughout the project from start to finish, with an emphasis on reconciling the differences between plan and actual. **As with the rest of the project closure processes, the performance reconciliation is an exercise for every stage and for the project as a whole.**

Lessons Learned

Lessons learned are crucial, and the project management team must capture them throughout the various project stages.

How to capture lessons learned?

With a mature project management culture within an organization and the right mindset, every team member must be aware of the importance of lessons learned. Therefore, any team member who discovers a new lesson should document it. One must not wait until the end of the stage or, even worse, the end of the project. At the end of a stage or project, the team should come together and reflect back to identify any other lessons they might have encountered but did not capture at the time. Furthermore, as the team assembles the performance report, other lessons learned might appear. An experienced team (and mature organization) recognizes that lessons are about the good, the bad, and sometimes the ugly.

Final note: Lessons learned can only work in mature organizations because lessons should be clear and open; in a culture of blame or scapegoats, one cannot honestly expect many valuable lessons.

Success Assessment

Measuring project success is another way that the organization can evaluate whether the project delivered the anticipated benefits. The only difference from current practices, or a narrow perspective that focuses on cost and schedule as the only criteria for measuring success, is that CAMMP™ includes four distinct dimensions, which we have already covered. At this time, it is worth noting that of these four dimensions, only two—product technical success and project management success—are possible to assess before closing the project, and the results must be documented in the close-out report. The other two dimensions—project delivery success and project objectives success—require a period of operation after project completion. In this case, the close-out report should include the dates at which these last two assessments would be possible, and the organization management would be responsible for performing these evaluations in due time.

Benefits Realization Evaluation

An organization authorizes a project because executive management recognizes that the output of the project adds capabilities to the organization, which, in turn, produces the outcome and desired benefits. Therefore, some of the questions here would be: Did the project add value? Did the organization realize the benefits that they expected when they authorized the project? Can the team even assess the situation at the time of closure? Benefits realization is a topic that is related to one of the dimensions of project success, which the organization can only assess in the future. On other projects, it might be possible to assess success and determine the realization of benefits with project closure, but these projects would not be the common type.

Organizational Records Update

Every organization that has adopted project management practices will have an organizational project management system (OPMS) in place that includes the organizational records (organizational knowledge base, OKB). The OKB contains data from completed projects, lessons learned, and reconciliation reports. All of these will be useful as a new team works on a new project. One can understand that some organizations are quite young or are new to project management, and do not have a history; therefore, they might not have kept their project records in an organized manner for ready access. In either case, it is necessary to establish a system for record keeping and maintaining information from past projects. The records are likely to include a methodology, procedures, guidelines, policies, and other relevant information. Also, the records contain information on past projects, such as cost and time data. Therefore, the team must incorporate the various closure reports into the organizational records. Furthermore, the project management team can summarize key factors from these reports to include, in a summary format, in a database for quick retrieval in the future and overall analysis of the organizational portfolio performance.

19.5 Closing the Chapter

This is a vast chapter, the largest in the book, so let us not make it any longer. Simply, in this chapter, we have briefly explained every stage and the major steps along the project life cycle of the CAMMP™ MP tailored approach. Just for reference, this chapter actually summarized nine chapters from *Project Management beyond Waterfall and Agile* (Ajam, 2017) and added the content that is relevant to megaprojects.

The project life cycle is the first dimension in the CAMMP™ three-dimensional model.* The next chapter, another substantial one, will cover the CAMMP™ second dimension, which is about the stage management processes and functions.

* Refer to Chapter 17 for the introduction of the three-dimensional model.

Chapter 20

Stage Management Processes and Functions

20.1 Navigation Message

I am sure you have noticed that my style is often to reflect back or remind you of something we have covered, giving you a heads up on what is to come. I follow this approach to help you navigate this book and the diversity of topics necessary to lead megaprojects successfully. This approach will allow you to pause, reflect back, analyze, understand where you are in the process, and continue on this journey of learning and exploration. So, let us do that again and look back at this section of the book.

After an introduction to the tailored approach, we established the critical success factors, then we discussed the need for an integrated approach. The integrated and applied project management approach views the management of projects (and megaprojects) as the act of integrating a project management methodological approach into the organizational project management system (OPMS), inclusive of project success and project control models. After these general concepts, we dove into CAMMP™ for Megaprojects (CAMMP™ MP), where we emphasized the need for an asset management perspective and provided an overview of the various streams needed for megaproject leadership. We also presented the proposed megaproject life cycle. In the last chapter, we briefly discussed the flow sequence of leading a megaproject from concept to success, covering each stage, with its deliverables and gates.

Process Groups Are NOT Phases

Part of the confusion among many practitioners, especially those who follow the PMI® school of thought, is thinking that the process *groups* are project *phases*. This is not a correct understanding of the *PMBOK® Guide* or ISO 21500,* because both guides clearly state that (1) the process groups are not project phases, and (2) the processes repeat for all phases.

Consequently, and as explained earlier, we refer to these processes as *stage* management processes instead of *project* management processes.

* ISO, 2012.

Now it is time to look at megaprojects from a different angle and discuss our approach for the stage management, the processes for managing a typical stage, and the relevant organizational project management functions (OPMF).

20.2 The Phase/Stage Management Process

Here we will elaborate on the phase or stage management process. We touched on this in Chapter 2 and other mentions in this book, but we need to go into a bit more detail. I say a bit more, because we do not want to rewrite *Project Management beyond Waterfall and Agile* (Ajam, 2017).

20.2.1 The Knowledge Foundation

We had mentioned earlier that to manage a stage of the project, we need the stage management processes, such as those proposed by ISO (International Standard Organization) and PMI (Project Management Institute). ISO calls these *process groups, initiating, planning, implementing, controlling,* and *closing,* per ISO 21500.* Please see the textbox "Process Groups Are NOT Phases" to understand and clarify a critical point and a common confusion in the project management community.

20.2.2 The CAMMP™ Alternative View

Here are a few points explaining the CAMMP™ deviations from these guides.

- The first part of this alternative view is that we think of those processes as stage management, not as project management processes, as we have already said more than once. This point is not about semantics or terminology; rather, it is about when or where to use these process groups.
- Another change, mostly a terminology change,† is what PMI/ISO call a process group we call a *process* or the *main process.* Also, what these organizations call a process, we call a *process step.* For example, what PMI/ISO call the initiating process group we call the *Authorize Process*; and what they call the Identify Risks process we call the *Identify Risks process step.* Therefore, to manage a stage, we have six processes—authorize, plan (management), plan (details), implement, control, and close. Please refer to Figure 20.1, which is a repeat of Figure 2.6 from Chapter 2 for these processes.
- The third variation, and the most important one, is splitting planning into two processes instead of one. If you study the PMI and ISO guides carefully, you will notice they also split the planning processes, but not explicitly. For example, the *PMBOK® Guide* (PMI, 2017) presents that the management planning process in each knowledge area leads to the Project Management Plan and subsidiary plans (PMI, 2017). The other planning processes lead to "project documents" (PMI, 2017). We believe this is a scenario that would be better understood and applied if we split planning into management planning, leading to a stage management plan, and detailed planning, producing a Stage Detailed Plan.‡

---

* The PMI terms are *initiating, planning, executing, monitoring and controlling,* and *closing* (PMI, 2017).

† One reason for the terminology change is to avoid using the PMI or ISO process names.

‡ If you would like to get the detailed explanations with graphics, we refer you to *Project Management beyond Waterfall and Agile* (Ajam, 2017).

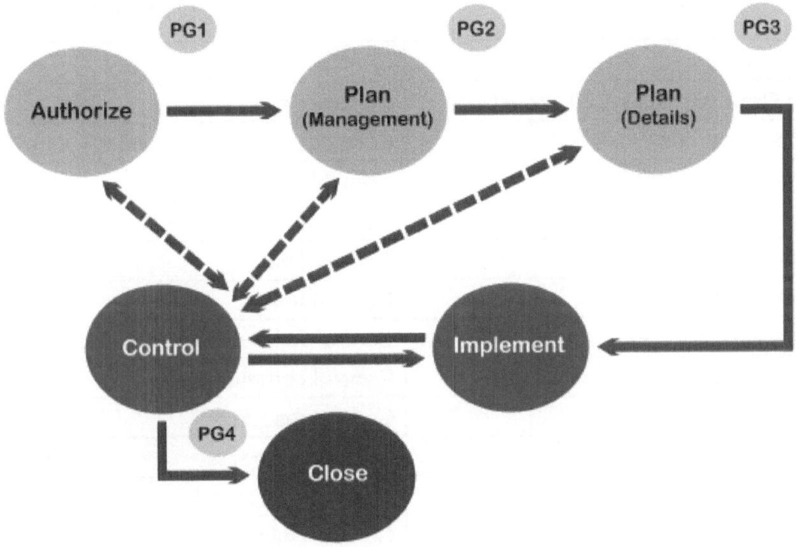

Figure 20.1 The Stage Management Processes Per CAMMP™

- In addition to splitting planning, we expanded the Authorize Process (initiating) and the Close Process significantly. CAMMP™ offers quite a few process steps in each of these processes, in contrast to PMI's and ISO's one or two processes for each process group.
- The above variations lead us to 64 process steps spread among the six processes. Here it would be useful to share a table from *Project Management beyond Waterfall and Agile* (Ajam, 2017)—Table 20.1 in this book.

Table 20.1 CAMMP™ Processes and Process Steps

Authorize Process (4 Steps)	**Implement Process (9 Steps)**
Develop Stage Authorization Document	Complete work packages
Establish stage success criteria	Mobilize implementation team
Mobilize project manager	Manage team
Identify stakeholders	Manage stakeholder engagement
	Manage communications
Plan Management Process (7 Steps)	Acquire resources
Develop Stage Management Plan	Select and mobilize service providers
Develop Change Management Plan	Manage contracts
Develop Control Plan	Lead and manage stage work
Develop Configuration Management Plan	
Develop stage initial schedule	**Control Process (16 Steps)**
Develop stage initial cost estimate	Control scope
Mobilize management team	Control schedule

(continues on next page)

Table 20.1 CAMMP™ Processes and Process Steps (cont.)

	Control costs
Plan Details Process (20 Steps)	Control risks
Explore for expectations	Carryout quality assurance
Create stage PBS and WBS	Control quality
Develop stage scope of work	Carryout HSE assurance
Develop stage schedule	Control HSE
Optimize stage schedule	Control team
Develop stage cost estimate	Control stakeholder engagement
Determine stage budget	Control communications
Identify and clarify risks	Control resources
Assess risks	Control contracts
Plan responses to risks	Control stage work
Define quality standards	Control changes
Define process improvements	Assess performance
Define HSE compliance requirements	
Mobilize detailed planning team	**Close Process (8 Steps)**
Determine implementation roles and staffing requirements	Validate product
	Demobilize team
Develop stakeholder engagement plan	Close-out report
Determine communications requirements	Close resources
Determine resources requirements	Close contracts
Develop procurement documents	Document lessons learned
Outline stage detailed plan	Reconcile stage work
	Assess stage

Please note that some will say there too many steps. ISO 21500 has only 39, and the *PMBOK® Guide* has 49 in the sixth edition, so why does CAMMP™ promote 64 steps?

- First, we reduced them from 72 (you can imagine a smiley at this point).
- Second, each of the project stages on large-complex and mega projects can easily be classified as a standalone project, in terms of the effort required and the complexity.
- Third, remember Merrow's (2011) findings: organizations do not take the necessary time and do not put the required effort into megaprojects in comparison to the other projects.
- Another point to stress here is that some of the process steps might not apply in every stage. For example, if you appoint a project manager (a project director) in the Feasibility Stage or Concept Stage, and you keep this person on the project, you do not need to go through that process step again. Similarly, we mobilize the initial project management team early, then we add to it in subsequent stages. In other words, if it does not apply, well, it does not apply—that is, it is not needed. Therefore, I would rather have more steps and tolerate a few N/As (not applicable) than forget or ignore a critical step.
- Finally, and most important, see the highlighted message.

The level of effort for a given project is a function of the size, complexity, and importance of the project, not a function of how many processes or sub-processes it contains!

- In other words, whether you have 5, 6, 39, 49, or 64 process steps should not change the level of effort required to manage a stage of the project. We can have the 64 processes done in hours or days on a small stage, or it could take weeks or months to apply the same 64 processes on a large stage.

20.2.3 Stage Management or Phase Management?

Is it *stage* management or *phase* management?

If you recall the earlier discussion on project classifications, we clearly stated that for medium-moderate complexity projects, we can apply the six processes and their 64 steps at the phase level. We do this because it would be the appropriate level of detail, and going to the stage level could be like using a sledgehammer to drive a nail—you just break the table.

On the other hand, for large-complex projects, we would consider going down to the stage level. However, for megaprojects, we say do not "consider," we say, "do" go down to the stage level—it will make the difference between success and failure, especially in the early stages of the project. I believe that many of the challenges we addressed in the earlier sections of the book, such as lack of clarity of objectives, not progressing the front end well enough to get a good level of definition, etc., are due to not taking the time to manage the front end. Both Oberlender (2016) and Merrow (2011) stressed similar points in their respective books. Further, the very first stages (in the Discovery Phase) are done by businesspeople, not project management, which contributes to the challenges. Therefore, **for megaprojects, we must manage each stage to manage the whole project; project management must lead (or co-lead) the early stages; these actions are necessary if your organization wants to deliver on the promise of concept-to-success**!

20.3 The Processes Outputs

It is clear now that to manage each stage, we apply the six-stage management processes. Here is a brief outline of how that works.

20.3.1 Authorize

The Authorize Process authorizes the stage work, which could be accomplished via a dedicated document that we call the Stage Authorization Document. For the first stage of the project, this would be the first document from management to authorize the Concept Stage. However, for the subsequent stages, the stage gate approval provides the input to authorizing the next stage. For example, approval at Stage Gate 4 (SG4) would authorize the Definition Stage, and so on. The main difference is that approving SG2 will lead to the overall Project Authorization Document.

If the stage gate is there, why do we need a stage authorization document?

It is possible that the OPMS mandates approval at a stage gate automatically authorizes the subsequent stage. This is possible and can be the preferred option. However, keep in mind that approving a stage gate might happen at an executive or even board level, basically saying, "We are OK to go ahead." On the other hand, a Stage Authorization Document might provide more specific information to the leader of the stage work, the project director, or a designee.

Table 20.1 shows the four process steps that are required to complete stage authorization.

There is a process gate at this point (PG1), but this could be just a project director or project executive decision. Process gates do not require the same high-level scrutiny that stage gates do.

20.3.2 Plan Management

The Plan Management Process is to develop the Stage Management Plan. It is important to remember that each stage can require many months and numerous resources. Therefore, the project director must ensure that every stage has its Stage Management Plan. Also remember, most of the management plans for the various stages could be quite similar and have a lot of copy–paste–modify (CPM). However, the CPM can reduce the work, but the team should still give the plan the necessary focus.

Table 20.1 also shows seven process steps that are required to complete the stage management plan.

There is a process gate at this point (PG2), but, similar to PG1, this could be just a project director or project executive decision.

20.3.3 Plan Details

The Plan Details Process is to develop the Stage Detailed Plan (SDP). The SDP for the various stages has a similar structure (covering all functions) but is specific to the type of work of the given stage. For example, the detailed plan for the Feasibility Stage would be quite different and possibly significantly abbreviated in comparison to the detailed plan for the Engineering Stage.

Table 20.1 shows the 20 process steps required to complete the SDP. It is the process with the most significant number of steps.

There is a process gate at this point (PG3), and once again, this could be just a project director or project executive decision.

20.3.4 Implement

The Implement Process is to implement the stage work. For example, in the Requirements Stage, the stage work is to develop the Project Requirements Document. In other words, every stage has a unique scope of work; we plan it in the Stage Detailed Plan, and that is the time to implement it.

Implementation is supported by nine process steps.

The implementation's process gate (PG4) is linked to the Control Process.

20.3.5 Control

The Control Process is to control the stage work. Control starts with the Stage Authorization Document and ends with stage closure.

Control is supported by 16 process steps.

The control actions must be ongoing and cover all functions within the stage. There is a process gate at the end (PG4) to ensure all work is implemented per the detailed plan, which triggers closure.

20.3.6 Close

The Close Process is to close the stage work. The various actions and content are like the project's Close Stage but limited to the stage work. In other words, we will have reconciliation, lessons learned, success assessment, and archiving project records.

There are eight process steps for the Close Process.

20.3.7 The Templates

The simulation section (Part B) will partially present the templates as we present the steps, but this is in a text format, not a template format. In the Appendices, we will show a few sample project and stage templates. However, for the detailed templates and instruction sheets, we will include them via a dedicated section on the SUKAD website.*

20.4 The Project Management Functions

20.4.1 Overview

First, a definition. Throughout this book, we have talked about the need for organizations to build their organizational project management function (OPMF). In that context, the focus is on establishing a formal structure, division, department, or other formalized business structure to treat project management as an essential and independent function, equal if not more vital than the other organizational functions such as IT, HR, marketing, etc. In the context of this chapter, when we use the term *project management functions* (plural form) we are referring to cost, risk, procurement, and all of the other specialized functions required for the management of projects.

With the definition out of the way, it is time to cover some of these project management functions in this book. Please note that many references and books discuss these topics in detail. PMI's *PMBOK® Guide* explains most of these functions, which it refers to as *knowledge areas.* Other references go deep into some of the topics. For example, AACE® International† has guides for cost engineering and even scheduling. Quality associations offer extensive coverage of quality-related topics. There are other associations and guides for these topics as well.

If all of the above exists, then why this section?

This section focuses on these functions in the context of (1) project life cycle model, and (2) megaprojects. For example, some references might offer us knowledge about estimating techniques (top down, bottom up, etc.) but do not provide enough detail on when or how to use them along a project life cycle (across the stages) and do not address the complexity of capital projects in general and megaprojects in particular. Once again, in this section, we summarize a few chapters from *Project Management beyond Waterfall and Agile* (Ajam, 2017) and incorporate the megaprojects context.

---

* A reader may have to show a proof of purchase to share these templates for free. In addition to the templates, the dedicated sections will include flowcharts and mind maps. We also expect that soon after publishing this book, the Uruk PPM Platform will be under development, which will be of great value to your organization if you want to apply CAMMP™ MP in your organization.

† http://web.aacei.org/resources/publications

20.4.2 What Are the Project Management Functions?

PMI offers 10 knowledge areas; ISO offers the same but calls them *subject* areas. The CAMMP™ standard model provides 12 functions. However, since publishing CAMMP™ version 3, we believe it is better to split the Overall Management into Integration and Performance Management. We also group these functions into five groups. Therefore, the functional groups and functions are:

1. Core Functions
   1.1. Scope management
   1.2. Schedule management
   1.3. Cost management
   1.4. Risk management
2. Compliance Functions
   2.1. Quality management
   2.2. HSE management
3. People functions
   3.1. Team management
   3.2. Stakeholders management
   3.3. Communications management
4. Support functions
   4.1. Resource management (physical)
   4.2. Procurement management
5. Overall management
   5.1. Integration
   5.2. Performance management

20.4.3 Linking the Functions to Processes

Remember, the stage management processes are Authorize, Plan Management, Plan Details, Implement, Control, and Close. Therefore, for the management of each stage and the overall project, these functions apply along with the stage management process in a similar fashion to what ISO:21500 (ISO, 2012) and *PMBOK® Guide* offer. What is shared between the project management functions and project management processes are the process steps shown in Table 20.1, which presents the process steps for each process. We will re-represent the same information in Figure 20.2, which shows the process steps mapped to a given function.

- We realize there is a great deal of information here, but keep in mind that leading megaprojects requires considerable effort and adequate staffing. This is why I choke when I hear that a project owner is leading a megaproject with a handful of people, and everything else is outsourced. The bad news is that these process steps would be repeated for every stage. The good news is that (1) different people might be doing some of this work, and (2) some of the general items could be CPM'd from one stage to another.
- For example, the Stage Management Plan from one stage to another has the same structure and even a lot of the same content, but with some variations that are specific to the given stage. The Stage Detailed Plan also has a similar structure from stage to stage, but the content is unique to the stage work.

In the following parts, we will cover the various functions.

Figure 20.2 The CAMMP™ Process Steps Mapped to the Project Management Functions

20.5 Expanding on the Functions

I think it should be clear by now that we will not cover every function or go into extensive detail.

20.5.1 Scope Management

I mentioned earlier that the word *scope* can be misleading, because depending on where you sit, you might see it from a different angle. If you are business or operation, to you, the scope is the *product* that the project will deliver. A project management purist perspective would see scope as the *project* scope, and even a narrower view, the *stage* scope. If you are a contractor, then you view things from your angle, which is the *contract* scope.

How do we develop the [___] scope?

The product scope is what the business needs, and the business assesses the product based on their understanding of the market conditions, economics, and other factors. That helps decide what to build. Other factors affect the capacity that the organization needs. Barshop (2016), Oberlender (2014), Construction Industry Institute (CII, 1994), and others have enough information about this and how to do it. We often link scope to change, but we will address change as a separate topic. In a typical project life cycle, the high-level scope is included in the Project Brief. However, the Project Requirements Document (PRD) is the main document that clarifies the product scope.

Project scope is mostly the project life cycle that we present that identifies all of the work that has to happen, from idea to closure. Within this project scope is the scope of work for each stage, which the team will extract and elaborate on. Similarly, a contract scope covers the work that a project owner wants to have accomplished through a service provider.

We think this enough on this subject. The key is: take it one stage at a time. We start with the high level and elaborate as we progress.

20.5.2 Cost Management

Cost management includes the actions of estimating, budgeting, and controlling. We will address controlling as an independent topic. This is a critical topic, so we will expand a bit.

Estimating

One of the challenges faced on megaprojects is the ability to generate a reasonable cost estimate that provides a predictable outcome, even from the early conceptual estimating. Another challenge we raised is related to some owners' understanding or lack of understanding of the links between uncertainty and estimating. I believe all would agree that as the project progresses, scope definition is elaborated, risks and uncertainty are clarified and reduced, and the accuracy of a project estimate improves. A good and detailed estimate, with a high level of accuracy, might not be possible until we complete engineering. However, waiting for that point—or even worse, waiting for bids—is too late and will cost a lot of money. Pre-engineering or the end of the preliminary engineering stage is a better point. In the capital project industry, it common to use this point for the final funding of the project, because the accuracy of the estimate at that point could be approximately ±10%. To get to that point, management would have spent about 3% to 5% of the total expected cost.

The above sounds good, except that management might not be willing to spend so much money before the final go or no-go decision. Management wants earlier indicators. As a result, the capital project process should include more than one estimate in the front end. It is common to have a project estimate in the Feasibility Stage that is a rough order of magnitude with high uncertainty, justifying wide-range estimating. This estimate is good enough for management to decide to grant the project an initial approval, justifying spending more money for the upcoming stages, until the next point. Then, another estimate would be developed with the Strategy Stage that management can use for the business decision, advanced funding, and the decision to move into the preliminary engineering stage.

Therefore, based on the above, we could have three estimates in the front end. The critical message is that these are estimates for the whole project! Also, there will be at least one estimate for each stage that could align with the overall estimate or lead to adjustments. The simulation (Part B) will include a further explanation on this point.

The biggest challenge, which can cause the lock-in mindset, is that management is often stuck on the numbers from the initial estimate as a single point estimate instead of a range. Therefore, it is crucial that the first estimate have a wide range to be predictable. This requires two things:

1. Management must understand project management; refer to the earlier discussions in Section III. In other words, they need to understand uncertainty, estimating ranges, and accuracy, among other factors.
2. The organization must have an excellent OPMS, inclusive of the cost estimating processes, historical data, and competent cost engineers.

If the above two conditions are absent, then organizations will continue to encounter challenges and disasters on megaprojects. On the other hand, with the above in place, all three estimates should provide us with a total expected cost that would be similar or within acceptable parameters.

How can we achieve the above when there is not enough definition of the project and product scope? Figure 20.3 provides the answers, and here is how to understand it.

- First, the estimate class numbering that we are using here is different from what AACE® International or other references offer. The CAMMP™ sequence is: Class 1 Estimate represents the first estimate developed during the Feasibility Study; Class 2 Estimate is prepared with the project management plan; and Class 3 Estimate uses implementation planning and the preliminary engineering information as input.
- The percentages on the y-axis are for demonstration purposes to relatively highlight the level of the scope definition from one estimate to another.

Based on the above, we can say the following:

- We do not know the final cost until all of the work is defined, known, and even completed. That is what we show via the right bar, as the final cost. The challenge is to predict this final total cost when we develop the earlier estimates.
- If the organization has an excellent organizational project management system and historical data, as we stated earlier, and has competent cost engineers, then we should know that when the team develops the first estimate, uncertainty is high. It is possible that only 60% of the final cost can be

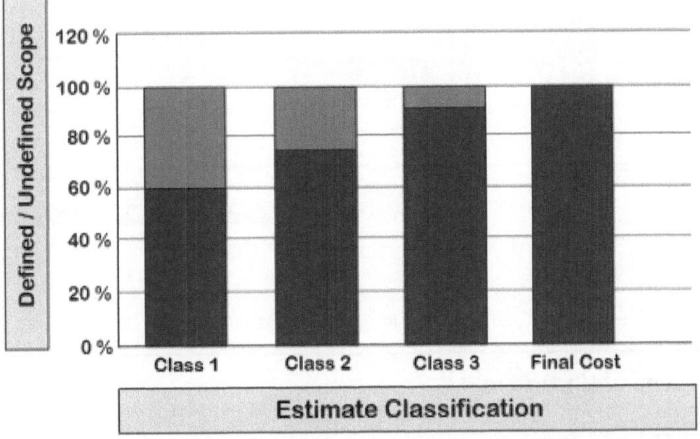

Figure 20.3 Linking Estimating to the Level of Scope Definition and Predictability

predicted. Alternatively, we can say the final cost could be –50% to +100% (or some other range) of the base estimated cost. A good OPMS and historical record can also tell us that in this case, we may need a robust contingency amount to cover the gap*—let's say we will have 40% of the total cost as contingency allowance. The key is to predict the final outcome.

- Similarly, for Class 2, we now know much more about the scope details than we did at the Feasibility Stage. For example, the project requirements have been done after Class 1. Therefore, uncertainty is lower, but we still do not see the full scope. Notice that in the image the dark part of the bar is higher than in Class 1, indicating a higher base cost (what is known). We cover this increase from the contingency allowance of the Class 1 Estimate. In other words, some of the contingency from the previous estimate is now base cost, but we may still need a contingency of about 25% (again the numbers are for a demonstration, not to be used as rules). In other words, we have converted 15% of the 40% of the Class 1 contingency into base costs and keep 25% as a contingency.
- Finally, in Class 3, with definition work done, we could reach a narrower range, possibly –10% to +15% (again, the numbers vary). As before, the increase will partially or fully come from the contingency of the Class 2 Estimate. However, we will still need some contingency at this point.
- All of the above is based on a critical assumption, which is that there are no objectives or major product scope changes, and that what the organization envisioned in the Concept Stage remains the same. What we mean is that the objective is relatively constant, but the details, of course, will grow significantly through design development or progressive elaboration. If there are changes, they must be addressed and reconciled for learning and future reference. This is why we advocate implementing change management events in the front end; more on this later.

Another way of looking at this situation is through the cone of uncertainty (see Fig. 20.4). Once again, do not fixate on the numbers—these might be true for one organization but not all. This image shows that the first estimate has the highest range, the second estimate is a narrower range, and finally, the third estimate is the narrowest. In a perfect world, the mean is middle (0%), which is the target cost. The image shows steeper lines in the front end and shallower lines during implementation, which indicate that most of the uncertainty has been identified and resolved in the front end.

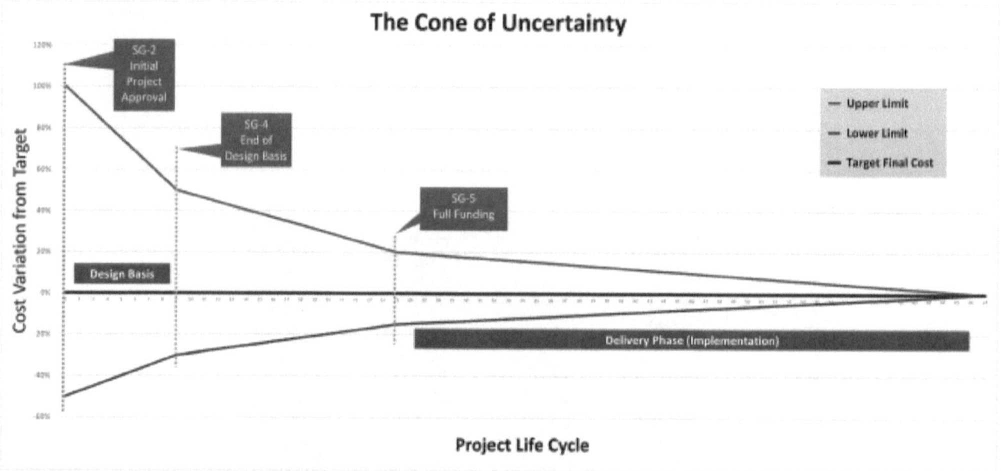

Figure 20.4 The Cone of Uncertainty

* A series of videos on the SUKAD Applied Project Management Channel offer an elaboration on these concepts; here is a link https://www.youtube.com/playlist?list=PLiXup1lJ-_TOxc-e3K2bZqdEWmtolzh41.

A closing comment: Scope definition and estimating are directly linked and are iterative. With every stage, we elaborate on the product scope to increase the level of definition, reduce uncertainty, reduce risks, and improve the estimating accuracy.

Contingency

We have mentioned contingency more than once in the previous discussion. So what is contingency? A contingency (contingency allowance, contingency amount) is the funds needed to cover risks and uncertainty. In the early stages of the project, we estimate the contingency amount based on the historical norms from the organizational knowledge base. Many organizations use the historical patterns even for the Class 3 Estimate; a typical value is 10% at that time (with Class 3). Some of the research (Barshop, 2016) indicates that, on average, this might not be enough. To repeat: you need to find out the numbers in your own organizational context.

A more academic way to estimate contingency is to identify all risks, determine their expected monetary values, then calculate the cumulative sum, which would be the amount of contingency. This is alright for the major risks, but there could be thousands of risks on megaprojects, and most would not be identified or assessed. Therefore, using historical norms would be a good approach, but we must consider any unique project challenges, such as remote locations, new technology, congested sites, and brownfield versus greenfield; all of these factors could require additional allowances.

One final statement here: It is common to have two types of safety factors, or allowances. This is the contingency reserve that we have been discussing here. This amount would be added to the estimated cost to determine the approved cost under the project director's control. The board of directors or the organizational system could also have a standard management reserve in the financial system above the project cost. This additional amount would be under the control of the project executive or the steering committee; this would be a management reserve.

> **A Sad Reality**
>
> Albert O. Hirschman's Principle of the Hidden Hand (Hirschman, n.d.) states, "*If people knew the real costs and difficulties from the start, nothing would ever be approved. Therefore, it is good they don't know, because we need to get things going, to start digging.*" (Flyvbjerg, 2017)

Other Costs

Everything that we have discussed so far is for facilities construction—project capital expenditures. There will be additional costs, such as working capital and other items. Those will be addressed as part of the financial package and funding request. These costs would be estimated by the operations team using similar estimating approaches to what we discussed for the facilities.

Estimating Pitfalls

There are many pitfalls related to estimating. We will touch on a couple of them here.

1. Playing games. Playing games might consist of inflating the estimate to get a higher budget, which could help avoid an overrun but would lead to higher costs than industry.* Another game is to underestimate in order to sell a project that might not be approved otherwise. This is unethical if done on purpose, but we know it happens, maybe more so in infrastructure and government projects, where the taxpayers suffer the consequences.

* We discussed these in Section III.

2. Giving too much power to project managers/directors to influence the above. In mature organizations, estimating is part of project management but is kept outside the control of a project manager to avoid a conflict of interest.

20.5.3 Schedule Management

A great deal of what we have discussed for cost applies to schedule, because a schedule is a time estimate. In Chapters 5 and 6, on challenges, we also touched on many of the challenges related to scheduling, and it would be useful to refer back to review that content. The vital information to re-share at this time is taking the necessary time to adequately define the project before final approval and the start of implementation. The CII's Project Definition Rating Index (PDRI) and IPA's FEL- (Front-End Loading) Index are both good references to help in this regard.

I also want to refer to the case studies. For the failed project (the North American project), one of the things that were done right was insisting that the schedule must be submitted with the invoice, and if no schedule, then no payment. A monthly invoice on a megaproject could be tens of millions of US dollars, so it is a significant incentive to require schedule updates with the invoice. However, even with this condition, the project ended up nine months behind the original date; if they were not linking scheduling to payment, the team would not have been able to find out the hidden problems that were happening. Unfortunately, the discovery came late,* and the damage was done, but it could have been much worse.

20.5.4 Planning

I am separating planning from scheduling because they are different. *"Project planning is the heart of good project management because it provides the central communication that coordinates the work of all parties"* (Oberlender, 2014, p. 66). It is important to remember that megaprojects are fragile; they are not flexible, and as Merrow (2011) said, they are not for filling a gap or market window. In other words, organizations must view megaprojects as strategic initiatives and not push the planning or schedule development to fill a short-term increase in market demand for a given product. If organizations try to accelerate a project for a market window, they are putting the project at considerable risk.

Project management, including the core aspects of estimating, scheduling, and overall planning, should be viewed as strategic imperatives. Management must learn to acknowledge project management and respect the plans that this function produces (assuming the organization has not already killed this function). We realize that many organizations now depend on project management consultants and contractors, but these cannot fill the shoes of the project owners, and there are often many opportunities for conflicts of interest.

View on Risks

In the project management community there are two schools of thought on risk management. One of them views risks as *threats and opportunities*. Therefore, risk management in this context is about managing the threats and opportunities. The other views risks as *threats only*. Consequently, followers of this school talk about managing risks and opportunities.

* Although the contract had that good clause, the owner did not have the personnel to manage it or even monitor it until the damage was done. Even then, the project owner felt that having a scheduler visit the project for three days per month was enough. In other words, there was no adequate staffing.

20.5.5 Risk Management

Lenfle and Loch (2017) stress the need for higher flexibility in managing uncertainty for innovative projects. They also state that, *"underestimation of, or refusal to acknowledge, uncertainty"* is one of the three common causes of project failures.*

The critical message here is the need to manage a megaproject across the stages, concept to closure, and to manage each stage. We have said this more than once, and we may repeat it a few more times. Therefore, we must look at risks (threats and opportunities) for the project and for each stage as well. With every project, such as business, marketing, safety, planning, etc., there are different types of risks. However, in the context of the project life cycle, risks are divided into two categories. The first category is the set of risks that are unique to a stage, meaning they will not be relevant when the stage is complete. Therefore, the team must treat these risks during the given stage or lose the opportunity for effective risk management. For example, risks that could affect the development of the project Feasibility Study will not be relevant once the study is completed and accepted. The same applies to the other stages. These risks are addressed during the stage detailed planning work.

The second category is the set of *project* risks; these are risks that affect the project's overall objectives and are not particular to a given stage. Let us view risks from these various perspectives.

What is a project?

A project is an opportunity that an organization is considering, such as a new product, a new market, or a new revenue-generation idea. A project could also be a response to a threat, such as a competitive threat, regulations, or safety/security matters. To elaborate, we will consider that the project is for a new product—a business opportunity—because if the organization delivers the project successfully, it will create the capabilities that enable the realization of benefits or positive consequences. If the team fails in delivering, the consequences will be negative. Within the above context, a Feasibility Study is a form of risk analysis in which the organization assesses the organizational decision (the project/opportunity) and determines whether the benefits justify pursuing the possibility. Through the Feasibility Study, the organization decides whether to exploit the opportunity presented to them.

What are the project risks?

During the development of the Feasibility Study, the focus is on commercial risks, market risks, and financial risks, among other high-level, project-wide risks. In the megaprojects context, other risks could be significant, such as partnering risks, political risks, corruption, security risks, and instability. These are risks that the team should address during the work on the Feasibility Study and the Shaping Stage. If management considers that the threats from these various risks are significant, the organization would likely decide not to exploit the opportunity and would discontinue the project.

However, let's say that at Stage Gate 2 (SG2), executive management decides to go ahead with the project; in that case, they are indicating that those risks (threats and opportunities) are manageable, and the team has considered those risks as part of the project's economic calculations. Then again, at SG4, when it is time for the business decision, they will go through the analysis again. However, when the decision is a go, it means that management has either managed or accepted the project risks. Alternatively, they could have accepted the risks at the time of authorization, on the condition that the team addresses those risks when appropriate, in the upcoming stage(s).

What are the stage risks?

As mentioned earlier, in addition to project risks, there are stage-specific risks that the team must deal with during the planning for each stage. Some of these risks can be people related, such as human resource

* The other two are "stakeholder neglect or mismanagement" and "inflexible contractor management" (prominently, awarding work to the lowest bidder).

availability and competence. Other risks can be time related—for instance, not enough time to do the stage work. Other risks can be a result of a lack of organizational systems and standards. Then there are the possibilities of poor planning and known distractions during the implementation of the stage work.

Because these are risks that only affect a stage, once the stage is over these risks are gone, whether they materialized or not, unless they are treated and managed. How does the team manage these risks? Like any other risk, the team has to identify them, analyze them, and respond to them.

Why is this important?

Consider the following scenario, using the Feasibility Stage as an example. Management makes a decision to spend millions of dollars (billions in the case of megaproject) when they approve a project based on the output of the Feasibility Study. Such a decision builds on the understanding that the Feasibility Study and conclusion is sound. Imagine the consequences if the Feasibility Study conclusion proves to be faulty—say, if it was based on poor analysis or by personnel who did not have the right level of competence. These are risks specific to the Feasibility Stage and, if not addressed, then all else is questionable, and the decision to approve the project lacks the proper due diligence. The findings from IPA (and others) clearly show that a common error is an underestimation of cost and overestimation of benefits (Flyvbjerg, Holm, and Buhl, 2002). If this is not an error, but a deceptive practice, then that is a massive risk in feasibility studies and other stages of the front end and must be managed.

A suggested resource for further study of this topic is ISO 31000 (ISO, 2009), which is a guide for enterprise risk management. This guide has recently been updated.

20.5.6 Quality Management

Quality of What?

Capital megaprojects require engineering and construction. Therefore, it is common to have the majority of the staff come from an engineering and construction background, and consequently, when they think of quality, they think about the quality of engineering and construction. The team can monitor this type of work via inspections and testing. However, quality is not limited to products, and it must include the management and project management process.

Why?

Because the ultimate goal of a project is to gain added value, which is based on production and revenue. If we fail to accomplish the expected value, we must look for the root cause. Let us consider this scenario.

Why can't the facility reach the production capacity? Technical quality issues.

Why technical quality issues? Installation issues.

Why installation issues? Engineering errors and omissions.

Why engineering errors and omissions? The front-end development was cut short or not enough details.

*Why was the front-end development cut short or not enough details?**

Here we get to the root cause, which is related to the challenges we presented earlier: lack of clear objectives, lack of leadership, lack of competence (organizational and individual), etc. We cannot blame Murphy for everything. So quality is in every step, every stage. If we fail to do proper feasibility, the decision-making process is jeopardized. If the team fails to define the scope correctly, all other functions suffer.

In closing this part, I am willing to say that the quality of the *products* is easier to manage because there are precise specifications, often that we can test and measure with physical means, and often undebatable.

* Do you see what we mean by the cycle of doom?

However, management of *process* quality—the technical and project management process—is highly opinioned and debatable. Sure, there is a scientific foundation for project management, but it is mostly an art and a lot of psychology (human behavior).

How do we enhance the chance of achieving quality in managing and leading projects?

Ongoing Project Reviews

The field of quality and quality certifications is vast and offers us the concept of quality planning, quality assurance, and quality control. We have Pareto* and Ishikawa,† control charts and histograms, and many other tools and techniques that we can use. However, shifting from the tools and techniques, let us consider this strategic model that would help in avoiding the cycle of doom on megaprojects. I have touched on this model in an earlier chapter, but it is worth revisiting it. This is not something we invented; I learned it as a young project professional. This model can be used to reinforce the stage gates, but it does not have to be limited to clearing these stage gates.

The core concept is to embed the necessary expertise throughout the project life cycle and project development—a multi-disciplinary expertise. The idea is not to close a stage unless the team (inclusive of management) has a high degree of confidence that there are no gaps or hidden holes in the work and output of a given stage. One way to do this is to look at the stage gate again and consider the question, *"Are we ready to move ahead?"*

Think about it: who can answer this question best? Experts from the next stage, because they are the ones receiving that output to use as input for their work. In other words, have experts from a given stage involved in the preceding stage or even two stages earlier. What we are saying is, *do not depend on management alone or the steering committee decision to clear you through the stage gates* unless experts from the succeeding stage are willing to accept the work of the preceding stage.

To clarify further, consider this simplified sequence of stages: preliminary engineering → detailed engineering → construction → startup.

The proposal is that during the preliminary engineering stage (closer to the end of the stage), and in preparation for the gate review, have a team of experts (could be from the same organization, but from outside the project) review the progress of the stage work. This team could include a project manager, project control, and an engineering expert (or two). The team can perform a review, like an audit, to gain comfort that work is progressing per the company guidelines, and it should be ready to move to detailed engineering. If there are gaps, they must be fixed before proceeding. In this approach, the project manager (or director level) would focus on the project management aspects, the project control expert will review matters related to control, and the engineering expert will examine the technical work.

The organization should consider repeating this process during engineering, with a similar team but instead of engineering expert(s) have construction expert(s). The project director and project control will have actions related to the prior review, whereas construction expert(s) will focus on construction readiness. Then, once more, during construction (late in the stage work), repeat the process but with operation and maintenance experts in the review team, to assess readiness for handover.

The above practice might seem exhausting, but it should not. In our view, it is a must on all large projects and megaprojects.

The above model can be readily applied if the organization has maintained a robust organizational project management function with in-house expertise to conduct these reviews. An alternative, not as effective but can be workable, is for the project owner to build a strong partnership with project management

* The Pareto Principle basically states for many events, roughly 80% of the effects come from 20% of the causes.

† Ishikawa diagrams (also called fishbone diagrams, herringbone diagrams, cause-and-effect diagrams, or Fishikawa) are causal diagrams created by Kaoru Ishikawa that show the causes of a specific event.

consultants, engineering firms, and construction contractors that would be able to help. However, this alternative might not be feasible if there are proprietary technologies, and any of the reviewees could object, opening up for potential competitors. This is why it is crucial for the project owners' organization to build and sustain their project management and project development functions.

20.5.7 Procurement Management

The Focus

This is a massive topic because it can cover many subjects, such as the various execution approaches or commercial variations of the contract types. Also, we can discuss purchasing and contracting. However, we will touch on the critical contracting aspects only. First, a few clarifications and definitions.

- We are focusing on contracting, rather than purchasing and contracting.
- We cover major contracts—those for the delivery phase of the project, in particular, the engineering and construction work. Therefore, it is critical to note that many other services are covered via contracts, and all are important. However, their cost and schedule impact could be negligible in comparison to implementation work.

Types of Contracts

For the implementation work, we need to consider two concepts that we can use to categorize contracts:

Delivery Approach

This is a way to look at contracts from their scope of work in the delivery life cycle. In other words, are we talking about design–build (DB) or design–bid–build (DBB)?* Another way of looking at this would be a contract for engineering–procurement–construction (EPC) or separating the EP from construction (EP & C).†

Keep in mind that these are not the only types. Another related type would be turnkey, in which the contractor handles the commissioning work. We can also see EPCM, in which a client hires a contractor to handle the engineering and procurement of engineering items and manages the construction contractor. In this case, the construction contractor could be a subcontractor to the EPCM contractor or have a separate contract directly with the project owner.

In megaprojects, most contracts for the implementation work are typically EPC or EPCM, regardless of the commercial aspects (Merrow, 2011).

Commercial Perspective

From a commercial perspective, the different types may include:

- Cost plus, also known as *reimbursable cost contracts.*
- Fixed price, and we often hear the *firm fixed price.*

* DB often means that one company would design and build the facilities; whereas DBB often refers to two companies—first design, then bid for building work. These terms are often common in building and infrastructure projects.

† Similar to the previous comments, EPC implies one company for the whole work, whereas EP & C means a separation. These terms are common for industrial projects.

- Lump sum; on its own, this term might be for short-term services for which the price is fixed and payment is made as one lump sum payment. However, it is common to hear *fixed price lump sum,* which is not accurate because, in a fixed price contract, payments would happen throughout the project, based on certain milestones.

Based on these first two sections, one can expect to define a contract by using one term from each category. For example, an EPC Cost Plus contract: this tells us that we have one contract responsible for the full implementation on a cost-plus basis. The varieties could be numerous, and that is without listing the various types of incentives that can considered.

Other Types

In addition to the above, there are different types of contractual arrangements such as the public–private partnerships. These would be contracts, often between a public entity and a private company, that would likely include a form of operational agreements. Then we have the concepts of BOT, BOOT, and their variations. BOT is Build–Operate–Transfer, and in this case, the Operate could be for a limited period, say, one to five years, or extended. BOOT would be to Build–Own–Operate–Transfer. In this case, the contractor would own the facility for an extended period, then transfer it at the agreed time and price.

There are numerous other variables, but we believe the above is good enough for what comes next.

Contracting and Megaprojects

It is clear, and from more than one reference, that the type of contracting relationship does have a direct impact on project success (Merrow, 2011), and project owners' organizations can benefit from innovation and flexibility in the contractual relations of megaprojects (Davies, Dodgson, and Gann, 2017). Merrow has shown that globally, about 53% of industrial megaprojects use EPC fixed-price contracts,[*] about 25% use EPC or EPCM on a reimbursable basis, 12% use a particular form called *alliances,*[†] and 10% use mixed—that is, use reimbursable for engineering and procurement and fixed price for construction.

Is there a best type?

Personally, I favor the concept of cost-plus incentive, but only for project owners who have maintained an excellent organizational project management function, and they can also handle most of the front-end work internally. However, this is an opinion based on personal observations and experience. If we fall back on research, we find that Merrow (2011) clearly shows a link between performance and the type of contract. The EPC fixed-price and EPC/EPCM reimbursable-cost contracts produced similar results. In other words, the differences in their performance and outcome of success and failure are close, with subtle differences. Alliance contracts were quite poor and should be abandoned. The star was the mixed approach. The majority of megaprojects performed under a mixed contracting strategy were outstanding—they experienced a reduction in the final cost and schedule targets, and production failures were almost nonexistent (Merrow, 2011, pp. 259–263). We are convinced, even though our preference is cost plus all the way.

Once again, what is the mixed type of contracts?

It is using a cost-plus arrangement for the engineering and procurement[‡] work, then awarding construction on a fixed-price basis to a different contractor.

Why would this make sense?

---

* Although the global average is 53%, it is almost 100% in West Asia (Middle East).

† This is not about the partnering best practice. It is the worst performing type per Merrow.

‡ What would make sense for engineering to procure.

> **Why the Preference for Fixed Price?**
>
> There are a few misleading concepts that drive organizations toward a fixed-price culture:
>
> - There is a belief that they are low-risk options or that they transfer the risks to contractors. *"Often, the underlying purpose of using [the] EPC lump-sum approach to contracting for megaprojects is not risk transfer but blame transfer"* (Merrow, 2011).
> - Bank financing favors and even requires fixed-price contracts as a result of the lack of understanding by the finance industry of the megaprojects industry.
> - Governments often require this type of contract for many reasons. This is especially the case in West Asia (the Middle East).
> - The situation is even more complicated with governments mandating lowest bid. *"All of the EPC lump-sum contracts that were won on significantly low bids . . . failed"* (Merrow, 2011).
> - There is a perception of lack of corruption in these contracts.

One of the main challenges with EPC fixed price is that there is too much uncertainty, especially with the construction cost. The uncertainty is magnified as a result of:

Construction work is one to three years into the future from the point of budget approval or initial contract, which leads to exposure to changing economic conditions (especially in a VUCA-driven world*).

Potential changes in design would impact construction and would not be known at the time of bidding.

Therefore, unless the scope is genuinely fixed and stable, the mixed approach is preferred. Once engineering is close to being done, uncertainty caused by changes is significantly reduced or eliminated, plus the timing is close to construction, diluting the effect of economic conditions. Therefore, that would justify the fixed price for the construction work.

I almost forgot—the project management for both contracts (EP & C) should be by the owner management. If you must outsource project management, then outsource it to other than the EP contractor.

Then, why are only 10% of the contracts for industrial megaprojects mixed contracts?

We do not have an absolute answer, but the assumptions that we share here could be the leading causes. First, maybe not many companies know that the research shows that this type performs the best. The second reason could be that project owners do not have the staff, competence, or capacity to manage two separate contracts—engineering and construction—especially with engineering being done on a cost-plus basis. The third assumption would be the reasons we listed for the preference for a fixed price (earlier textbox). We know there are other reasons, but maybe it would not be politically correct to mention them here.

Potential Pitfalls

The first potential pitfall is related to incentives. Incentives could have negative consequences under these conditions, collectively: (1) it is in a fixed-price environment, (2) the owner does not have enough project management capacity or competence to manage the work, (3) the owner depends on a FEED consultant (preliminary engineering) to develop the implementation plan and cost estimates, and (4) the FEED contractor is allowed to bid (Merrow, 2011). If this does not present a massive potential conflict of interest, I'm not sure what would. We have seen this happen too many times, and the consequences are project cost significantly higher than the industry average—it could be as much as 50% higher.

Let us also list a few points on the risk of a fixed price, because this is one of the significant challenges facing megaprojects:

- Fixed-price contracts always cost more than the other contracting approaches.

---

* VUCA: Volatile, Uncertain, Complex, and Ambiguous.

- Projects could achieve their cost and schedule target and still fail on the IPA competitiveness indices: *"Some of our [failed] projects in the Middle East came in on budget, on schedule, and started up appropriately but paid 40 percent more for the project than they should have."* (Merrow, 2011, p. 278)
- If the market conditions allow the contractors to be conservative, they have many risk factors that they include in their bids; each of those factors could be 10% to 40% of the base price (Merrow, 2011).
- The analysis shows that incentives might not work well. On this last point, we have seen cost plus incentive contracts backfire because the contractor felt that the owner controlled the budget and was not taking responsibility for the owner's problems. In other words, the contractor believed that the owner was taking advantage of them.

*20.5.8 Project Control**

It should be clear by now that control, like all other functions, is at two levels: the project level following the project life cycle, and the stage level following the stage management processes. Furthermore, the concept of *baselines* is established in project management; however, in this part, the elaboration provided will offer a different perspective.

Baseline

The common definition of a baseline is, *a line of reference, the basis against which we compare project performance.* In other words, the baseline is essential for project control because it reflects the "approved plan." Then the team performs project control by comparing actual performance against approved plans.

Where is this line of reference? What plan should the team use for control and measuring performance?

As stated repeatedly, the traditional view, or common practice, is that project control is about comparing actual performance to the plan, which is typically the plan used to obtain the final approval. In the CAMMP™ MP, this is at SG5. Many of the references we have reviewed measure the project performance in implementation, in comparison to this point. On the stage level, control (assuming there are controls at this level) would be against the Stage Detailed Plan. However, this concept contradicts a fundamental aspect of project performance. As project owners, the interest in control should be along the whole project life cycle and not limited to a detailed plan, project, or stage. In other words, control happens almost end to end.

How is that?

Let us elaborate.

Stage-Level Control

If one considers the CAMMP™ processes, then control starts as soon as the team begins the Authorize Process, with the control reference being the statement of work. Furthermore, control activities continue during all of the other processes. Figure 20.5† shows the six CAMMP™ processes in a sequential style. The sequence shows five of the six processes, with the Control Process below them, since it is linked to the other five processes. If you refer to Figure 2.6 (on page 21), you will notice that control is connected to the first three processes via the dashed lines and to implementation and closing via solid lines. These dashed lines indicate that control is active throughout the stage. If you study Figure 20.5, you will

* The original content is from *Project Management beyond Waterfall and Agile* (Ajam, 2017), including some modifications.

† In the final version I need to update this graphics to use the same term as CAMMP™ version 3.

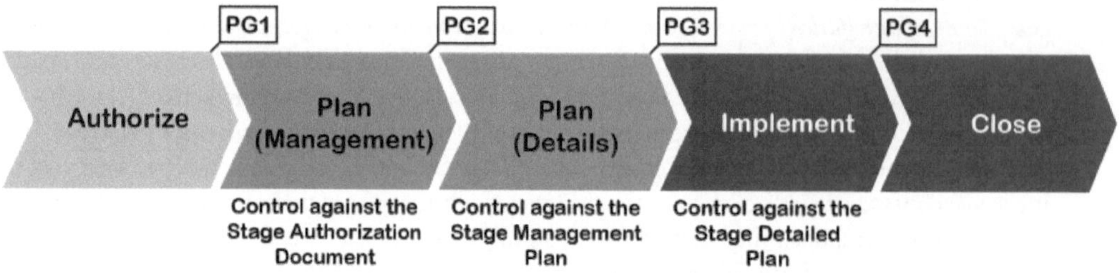

Figure 20.5 Stage Control and Stage Control Reference Points

notice that, as the work proceeds from one process to another, the control reference point is superseded by the new control reference. For example, during the Authorize Process, the control reference is per the idea statement or Project Brief. Once the stage authorization is issued and approved at Process Gate 1 (PG1), that document will become the control reference during the Plan Management Process work. In other words, as the project management team develops the Stage Management Plan (SMP), they must continue to compare it to their control reference—the stage authorization.

Similarly, once the SMP is published, it becomes the control reference point during the Plan Details Process. Next, during the Implement Process, the control reference for validating the work is the Stage Detailed Plan (SDP). Finally, the output verification for the stage is at PG4, just before closure. This concept might be implied in various references, but I have not found an explicit reference.

Project-Level Control

Many organizations that have a reasonable level of project management maturity use one form or another of the concepts that are presented in this section—but possibly implicitly, not necessarily explicitly. The proposed approach is that the baseline is not fixed, and it is not limited to the Project Management Plan (PMP). The baseline moves as the project progresses from one stage to another (as shown in Figure 20.6). The shifting baselines, or what CAMMP™ calls *control reference points,* start with the Project Brief.

It is important to note that *moving baselines* does not mean *changing the baseline.* In this context, changing the baseline refers to the act of re-planning the project and changing the approved plan's target of cost and schedule. Unless the project objectives change, changing the baseline (budget or schedule) is not a proper project management practice. However, the term *moving baselines* means that, as the project progresses from one stage to another, a new control reference point will supersede the previous one.

Well, is not this an example of changing the baseline?

Not really—read on as we explain The Four Control Reference Points™!

First Control Reference Point

In Figure 20.6, the lower arrow touching the Project Brief upward arrow on the left represents control in comparison to the Project Brief—the idea and business case of the project. This is the first control reference point (CRP).

Can a team control against the idea?

Yes, they can, although control in comparison to the idea is conceptual and not quantitative. The phrasing of the Project Brief is critical, and a slight variation can indicate different projects. For example:

- **Scenario 1.** The project's purpose is "to get drinkable water from city A to city B."
- **Scenario 2.** The project's purpose is "to build a pipeline to get drinkable water from city A to city B."

Project control must start early, from SG1

Clearly, Scenario 2 is more specific, and one would expect that the project manager will work on building the pipeline. On the other hand, in Scenario 1, a good project manager will look at all options for moving water and choose the optimal one. To build on the above, if the project manager in Scenario 2 starts considering options other than building the pipeline, then he or she is deviating from the project mandate, which would be outside the project objectives. The role of project control is to stop or eliminate such deviation. However, in Scenario 1, a project manager who is not evaluating the various possible options for moving water between these cities is not performing per the project mandate. Meaning, the project manager is deviating from the project objectives. Therefore, the highest-level control on the project, during the first phase, is working per the Project Brief. In other words, this initial control action is to ensure that the Feasibility Study is developed per the Project Brief and does not deviate from it.

Second Control Reference Point

Control is an ongoing effort that starts with the Project Brief, as already stated. The second CRP is in comparison to the Project Authorization Document (PAD), which is the second arrow from the bottom

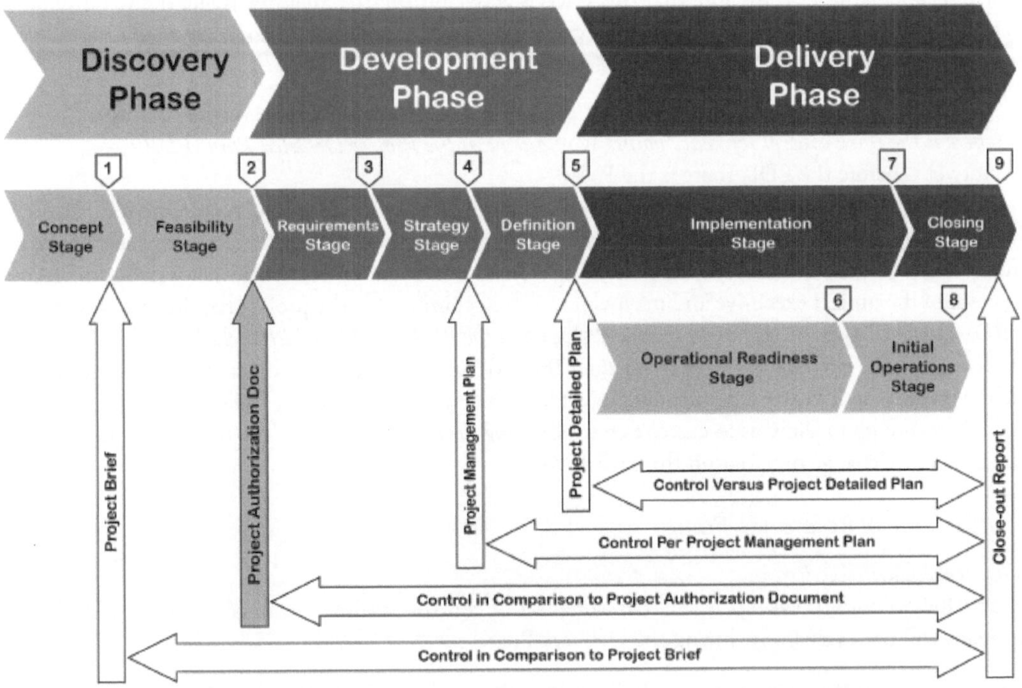

Figure 20.6 Project Control with the Concept of Moving Baseline (Control Reference Points)

of the figure. Assuming the team has performed a proper Feasibility Study and the sponsor developed a clear PAD, the second CRP supersedes the first because it will encompass its purpose. Stated differently, the PAD incorporates the Project Brief and expands on it.

One may ask a similar question: "How can the team control against a brief, high-level document, such as the PAD? There is not enough information to control against at this level; there is no baseline yet."

The proper response is: there is enough to exercise control! The team has the project description, justification, and objectives—all part of the PAD—and these are the ultimate control points. They also have a rough-order-of-magnitude estimate, the Class 1 Estimate.

Aren't these items vague?

Agreed, they do lack detail, but if the PAD's brief contents cannot clearly define the project objectives for the team, then it has a deficiency, and there is an issue that is likely to impact the project's performance. Therefore, control must continue as soon as the project executive issues the PAD. The team might not be able to use the full, earned-value management approach at this point, because they do not have quantifiable means for control of the entire project, but that is perfectly acceptable. The main concern during the Development Phase is to ensure that the team is developing the project per the project objectives as defined in the PAD.

What happens if there is a need to change the objectives?

This can be done via two mechanisms:

1. The project executive can amend the PAD.
2. The organization has a *project change control system** that includes the policy and guidelines on how to initiate and approve changes to the PAD. These are project objective changes.

The second approach is recommended because it documents a change for historical purposes and project reconciliation. Please note that this CRP is active during the Shaping Stage, the Requirements Stage, and even the Strategy Stage. This is a critical period, and a foggy one; *refer to page 226.*

Third Control Reference Point

So where is the third control reference point? Would it be at the Project Detailed Plan (PDP)?

Not yet—before the PDP, there is the PMP.

After the PAD, the project manager and the team start working on the project requirements in line with the project objectives. Subsequently, the team develops the PMP. As stated before, during these stages, the project director must use the PAD as the guiding reference and cannot deviate from it without the formal approval of the project executive and maybe the steering committee. The project requirements document (PRD) and PMP must be complete per the PAD. *Once the PMP is ready, which includes the updated Class 2 Estimate, the team has the third CRP or baseline.* The PMP includes various project management elements, such as procurement strategy, confirmed project life cycle, updated schedule, and cost estimate. These are more than enough to allow us to exercise control *during the Definition Stage.* The team must monitor and control for any change or variation from the PMP.

Fourth Control Reference Point

What is the fourth CRP?

This fourth point is easy! It is the traditional set of control baselines established by the PDP, which the team will use extensively during the Delivery Phase.

---

* Part of the overall OPMS.

Why Four Points?

It is vital to note here that the team is not necessarily, and concurrently, controlling against these four CRPs, which, in practice, will supersede each other.

Why is there a need for four points of control? Is this approach necessary? Is it not enough to control against the baselines we defined per the PDP?

To answer these questions, take a step back and reverse the question, *"Is one point sufficient?"*

One reason that the traditional and shared view of control starts with the PDP is its linkage to project funding at final approval. Some practitioners think that unless there is an approved budget, there is no need for control. In other words, they link control to the money in an approved budget. In other cases, there is no formal control or change management unless there is a contract for services, because the view is that internal changes are part of the development and so are not real changes—changes only exist in the context of a contractual relationship.

Project control is not only about money or contracts. Project control is about variations from the plan, including variations in scope, time, cost, quality, and human resources. Notice the use of the word *plan* and not *Project Detailed Plan*: the plan in question is an official reference (the last approved deliverable or estimate) that the team can compare against, such as the PDP, PMP, and even the PAD. Therefore, the team must monitor for variations and changes throughout the project.

Project Management Maturity and Control

OK, clear, but still, what is the significance of project control during the Development Phase? Wouldn't the PDP absorb all of these variations and changes and become the document that supersedes all before it?

Well, yes, but is that enough?

Let us put on our organizational maturity hat.

The assumption here is that during the Development Phase, the team would expand the level of detail in the project scope from the Discovery Phase. This development of concept action—progressive elaboration—will likely introduce details that were not listed earlier. Some of these could be deviations, whereas others could be changes or scope creep. By "absorbing" these variations and changes in the PDP without a formal project control and approval mechanism, the team is essentially burying or hiding these changes in the project details. If these variations and changes are due to poor planning, lack of experience, or an inadequate OPMS, then how can the organization know, how can it learn, and how can it seek continual improvement? Alternatively, the changes could be due to environmental factors, changing economic conditions, or numerous other factors. It does not matter what the reasons for variations and changes are—the team must know them and understand them. They also must document them so that the organization can learn from them and build toward higher levels of project management maturity.

If we reflect back on the literature review and identified challenges, we see that significant growth could be happening in the early stages of the projects. The Australian Project study showed that about two-thirds of the overall growth is in these stages, and only one third happens during the Implementation Stage (Terrill and Danks, 2016).

The Challenge with On Time and Within Budget

We have touched on this subject earlier and in our other books and blog articles. The concept of "on time and within approved cost" can be taken dangerously out of context. To avoid repeating, we advise you to pause and refer back to paragraph 12.7.3, starting on page 109.

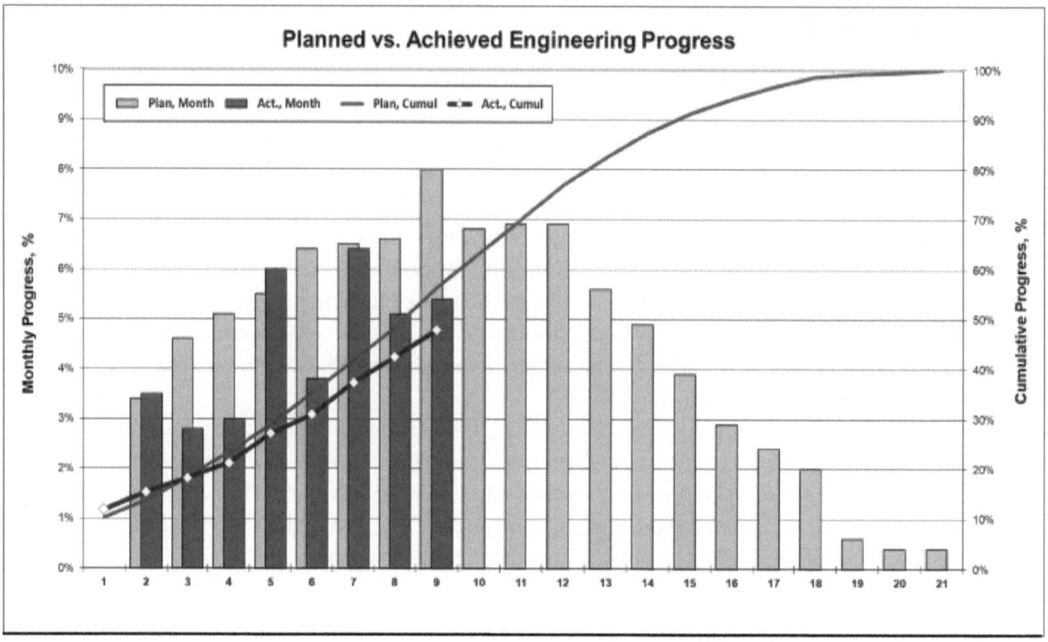

Figure 20.7 Engineering Progress, Actual Versus Plan on a Real Project

Schedule Control

Although all of these discussions are about overall project control, which includes schedule control, it would be good to address a few points on schedule control.

Schedule performance (delays) is one of the common challenges of megaprojects. There are many reasons for this, and it can be detected early if competence and discipline decide to partner. For example, it is vital that project owners mandate a resource-loaded schedule with the bids. However, the owner team needs to have the discipline to check it and not to treat it like just another tick mark on the checklist. Further, delays in engineering would be an *"excellent indicator that the project is in the process of failing"* (Merrow, 2011, p. 311). Therefore, if the engineering is slipping, the project executive must be willing to accept the delay and rework the schedule rather than force mobilization of the construction contractor per the original schedule, which would be premature if the delay has happened.

Project control must be in real time, and accurate.

Let us consider Figure 20.7, which is from a real project.* Please notice the following:

- The *x*-axis is for the month, not the actual date.
- The project started much earlier than shown in the image, but there was no progress data for that period (an indicator of delayed setup of the control system).
- Engineering is slipping, and the gap is widening, yet no action was being taken.

* I have masked the dates.

- Construction started on the equivalent of month 6 (in the image), which was at the point of 30% engineering progress.

What were the consequences?

The net result was that engineering was delayed, and construction productivity suffered as well, because the construction team had mobilized but did not have enough work to keep them busy; the ultimate result? At least nine months' delay.

Closing Comments

I will fall back on the literature review to close this part.

"Project control is difficult because it involves a quantitative and qualitative evaluation of a project that is in a continuous state of change. To be effective, a project control system must be simple to administer and easily understood by all participants in a project" (Oberlender, 2014, p. 217).

Project control must start early because *"very few projects actually fail due to problems that originate in execution. If a project has been prepared well, execution is about maintaining the value that has already been created."* (Merrow, 2011, p. 305)

Project control is critical for project success; it is the project navigation system.

20.5.9 Change Management[*]

Project change management and project control have a similar foundation. The four control reference points discussed earlier apply to change management and are equally important. Although it might be clear to many readers, it is of value to differentiate between control and change management; see highlighted text.

Project control is about variations and deviations from the plan,
which happens as a result of performance-related matters (doing work).
Whereas (project) change management is about the conscious
decisions to change the plan, or part of the plan!

Why Change Management Is Critical

In the capital projects world, especially for megaprojects, change management is critical and a significant contributor to project success or failure. In general, changes to projects are disruptive and are leading causes of threats to the project's objectives. Studies by the CII[†] for capital-intensive projects—industrial and engineering/construction projects—find that poor change management results in poor productivity, and in turn, over-budget and behind-schedule performance.

[*] The original content is from *Project Management beyond Waterfall and Agile* (Ajam, 2017), including some modifications.

[†] http://www.construction-institute.org

On the other hand, effective change management can lead to better-than-planned performance. We touched on this in Chapter 11, but it is always good to recall that information. If the team does not adequately manage change, starting with proper identification, documentation, and analysis before decision-making and implementation, that is a form of weak project governance. This lack of change management is a critical factor for organizations without a proper project management system and an absence of appropriate accountability. The impact of change is not always evident, but CII best practice does offer us a view of the threats triggered by project changes. These include:

- Impact on engineering and construction productivity resulting from a high level of changes.
- Impact to project performance caused by changes.
- Impact from hidden changes.
- Failure to identify the real impact of changes.
- Cumulative impact of changes, which the team cannot readily evaluate!

All of these are factors that confirm the need for proper change management, especially in the "foggy zone," from idea to SG5. In the following section, we will focus on three areas of attention concerning this lively topic.

The Hidden Changes—Scope Creep

One of the key advantages of the concept of four CRPs addressed in the previous section is to avoid *scope creep* (creep, as in *sneak in, creepy, sinister,* or whatever term the reader prefers to use); this is a significant challenge in projects.

Why does scope creep happen?

One factor is that the Project Brief and PAD are high level, lack detail, and therefore leave room for interpretation. Even the Project Requirements Document (PRD) and Project Management Plan (PMP) do not have the same level of detail as the project detailed plan (PDP). Another factor is that, as the project progresses along the project lifecycle, the level of detail increases. As a result of these factors, it is often difficult to decide what is (1) a natural variation of scope, (2) a conscious change, or (3) a natural expansion of the project detail owing to progressive elaboration. Although it is difficult, it is possible, with proper due diligence, to tell the difference between a variation, a change, and design growth.

Often, scope creep can be innocent, and things happen as a result of stakeholders' discovering missing items, so they add them during the Definition Stage. Missing things is often a sign of a weak system or lack of proper planning. However, assume these are innocent mistakes (although reality tells us that scope creep is often not innocent). Stakeholders use the ambiguity of the earlier stages to add scope—preference items—for things that would not be required to meet the project's objectives. These are hidden changes that get added (or sneaked in) without proper justification or documentation and might not be discovered until the project cost in the Class 3 Estimate shows significant growth over the Class 2 level.

Change Management in the Fog

Project change management in the fog is a term we like to use to refer to the part of the project life cycle during which control is mostly qualitative because of the ambiguity of the scope definition.

Why use the analogy of fog?

Well, imagine you are driving in extremely dense fog; you can't see anything, and driving is dangerous, so you should stop. Relating this concept to project management, a project in which the team does not know the scope or the objectives *is not a project and should be stopped.*

Now, imagine that you are driving in dense fog; you can see some things, but not very clearly. Driving under these conditions is dangerous, but if you must drive, extra care is in order. Even under these circumstances, you would be able to see a few meters ahead—not the details, but enough to stay on the road. This is similar to the period between Project Brief and PAD, during which the team does not see the details but will know if they sidetrack off the project objectives—assuming the Project Brief is clear and unambiguous.

Continuing with the above analogy, imagine you are driving in fog, but not as dense as before. You will be able to see things, not very clearly, but enough to pick up speed while maintaining caution. This is similar to the period between PAD and PMP or between PMP and PDP, depending on the density. In this period, there is more clarity, but still not enough to see the full details—that is, not quantitative.

Why the above analogy?

Because of the traditional view of project change management, some organizations do not implement a change management process until after full funding, from SG5 onward, and in some cases only concerning contracts. To these organizations, unless everything is clear quantitatively, there is no control, which is not acceptable. As stated earlier, the early stages, when things are not clear, is when scope creeps into the project and might not be discovered until too late. Hence, the need for careful and proper monitoring and control, concept to success.

Traditional Change Management

Traditional project change management refers to the common practices for managing project changes. In many organizations, management of the project's changes is limited to projects for which the work is under contract, and the changes are specific to the contracts. This approach represents control against the fourth control reference point, as discussed in paragraph 20.5.8 (page 219). With this approach, the team is skipping proper control between CRP 1 and 4. This period is the foggy, ambiguous area—the critical area.

Types of Changes

There are different kinds of changes that happen on projects.

1. There are the typical contract changes when working under a contract.
2. There are project changes that do not alter the project objectives. These could be changes in the plan or other changes. These changes could happen whether there is a contract or not. Usually, these changes are within the control of the project director and are covered by the project budget.
3. There are objective changes. These are defined as any change to the project objectives outside the control of the project manager—for example, moving the location of a facility project or altering the capacity of a production facility. Usually, these types of changes will require supplemental funds and an extension of time, because they are outside the power and influence of the project director.

20.6 Chapter Summary

This has been an extended chapter, which also summarizes approximately ten chapters from *Project Management beyond Waterfall and Agile* (Ajam, 2017). In this chapter, we focused initially on the project management processes required to manage every stage of the project. Then we shifted the focus to introduce the various functions needed for the project, stage, and project level. Finally, the most extensive section covered most of the project management functions in the context of megaprojects. What we did not discuss is the people aspects, which is the subject of the next chapter.

Chapter 21

The People Aspects of Megaproject Leadership

21.1 Introduction

In project management literature, we often hear that project management is about the art and science of managing projects. We extend that view and see that project leadership is about art and science, and more importantly, about psychology. It is about the behaviors of people and the culture of organizational project management—the culture of performance, growth, and excellence. What I am saying could be a cliché, and maybe it has been for years, since projects and megaprojects continue to fail at higher rates than they succeed.

I realize there are numerous books about people, management, leadership, and even politics in the project management context, but I do agree with Merrow when he stated, *"I don't find much of what the literature has to say very helpful in understanding project teams generally and megaproject teams in particular"* (Merrow, 2011, p.159). Some challenges are unique to megaprojects,* such as geographic dispersion, integration of partner personnel, and interface management.

Further, it is common to observe both the lack of formal project management in the early stages of the projects and the belief that a small team can handle a megaproject in the context of fixed-price contracting. Another challenge that we mentioned earlier is the over-dependence on outsourcing resulting from the dilution of project owners' technical and project management staff. The research shows that all of these factors, and more, are causing the current sad state of practice. We have covered these challenges extensively, so there's no need to repeat here, just remember that those who do not learn from history are doomed to repeat it (Clairmont, 2013). Let us close Part A of this book with a focus on the people aspects, so we can remember, learn, and genuinely start building the road to success. In this chapter, we discuss the essential recommendations for organizing the project to success, which include crucial concepts and a proposed structure. Just remember, what we offer are general concepts

* If not totally unique, then highly amplified in the megaproject context.

and a sample leadership structure, which is a useful reference but does not work in every scenario or organizational setting.

21.2 Essential Recommendations

We offer five vital recommendations in team formation to enable the successful leadership of megaprojects.

21.2.1 Build and Sustain the Organizational Project Management Function

We covered this topic extensively in Section III of Part A. Without this action, the other four recommendations might not be possible. Building and sustaining the organizational project management function (OPMF) includes:

- A robust organizational project management system (OPMS)
- A continuously updated organizational knowledge base (knowledge management)
- A proper professional development and leadership program, all the way to the top
- An approach for organizational learning
- The right cultural mindset for excellence in product development

21.2.2 Adequate Staffing

We know a familiar cry in business is, *"We do not have enough people,"* or, *"We do not have enough time."* We do not necessarily agree with this blanket statement, because it is often a mask for lack of competence. However, it is a real challenge in many situations. If you recall, one of the evaluation topics we address in the CAMMP™ Feasibility Study is about the people and resources that are necessary to help achieve success on the project. This recommendation is followed by a further one, which is having adequate and competent resources to lead the projects. What happens if these two conditions are not met? It would be prudent to reconsider your options and your strategy, maybe even deciding not to proceed with the megaproject. Perhaps gaining market share through mergers and acquisitions might be a better choice.

Let us stay with the need for adequate staffing. The project owner team size is crucial, and I think it is clear that larger is better—of course without crossing the line into a bloated team size. One of the key differences between the Asian Project and the North American Project (in Chapter 7) was team size. The AP had more than 20 professionals on the project management team, whereas the NAP started with six people and grew to about 10 (after some of the damage was done).

What is the proper size for the core project management team?

The team size is a factor of project size, but complexity is the most relevant factor. Therefore, 30 or more people could be a functional level for the average megaproject. Merrow (2011, p. 168) quantifies this concept and states that, *"Each member added to the core owner team increases the probability of having a successful project by more than 3 percent."* Another reference from IPA (Barshop 2016, p.97): *"Robust owner teams produce capital assets that are higher quality and less expensive than owner teams that are understaffed or that have outsourced much of the owner team role to contractors."* Barshop also states that, *"The difference between understaffed and adequately staffed is about 7 percent,"* per IPA's database.

Need we say more?

21.2.3 Team Competence

Let me start with a funny-sad story—a personal story.

In a four-year alliance contract for engineering–procurement–construction (EPC) services on a cost-plus-incentive basis, the situation was not good. The project owner hand ended the first four-year alliance contract with one company and brought in another company to handle all projects in this geographic region, in which the owner had a few operating facilities. In other words, all of these projects were small in nature, less than $50 million at that time, and all were within existing operating facilities. A few projects were adding a process unit or equivalent facility, but all were brownfield projects.

So, cost plus incentive and the budget is set by the owner, with some input from the contractor. The contractor, well known globally, was used to building major projects and even megaprojects. However, the situation was now different, because it was about managing multiple small projects in an operating environment. As a result, most of the projects in the first two years of the arrangement were over budget. The owner was obviously not happy with the overruns. The contractor was not happy because, as a result of the incentive nature of the contract, they were not making any profit.

At that time, the owner decided to bring in an experienced professional to set up a project control office (within the alliance management setup).* Before this step, the project engineers were functioning as project managers responsible for the implementation work (EPC), including project control. Further, because the contractor was not happy, management replaced the director and brought in a strict director with strong tradition and royalty to the contracting company.

The projects' performance was not good, primarily because, as we stated earlier, the contractor and its personnel were not able to adjust to the small-multiple projects environment, but they would not acknowledge it because they genuinely believed the owner (who had set the budget) was cheating them. The contractor project control team (a huge team) were not able to forecast correctly. It was not uncommon to see expenditures higher than forecast.

Finally, in one meeting, the showdown happened. The director from the contractor side literally exploded, blaming the owner and, in particular, yours truly, who had been on the job a few months and had been leading the owner's project control office and had identified many of the issues on both sides. However, the main problem was the contractor's competence as an organization and the individuals they had on the job. They were ready to pull out, and the owner was prepared to terminate the contract, or at least not renew.

After the dust settled and the noise subsided,† we had a heart to heart. We proposed a multifaceted solution to address the concerns of both parties, and once agreed upon, we moved to implement it. We conducted awareness sessions about the owner's estimating system and how it was different from the quantity-driven contractor's estimating approach. We explained the need to think

> **Leadership Competence**
>
> If the project director (and other directors) or other senior-level professionals do not have enough experience in project management, they should undergo proper training and not have their ego stand in the way. Another good idea is cross training, especially between the project management team and the operational readiness team.
>
> Further, the project executive from the project owner can choose to have a project management advisor or can depend on the project director for advice.
>
> In the story, unless the contractor director got over the ego of *"We are a global contractor and we know what we are doing,"* it would not have been possible for him to accept training for all of his senior staff, and we could not have come out of the storming stage.

* Although the owner and contractor teams were working side by side in the main office and on sites, they were not integrated.

† It took a few days.

small-multiple instead of large and major. Then the contractor's director—recognizing the competence challenges of their project control staff and seeing that they were acting more like messengers rather than leading the control aspects—proposed to integrate the project control teams, the contractor team, with my team on the owner side. There was resistance from both sides (my management and his staff), but we pushed the idea, and it happened.

Now it was time to rebuild the team, which we did through training, mentoring, and coaching. We also formed small groups to identify all of the common challenges the projects were facing, developed solutions, and trained everyone on these solutions. As the team (and the whole alliance arrangement) was coming out of the storming into the norming stage, the project owner renewed the alliance contract for another four years, and it was time for me to leave.

Next, we will discuss team integration, but before that I want to reflect on the above story. It was not a mega project, but the size of the portfolio of projects was more than $100 million in 1990s value of money. In a cost-plus, incentive-based arrangement, the first critical challenge was to build trust among all of the involved parties; without trust, it would not have been possible to come out of the storming stage, which would lead to project failure. This was one of the contributing factors to the failure of the North American project (NAP) in Chapter 7. Building trust is not about persons or personalities; it is about competence. If one party does not believe the other party is competent, it would not be easy to trust, even if Mother Theresa was one of the parties.

Finally, competence starts with organizational competence, with individual competence at the second level. We had a whole chapter on this earlier, so time to move on.

21.2.4 Team Integration

The previous story includes aspects of integration, at least for the project control. A similar approach was used on the AP project. However, let us explain integration first.

When you hear the term *integrated team,* you have to dig down to understand what it means.

- One aspect of an integrated team, maybe the most common, is to have an integrated project owner team—that is, the project management team needs to include the technical functions involved as full-time* members of the owner team.
- Another form of integration occurs in situations in which the project owner is actually part of a joint venture or a partnership. In these scenarios, it is common to have one partner lead, but the other partner(s) could have team representatives. Sometimes these representatives are there as oversight to keep their management apprised of what is happening—not a healthy scenario. It would be best to embed these personnel within the project management team and assign them actual areas to manage. This saves time, builds trust, increases efficiency, and reduces conflict.
- The third form of integration occurs when the project owner and contractor management decide to integrate their teams, either fully or partially, as we shared in the alliance story. This is the most difficult of the three types, and we do not think it is possible except in a cost-plus environment. If the organizations, especially project owners, can let go of egos and plan this well, it can be a highly effective way to manage the project, with benefits to all, because they would be able to see eye to eye and have a better understanding for the other side's perspective.

What is the value of integration?

Unfortunately, I do not have studies to substantiate a recommendation for all of the above, especially the third one, because of a lack of data and literature. However, for integrating owners' personnel, Merrow

* For megaprojects, full-time would be necessary.

does identify measurable benefits. For all of the five factors that IPA uses to assess project performance, integrated teams produce better results on all five (Merrow, 2011, p. 170). For overall performance, let us look at these numbers:

- Of the projects in the database, 35% were graded as successful.
- Only 22% of the projects that did not have integrated teams were successful.
- On the other hand, 51% of the projects with integrated teams were successful.

The bottom line—integration—does not guarantee success, but it definitely has a significant positive impact. If integration with the contractor is not possible, then consider this statement from Oberlender (2014): *"A team approach between the owner, designer, and contractor must be created with cooperative relationship to complete the project in the most efficient manner. Too often, an adverse relationship develops that does not serve the interest of anyone."*

One more quotation to close this part, *"All things being equal, projects with functionally integrated teams build assets that have 10 percent better technical performance, are completed 15 percent faster and are 10 percent cheaper"* (Barshop, 2016, p. 99).

21.2.5 Leadership Continuity

I think most would agree on the value of leadership continuity, so let me go straight to the numbers. About 50 percent of projects in the IPA database experienced change in the sponsor (project executive). What are the consequences? The measured effects included a 5 percent cost growth, 11 percent schedule growth on the full project life cycle, and a 15 percent asset performance reduction (Merrow, 2011).

What is important to remember is that projects are temporary and do not have the organizational memory of a typical department. Most of the knowledge, relationships, stakeholders' engagements, and team building reside with the leadership team. Changes in this team, especially at the top, have adverse effects, as the numbers from IPA have shown.

Also, it is common that many of the executives that initiate projects and may push for unrealistic budgets or deadlines are likely to be gone before the damage occurs—possibly promoted or parachuted out of the company. These are the challenges the organization must deal with via early scrutiny and maybe some form of compensation structure to encourage continuity, even among the contractor leadership.

21.3 Project Organization

There is NO single format or structure that works universally. Further, more than one of the references that we have reviewed address project organizations and the various models that could exist. Oberlender (2014) is a useful reference here.

I will present a high-level structure via more than one image.

21.3.1 Project Governance

Figure 21.1 shows the venture leadership, in line with the streams that we discussed earlier. The lower part of the image shows three directors, and their roles and responsibilities are in line with what we have discussed for each stream in Chapter 18. The project executive is what various references call the *sponsor*. The project executive is more appropriate for megaprojects, and that would be the person responsible for the whole project. That position is accountable to executive management or the board of directors,

Figure 21.1 Organizational Governance and Venture Leadership Team

whether through a steering committee or not. If there is a steering committee, its membership should include the top decision makers from the functions involved in the project. This is a must to ensure efficient work and quick action by all involved.

21.3.2 The Facilities Stream

The facilities stream is where most of the project management team operate, because most of the effort and cost are required for the product development work. As the project approaches the handover point, the production stream could start to take over. For the facilities stream, Figure 21.2 shows its leadership. Notice that we are staying at the managers' level. This image shows five areas, which is what we recommend, but depending on the nature of the project, it could be reduced or expanded. For example, management may decide to combine support with project control. Possible additions could include legal and government relations. Permitting could be outsourced or part of the support. Interface management could also be independent and at high level or under construction or engineering. Also notice that for engineering and construction we are showing director level positions, not managers, since these areas require a large number of resources, which would necessitate director-level leadership. What would affect this structure is also the capability of the project owner team and whether the owner would manage

Figure 21.2 Facilities Stream, Project Management Team Leadership

directly or through outsourcing. Quality and HSE are shown as assistant positions, because most of the staff from these two areas could be contractors' personnel. In some organizations, these positions could also report to the home offices to ensure that the project's managers do not influence them.

21.3.3 The Other Streams

I had initially planned to cover each of these areas, but they could vary significantly based upon the nature of the projects. For example, some building projects could be substantially less complicated than industrial or even airport projects. Dams and metro lines, especially undergrounds, require a lot of work in comparison to buses, rapid transit, or roads projects. Therefore, instead of bumbling through this content, we advise seeking professional help.

21.4 Chapter Closing

The management of megaprojects is highly complex, to say the least. There are too many organizational stakeholders and possibly hundreds of stakeholders who have direct involvement or influence, not to mention the thousands or millions of potentially passive stakeholders who would be impacted by the legacy these projects leave behind. Therefore, how can we lead these projects to success despite the complexity? Can we reduce or minimize the conflict among the various parties? Can we break the traditional tension among the various parties? These were some of the questions that we have been addressing throughout the book and in this chapter. The critical thing to remember is that the complexity of megaprojects should be taken seriously, and organizations must give them the necessary attention. Do not fall into the trap of understaffing your megaproject. Adequate and competent staffing affects all of the other challenges that we have raised in this book.

21.5 Closing Part A

As we close Part A, it is vital to remember this fact: **project success and failure are predetermined in the Discovery Phase of the project, with the Feasibility Study and the Shaping Process**. In the early stages of project development, the owner must develop the project strategy, then diligently manage its stage as if the organization's life depends on it. The current state of practice is not a good one. If we can trust all of the numbers, we realize that industrial megaprojects do significantly better than infrastructure projects, but there are still more failures than successes. As we write these words, we still see numerous reports about mega challenges facing megaprojects; the recent reports on the UK cross-rail or the California high-speed rails are very sobering.

The sadness in the current reality is that we are not learning our lessons, but we know there are solutions. Some of the references we have reviewed offer best practices that have been proven to work. In this book, we have tried our best to filter and distill the learning from these various sources, from the case studies, and from our own experience, which spans more than three decades of global projects.

What we offer is a model for leading megaprojects, a tailored approach that incorporates the learning. In this book, we provided what we know works, via a practical approach that covers a project, concept to success. As one would expect, we cannot cover every aspect because of the massive number of disciplines and fields of studies that are required in the megaproject world. What we offered, the sources of information that we have shared, are readily available and can be put to use immediately by those organizations willing to acknowledge the problems and put their collective egos to the side to drive the change or face the consequences.

We believe in The Possible, but possible is not easy.

Let me close by this statement: I know it does not apply to megaprojects, but we can extrapolate. If you think project management is simple, you are in the wrong field. If you think managing megaprojects is impossible, you are in the wrong field. We believe in The Possible, but possible is not easy. If you think that managing projects should be something you can learn in a day, or that a five-day certification training will qualify you to master project management, you might need to question those claims. What is necessary for project leadership? The three engines of project management—people, process, and tools or technology. The process is critical, but culture is the real differentiator.

Part B will shift toward the application of what we have covered in Part A. We will do so via an extensive step-by-step simulation of a hypothetical but realistic megaproject—concept to success!

Part B
The Simulation

Section I
The Project

Chapter 22

The Integrated City Megaproject

22.1 Disclaimer

The project presented in this book is NOT a real project. It is a hypothetical project that was inspired by and used in workshops to stimulate learning. We have used different types of projects in workshops, including large and complex projects. However, the Integrated City Megaproject was inspired by a series of workshops for a transportation ministry. It is important to note, it was *inspired by* these workshops and not *taken from* the ministry work. In other words, we are not sharing any confidential information.

This is a realistic but not a real megaproject.

The simulation describes how to apply The Customizable and Adaptable Methodology for Managing Projects™, most specifically CAMMP™ MP, on a realistic megaproject. The original writing was complete in 2015, and the content was based on CAMMP™ version 2. We will keep most of the original content and update it to align as much as possible to the third generation of CAMMP™, and the CAMMP™ tailored approach that we presented in Part A of this book.

Once again, none of the information presented here represents a particular company, organization, or individual. The usage of the Prime Minister's Office, Strategic Planning Authority, National Project Management Authority, or other similar references are just for illustrative purposes. This project was used in the workshops mentioned earlier and will be used again in future workshops.

22.2 Megaproject Overview

The megaproject in the simulation is a strategic initiative that could take place in any country. It is to develop an integrated logistics city to stimulate the economy, drive direct international investment, and facilitate community and social development. The Integrated City Megaproject (ICMP) consists of the following components:

1. An international airport and supporting facilities
2. A seaport that can accommodate shipping and cruise lines
3. Railroads and stations for interconnecting the airport, seaport, and other cities, in addition to connecting to the cross-country rail and regional links
4. Roads and highways, including modifications to the existing network and any additional facilities
5. City facilities, including the infrastructure, housing communities, schools, public transportation, medical facilities, and other services
6. Professional development centers, focusing on the skills needed to manage the port, airport, and other facilities

The government is the lead investor and developer but will work with partners. The investment may include different investment platforms, such as direct cash investment, partnering, BOT (build–operate–transfer), BOOT (build-own-operate-transfer), or any other mechanism. We will refer to this group of stakeholders as *the Investor.*

Figure 22.1 The Integrated City Megaproject and Components

The project is discussed from the perspective of the investor, as the project owner.

22.3 Why a Project, Not a Program

A strategic initiative, such as the ICMP presented here, would be considered a strategic program. Therefore, it is best managed as a program rather than a project.* However, for the purpose of this book, we will treat it as a megaproject,† to illustrate the tailored approach. Maybe in a future publication, we can revisit this project as a program and highlight the differences, which would be at a higher level in managing the program by managing each component. If you are still struggling with this concept, then do not think of the city, think that we are building an airport or a port. Therefore, the book presents a simulated example of this megaproject from ideation to benefits realization and success.

22.4 Depth of Coverage

The intent is to present the application of the concept from a project management perspective, not technical or engineering perspective!

ICMP is a massive project—a megaproject. The details of such an initiative would require thousands of pages, developed by hundreds of people working for many years. Because the level of effort is not available for the author, nor is it practical to summarize in a book, the depth of coverage in this book will be a fraction of what is required if the project were real, or if we were to include all critical documents. We are also confident there will be gaps in various areas because one person, the author, cannot have enough expertise in all of the necessary technical and management areas. On the other hand, this is not an engineering, construction, or technical book—it is a project management and product delivery book. Therefore, the sample project will touch on most of the deliverables and offer brief write-ups on the various steps. The coverage is sufficient to visualize what is required and how to apply CAMMP™ MP (CAMMP™ for Megaprojects), the tailored approach for leading megaprojects, from concept to success.

* In Part A, we shared some insights on this point.

† We do so to demonstrate the tailored approach.

Chapter 23

Guide to Part B

23.1 Part B

In addition to this first section, there are three more sections in Part B. Each of those sections is dedicated to one of the three CAMMP™ MP phases: Section II deals with the Discovery Phase, Section III covers the Development Phase, and finally, Section IV is specific to the Delivery Phase. Then, each section may have a chapter about the whole phase and a dedicated chapter for each stage. Consequently, most of the chapters in Section II to IV will have similar structure and topics, since most of these chapters cover a project stage, but in a truncated way.

23.2 Stage Process Sequence

If we were to include all of the information, even those that would repeat from stage to stage, it would be too much content. In general, each stage follows the processes presented in Figure 23.1, end to end. For some stages, the actual process might be somewhat different but will have a unified six-process flow. However, I must stress that the sequence presented is at a high level and not the detailed, step-by-step process. Therefore, the focus is on the main stage actions; the **stage management processes**. The specific process steps (for each process) are not listed here; we had listed them in Part A (paragraph 20.2 on page 200).*

* We have many stages in the CAMMP™ MP tailored approach. Technically, every stage will have a Stage Authorization Document, a Stage Management Plan, a Stage Detailed Plan, implementation comments, control and change management actions, and a close-out report. We will not repeat all of these documents in every stage, it would get boring very quickly and make this book too long without adding a great deal of value. Therefore, we will demonstrate these concepts so you know how to apply them, while limiting the repetition.

The project management process within a typical stage

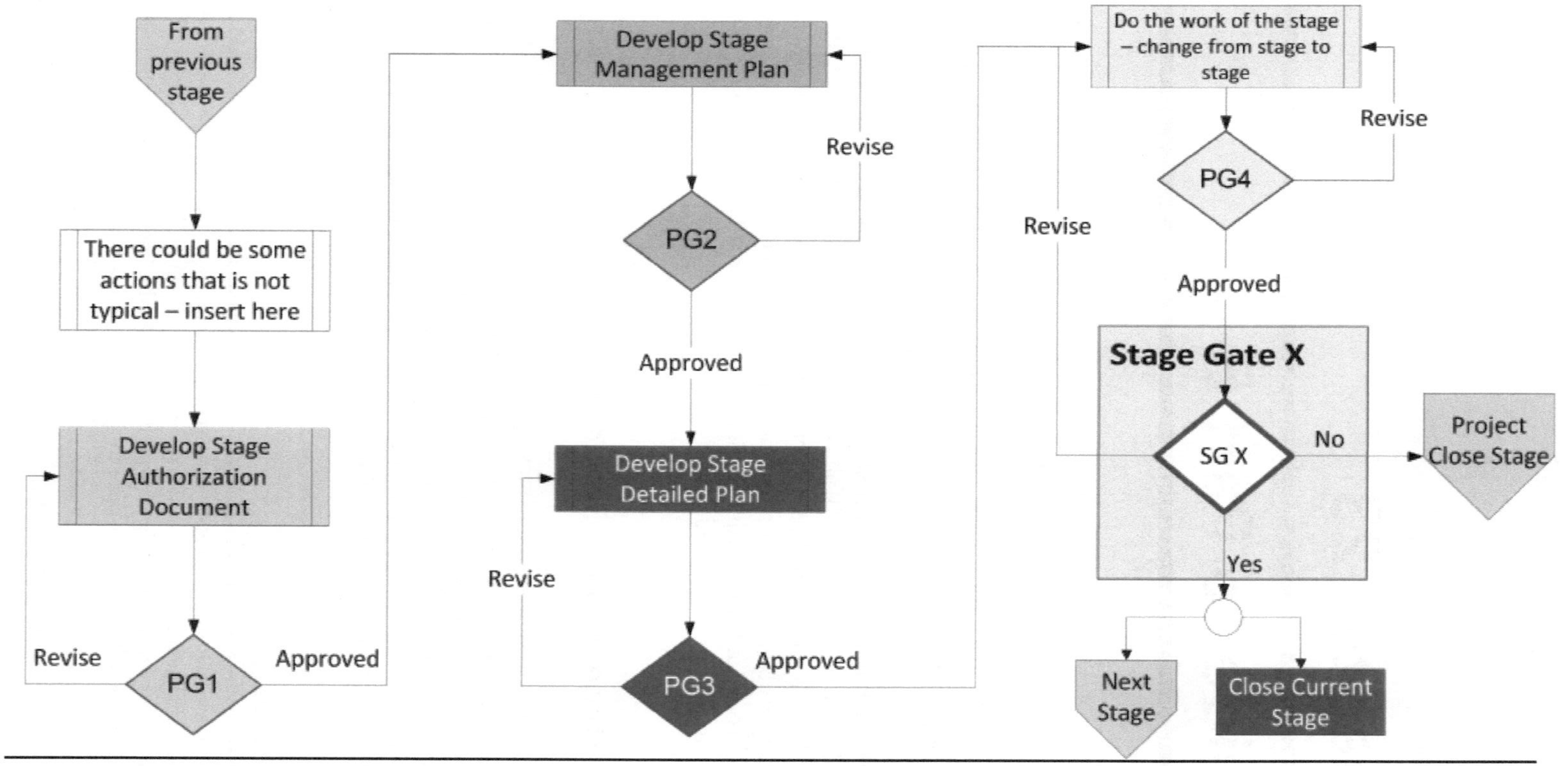

Figure 23.1 A Typical Stage Flowchart

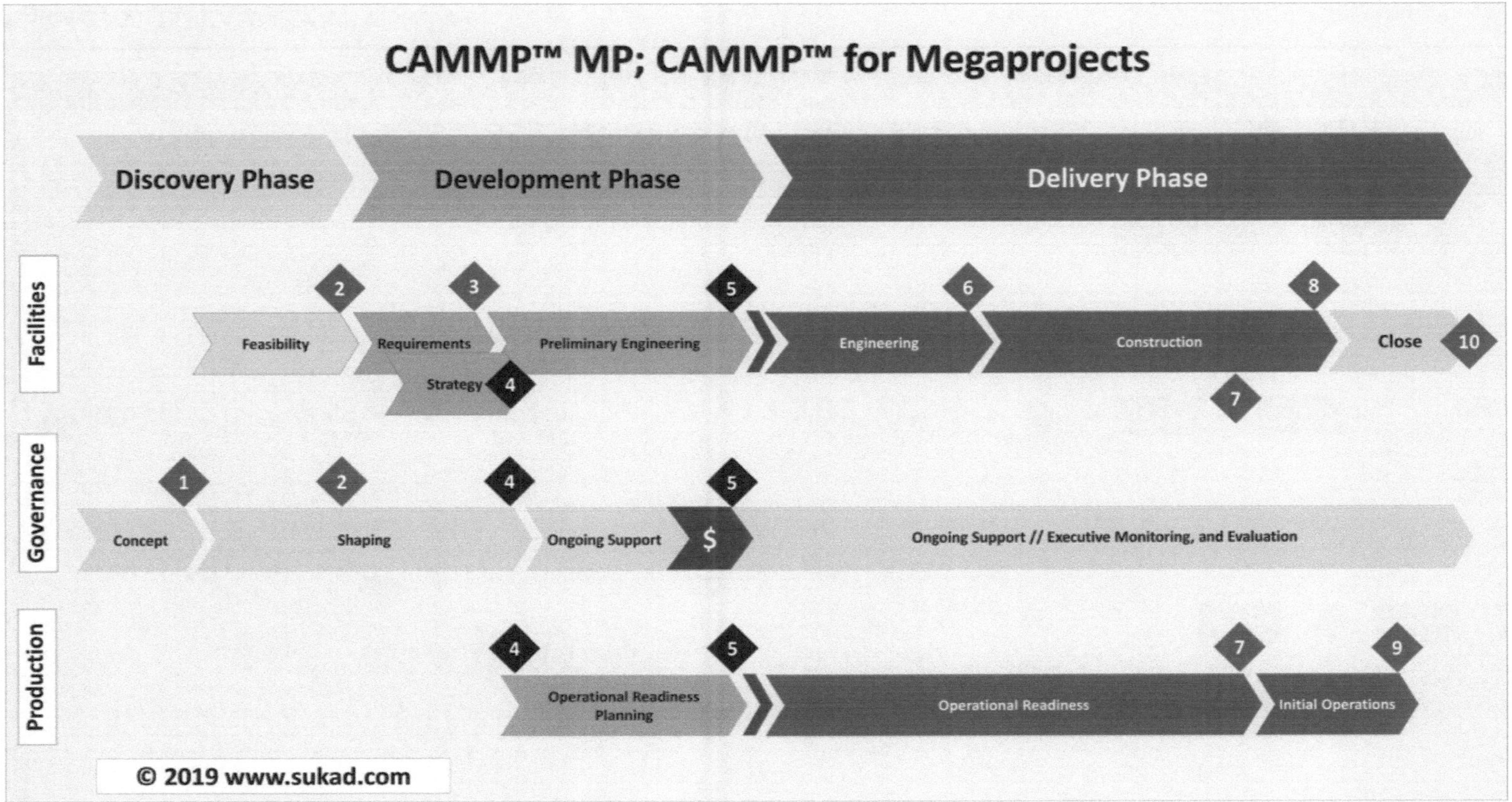

Figure 23.2 The CAMMP™ MP, Project Life Cycle

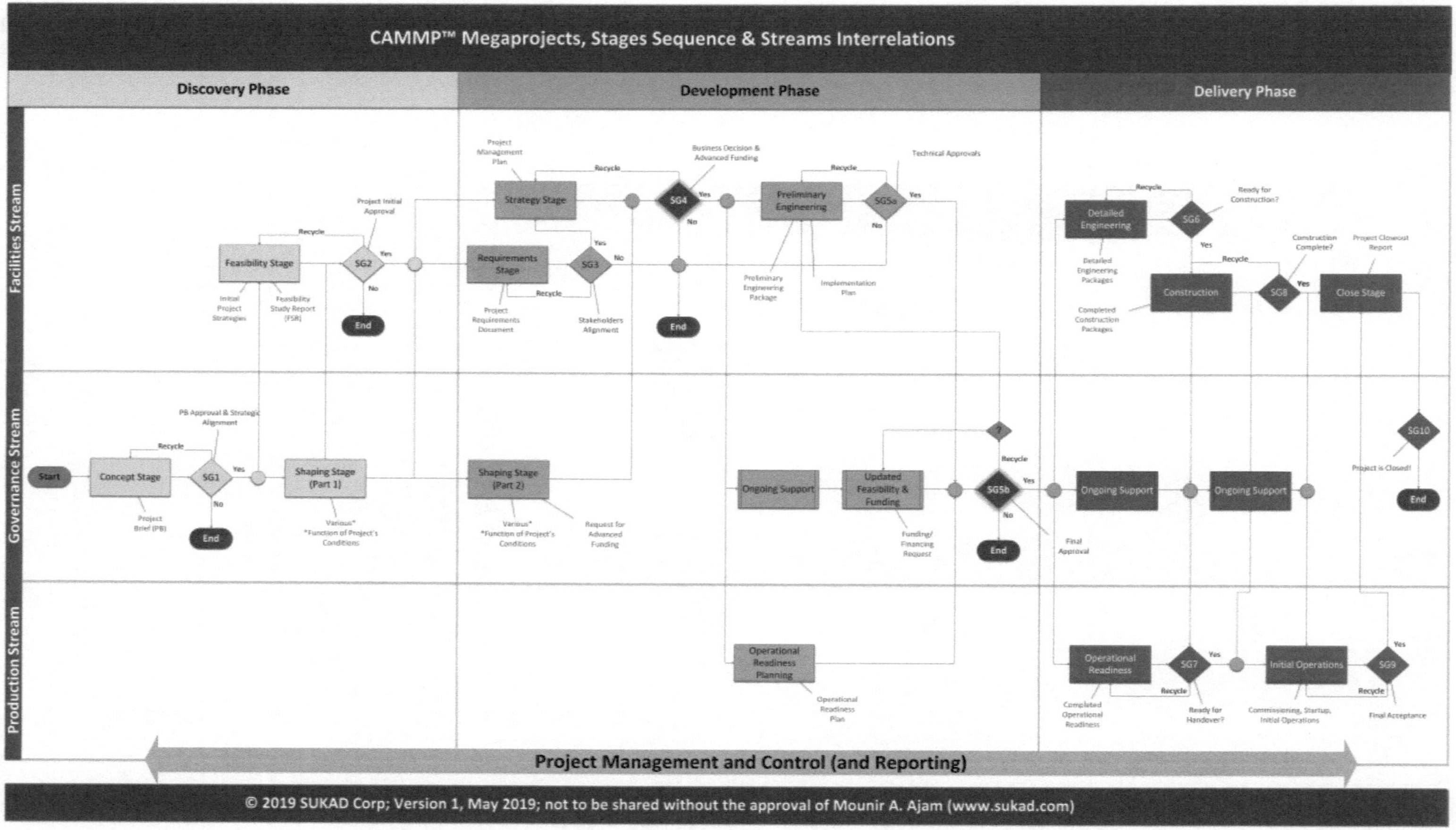

Figure 23.3 The CAMMP™ MP, Flowchart

Unlike the process deliverables and gates,
the stage deliverables and gates will vary from stage to stage.

23.3 Process Deliverables and Process Gates

Each of the stage sequence sections is an overview of the various processes within the stage; these are the CAMMP™ stage management processes that we just mentioned. Therefore, this sequence covers the stage, following the stage management processes. It includes descriptive information that is related to the case study. These sections cover the various process deliverables, including their process gates.

Most of the process deliverables and process gates of the stage management processes are similar from one stage to another. Only the implementation actions vary from one stage to another, because it is about doing the work of the stage, and every stage has a different focus and scope; the work of each stage is related to the purpose of the stage. In other words, the stage management processes are about managing the work, with the implementing process being about doing the product's related work. The plan details process would also be somewhat unique to the stage, although the structure of the stage detailed plans would be similar.

23.4 Stage Deliverables and Stage Gates

These sections will possibly be the largest in each chapter and cover the main stage deliverables per the CAMMP™ MP. These sections also include the stage gates. The stage deliverables are mentioned within the implementing process discussions, but the details are provided in these sections of each chapter.

23.5 Chapter Closing

We do not have much to add here. We will only repeat a couple of the images from before to use as a map for the rest of the book. Enjoy the simulation.

Section II
The Discovery Phase

Chapter 24

Introduction to the Phase*

The primary objective of this phase is to understand the concept behind the megaproject, validate the business needs, and determine its feasibility. Also, for megaprojects, the organization should start the shaping process for the opportunity. In the previous chapter, we presented the CAMMP™ MP project life cycle, the project flowchart, and the process flowchart for a typical stage. All of these images will be referenced here, as you continue to read.

The Integrated City Megaproject (ICMP) is a strategic initiative requiring a tremendous investment, on the order of billions of US dollars. Therefore, regardless of where the idea originates, to formulate the Project Brief and start the process, a strategic planning business unit will have to lead the effort. In other words, an initiative like this will be managed by strategic planning or strategic business development in one of the ministries or from the Prime Minister's office,† which would be the Strategic Planning Authority‡ (SPA).

The SPA personnel performs initial research and analysis to determine if it is worthwhile to consider this strategic initiative. If they think it is valuable, they will draft and issue the Project Brief, which is done in the Concept Stage, which we will discuss in the next chapter. The Project Brief will be presented to executive management, at least at minister or Prime Minister level. This will lead to the first stage gate, Stage Gate 1 (SG1), at which executive management, including the Prime Minister, could reject the idea, leading to terminating the megaproject. If they approve the idea and want to proceed, they will move on to the Feasibility Stage and start to develop and shape the opportunity, through the Shaping Stage. The feasibility and shaping will be discussed jointly in the Chapter 26.

> **SPA and NPMA**
>
> Here we share a few concepts that we believe would be great, but which are not necessarily specific to megaprojects. For example, in this simulation, we include the concept of a Strategic Planning Authority (SPA) and a National Project Management Authority (NPMA). These two authorities are at the highest level of government, with their executive directors reporting to the Prime Minister.
>
> [We touched on the concept of an NPMA in Chapter 14.]

* We have defined and briefly explained the phases and the stages in Part A, so we will not repeat the explanations. We will only include the necessary minimal information to set the scene in respect to the simulated megaproject.

† Remember, ICPM is simulating a megaproject, in any given country, that will be led by the government.

‡ This is not a real entity—it is presumed here for discussion purposes.

Chapter 25

The Concept Stage

25.1 Introduction

Ideas, concepts for projects, and especially megaprojects will not come out of thin air. For something like the Integrated City Megaproject (ICMP), the idea could come from the government or the private sector. We will say more as we go through the steps.

25.2 Process Deliverables and Process Gates

These are the CAMMP™ six processes to manage a stage and their associated process gates.

25.2.1 Authorize Process

The newly elected government wanted to look at the future with optimism and a focus on development and growth. Therefore, the Prime Minister's office, through the Strategic Planning Authority (SAP), led to the development of the 2050 vision. The vision included numerous development programs and megaprojects, such as upgrading the power infrastructure and major development projects in various cities, among various infrastructure initiatives.

Consequently, the central government decided to pursue the ICMP for the least developed region in the country. The Prime Minister's office instructed the SPA to form a team and study the needs and business case for this concept. The concept at this time was not specific about building an integrated city, but the team mandate was to explore various options for the development of the South.

The SPA requested the support of the National Project Management Authority (NPMA). Together, they formed an integrated team and appointed a project executive* to lead the Concept Stage, and ultimately,

* Per the discussion in Part A, the project executive is the role we recommend for the person who will lead the whole venture development and its three streams—the facilities, governance, and production streams.

the whole development of the megaproject, if it is authorized. The project executive issued a Stage Authorization Document, which included the appointment of a business director* to start the process.†

Educational Notice: *We realize that at this point in time, senior managers or managers could be enough for the level of effort required. However, we chose the director and executive levels to emphasize that the level of effort might not be great, but the spirit of CAMMP™ MP is that we need the appropriate level of executives for this critical stage.*

25.2.2 Plan Management Process

Upon receiving the Stage Authorization Document, the business director selected a CAMMP™ certified project manager to help manage the stage work. In this case, the business director leads the overall work; the project manager role is to be the director's deputy, with a focus on project management process. The project manager requested and mobilized an assistant project manager to help him develop the Stage Management Plan, which was expeditiously approved by the business director.

25.2.3 Plan Details Process

The project manager requested the addition of a team of top experts to help develop the Stage Detailed Plan, because some of this work is specialized and not project management work. The team was assembled and developed the plan. The Stage Detailed Plan identified the need for various specialists, developed a complete estimate and schedule for the stage work, assessed and planned responses to the various risks, along with the other detailed planning actions. With an approved Stage Detailed Plan at the third process gate, the team moved to apply the Implement Process and its steps.

25.2.4 Implement Process

Implement, for this stage, is to conduct the initial assessment and concept of the megaproject.

The business director and project manager mobilized the necessary specialists and other resources to conduct an analysis of the region and what could be done to stimulate the economy and declining infrastructure in the area. This region includes the only significant opening to open water and the ocean and global sea routes. The team considered investing in the existing major cities in the area but believed that a more strategic option would be to build a new city with integrated facilities, such as what we discussed in Chapter 22. The joint SPA–NPMA integrated team developed their report and proposed the vision of the integrated city megaproject, along with the justifications. The team also included information about an alternative megaproject, which consisted of various semi-independent projects, as a potentially lower-cost alternative.

25.2.5 Control Process

Control within the stage would take place from the stage authorization to closure. We will elaborate on this more in future chapters. In this simulation, because it is not a real case study, we do not have information on actual control actions, so we will discuss generically.

---

* The business director is the senior manager who will lead the governance stream.

† We are not including the Stage Authorization Document here; we will present one in the next stage. Please consider a Concept Stage authorization document was issued. The same for the other process deliverables for this stage.

The Control and Implement Processes share a process gate (PG4) to confirm that the team has completed the work of the stage and is ready to submit the work to the stage gate review, SG1 in this case. Refer to Figure 23.1 on page 244 for a visual.

23.2.6 Close Process

After the approval of the Project Brief at SG1, some of the team members work to close the stage by going through the CAMMP™ stage closure process steps. We do not include a closeout report in this chapter, but because we are following project management leading practices, a closeout report is done. While the closeout work is ongoing, the rest of the team will move to the next stage or be demobilized if their skills are no longer needed.

25.3 Stage Deliverable and Stage Gate

Moving back to the project level for the Concept Stage, there is one stage deliverable, the Project Brief. This first stage deliverable is subjected to the first stage gate, SG1.

25.3.1 The Project Brief

Following are the main sections of the Project Brief per the CAMMP™ template. However, a warning first. This document is very brief, by design. In the real world, it is possible that the Project Brief could be longer than we are showing here, but not necessarily, because the purpose is only to summarize the key points. However, the supporting data could be quite comprehensive and would be available to the working teams.

(Mega)Project Description

The project is to develop an integrated logistics city, which would include airport, seaport, railroads, roads and highways, housing community with adequate infrastructures, and professional development centers. The professional development centers will focus on building the skills needed to serve the needs of ICMP. These could be technical, engineering, management, and vocational skills.

The proposal is for the government to be the lead investor and developer, but to work with partners, which may include different investment platforms, such as direct cash investment, BOT (build–operate–transfer), BOOT (build–own–operate–transfer), joint ventures, or any other mechanism. In this book, this group of stakeholders is *the Investor.*

(Mega)Project Justification

The location of the ICMP is ideal, because it already has a seaport that can be expanded. It is also suitable in terms of proximity for ships carrying goods from the new manufacturing hubs of the world. Once at the seaport, ships can be offloaded, products re-packaged and moved by rail or trucks to the airport for re-export or transported by rail or trucks into neighboring and regional countries. It will also serve local import/export needs.

By integrating these facilities, and using modern technology, robotics, and automation, this megaproject can become a significant import, export, and re-export center, which brings tremendous benefits to the county and the communities in a region that has been deprived of modern facilities due to decades of neglect. This project will have access to the highly educated labor force at rates significantly below regional costs. This major center will be a vital economic stimulus, driving direct international investment, contributing income, offering excellent employment opportunities for thousands of people, and facilitating social and community development.

Strategic Alignment

The project aligns with the long-term strategic 2050 Vision of the government. Also, this project can alleviate poverty and underdevelopment in the country and this specific region, and help in eliminating, or minimizing, "separatist movements" that call for semi-independence from the central government.

25.3.2 Stage Gate 1: Idea Approval

☒ Approved ☐ Revise ☐ Hold ☐ Not Approved

Explanation of Decision

There is no need to add anything to what has already been stated in the Project Brief. The approval of SG1 is at the level of the Prime Minister's office. This approval will lead to closing the Concept Stage and trigger the start of the Feasibility Stage and the Shaping Stage.

Chapter 26

Feasibility and Shaping Stages

26.1 Introduction

Where are we in the overall megaproject process flowchart?

If you recall, the Concept Stage is part of the Governance Stream, which is why the leader of this stage was the business director, assisted by a project manager. In this chapter, we discuss two stages together to keep the book focused and avoid repetition. In the real world, your organization would be working on these two stages independently. Concerning the project streams, the Feasibility Stage is part of the facilities stream, whereas the Shaping Stage (part 1 in this case) is part of the governance stream.

26.2 Stage Process Sequence

26.2.1 Authorize Process

Because the Project Brief was approved at Stage Gate 1, the project executive appointed a project director to lead the facilities team. This person could have been the project manager mobilized in the Concept Stage or a different, more senior leader. At this time, the project executive could also have decided to keep the business director who lead the Concept Stage work or replace that person with a more senior leader. The need for seniority for these streams, facilities, and governance is due to the high-level complexity of megaprojects.

Educational Note: *It is highly recommended that we continue to use as many of the team members as possible from stage to stage, especially the seniors and leading positions.*

The project executive issued a Stage Authorization Document (SAD) authorizing the project director to proceed with the Feasibility Stage. The project executive also released a Stage Authorization

Document to the business director to start the Shaping Stage. You read it right, two authorization documents, one for each stage, and although they had similar structures, they would be authorizing different leaders to perform various work. We will cover the authorization document for the Feasibility Study in this chapter as an example.

26.2.2 Plan Management Process

Once again, remember: two stages, two documents, although the simulation might be discussing one. After authorizing the Feasibility Stage, the project director mobilized the necessary project management resources to help develop the Stage Management Plan (SMP) (for the Feasibility Stage), sought approval from the project executive at Process Gate 2, then managed the stage work to closure.

26.2.3 Plan Details Process

Next, the project director and other personnel (the stage detailed planning team) developed the Stage Detailed Plan (SDP) and sought approval at Process Gate 3 (PG3). However, keep in mind that before the detailed plan could be developed, it was required to identify and mobilize a planning team consisting of a few subject matter experts.

26.2.4 Implement Process

With the stage management and detailed plans approved, it was time to implement the stage work. For this stage, the Implement Process consisted of performing the **Project** Feasibility Study,* leading to management reviews and decision at Stage Gate 2 (SG2).

26.2.5 Control Process

The Control Process and its various steps were active throughout the stage and at every step. As demonstrated earlier (in Part A), there were two levels of controls at this point: *stage* control and *project* control. For project control, the central Control Reference Point (CRP) in this stage was the Project Brief. For stage control, we initially focused on the SAD, then the stage plans (SMP and SDP), once these plans were completed and approved in the set sequence.

26.2.6 Close

Once the Feasibility Study was approved, the project executive issued the Project Authorization Document (PAD),† and some of the teams‡ moved to close the stage, while others proceeded to the next stage, and some left the project if their skills were not needed anymore.

---

* Remember, a parallel document would be needed for the Shaping Stage.

† This document is for the whole project, what is known as *project charter* in the ISO and PMI worlds.

‡ Notice the term *teams* (plural) because we could have more than one team, such as the project management team, feasibility and technical team, and business team.

26.3 Process Deliverables and Process Gates

26.3.1 Authorize (Stage Work)

Authorizing the stage work started with the last stage gate documentation—in this case, SG1. It is possible the organization process would consider that that is enough. However, we do recommend formalizing the authorization of the stage work via a formal document. For this megaproject, we already had the project executive, the business director, and the project director assigned. It was time for formal authorization.

The CAMMP™ Authorize Process has four process steps, per what we listed in Table 20.1 (on page 201). These four steps are:

1. Develop Stage Authorization Document*
2. Establish stage success criteria
3. Mobilize project manager
4. Identify stakeholders

In this case, we had a project director already on board, so no need for Step 3. The following includes the results of the other steps and additional reference information.

Stage Authorization Document

The project executive developed an SAD authorizing the project director to form a team and a plan and conduct a Feasibility Study for this project. The following is the completed SAD† for the sample project (the Integrated City Megaproject).

Phase/Stage

Discovery Phase, Feasibility Stage

General Instructions

The project director is advised to refer to the attached Project Brief and is authorized to form a team to conduct a Feasibility Study for the ICMP. The Prime Minister's Office would decide on project authorization upon the completion of this study.

Further, because of the strategic importance of this initiative, the team must carry out this work with due diligence and without bias for or against the project. The team must not deviate from the Project Brief or proceed beyond the Feasibility Study without justification and authorization. It is vital to perform this work confidentially, with no publicity, and to restrict the communication to the Strategic Planning Authority in the Prime Minister's Office.

Duration

It was expected that the stage work would be completed within six months (the end of July of Year 1).‡

* As you will see below, this item has a few points to include.

† The content is included here as text in a book; in real life, this could be on a branded template.

‡ The project dates reflect a start (the idea) on January 1 of Year 1.

Estimated Cost

It was anticipated that the work of this stage, inclusive of the Feasibility Study and up to stage closure, would cost about $2 million US. This amount is to cover internal and external resources.

Assumptions

Because the Strategic Planning Authority (SPA) and National Project Management Authority (NPMA) were new authorities and still building their competence and capacity, they did not have enough resources and expertise to perform this study. The assumption was that they would not be able to mobilize enough resources from other ministries. Therefore, it was expected that some of the work would be outsourced.

Constraints

There were no available specialized consultants in this type of work within the country. Therefore, the team had to outsource to consultancies from the international market. Another limitation was location specific—the project's location was chosen because it is the only area with access to open water and ocean shipping.

Stakeholders

The stakeholders at this stage were limited to:

- SPA personnel and management
- NPMA personnel and management
- Prime Minister's Office and staff
- Project executives, directors, feasibility team
- Other ministries
- Feasibility Study consultants

Stage Success

For the stage success, we used two of The Four Dimensions of Project Success™[*]; these are the "product" technical success and project management success. In this case, the "product" is the product of the stage work, which is the Feasibility Study. Also, for project management success, a better term is *stage management success.* Therefore, the criteria for stage management success is the completion of the stage work within the expected cost and schedule parameters. These are the cost estimate and duration already identified of the stage work, with a tolerance of ±50%.[†]

The success of the product—the Feasibility Study—is a challenge to assess quantitatively. However, we can include a couple of qualitative metrics, such as the Feasibility Study must be thorough and stand the scrutiny of the executive management team. The study must consist of a robust, comprehensive, and convincing case to defend the study conclusions and include any credible alternative.[‡]

Please note that the ultimate success assessment will come as the megaproject progresses into the upcoming stages. We realize that discovering any issues or problems later might be after the fact and might not help the current project. However, in that case, there will be lessons learned for future projects.

---

[*] Covered in depth in Section 17.5, beginning on page 148.

[†] The ±50% is arbitrary; in your organization and for your context, you could have more accurate or representative range. For this early estimate, we expect to have a broad range.

[‡] Because I am not an expert in this type of work, there could be other metrics to use to determine success.

Process Gate 1

☒ Approved ☐ Revise ☐ Hold ☐ Not Approved

Because this was a strategic initiative, the project executive escalated the decision to the Prime Minister's Office. It was determined that the project executive signature was enough for the process gates. There was no need to get approvals from the prime minister until SG2—the initial project approval.

> **Process Gate**
>
> Notice this is a process gate, which is at the end of the Authorize Process for the Feasibility Stage, and not a stage gate. Stage Gate 1 was before this work and SG2 will be after the Feasibility Study.

26.3.2 Plan Management

Upon receiving the Stage Authorization Document, the project director requested (and gained approval) to mobilize a couple of project management specialists to develop the Stage Management Plan. For this plan, the project director used a similar structure to the Project Management Plan that we discussed in Chapter 19.

The following is a sample stage management pan, specific to the ICMP.

Stage Management Plan

Phase and Stage

Discovery Phase, Feasibility Stage

How to Develop the Stage Detailed Plan?

The project director, project management team, and other project personnel, as defined below, formed the stage planning team (detailed). This team consisted of only internal resources initially and developed the Stage Detailed Plan per the organizational guidelines.*

The Stage Detailed Plan included all of the functions, such as scope, quality, time, sustainability, and risk. Therefore, the planning team defined the scope; defined the applicable quality standards; developed the cost and time estimates; described the roles and responsibilities and the staffing plan; determined the stakeholder and communication management requirements; identified any safety, health, environment, and sustainability requirements; performed risk assessment; and developed the procurement plan, as required. All of these were for the project's Feasibility Stage, only.

How to Implement and Manage the Stage Work

The stage management team managed all personnel and project work from this point onward, all the way to stage closure. The team, which we will define below, followed the project management organizational processes that existed within the NPMA. Management of stage work included all aspects

* We will mention "per the organizational guidelines," "per the company organizational project management system," and similar terms every once in a while to remind the reader that it is vital that such organizational systems exist for project management to be effective and deliver success. However, even when it is not mentioned, it is implied, because following the organizational system is required for all activities and processes.

> **Control at Two Levels**
>
> Control for the project focused on ensuring that the team did not deviate from the objectives of the project. At this point along the project life cycle, the *project* Control Reference Point (CRP) was the Project Brief.
>
> Control of the *stage* was to ensure that the team stay focused on the stage work and not become over-anxious and jump ahead into the work of other stages. At this point within the stage, control was per the Stage Detailed Plan.

and functions, such as managing the delivery of stage work packages, meeting and communicating with the various team members and stakeholders, leading the team members, and managing contracts, if applicable. The final product of this **stage work** was the completed Feasibility Study and Project Authorization Document.

How to Control the Stage Work

The stage management team monitored the ongoing stage work to identify potential defects, deviations from the plan, or any areas for improvements, then took the necessary action. This team followed the organizational processes, which included performance management, quality assurance and control, control of the various team members and stakeholders, monitor of communication, monitor and control of risks, and control of contracts. It is worth noting that there were overlaps between stage and project control, although stage control was more detailed than project control. Refer to the textbox.

How to Manage Change During the Stage

The project management team followed the organization change management system, which emphasized the need to review and assess all changes before approval and implementation.

The organizational policies required the team to manage all types of changes, whether internal to the organization or external with the consultants. Further, the team managed change from the Project Brief, despite the ambiguity. If in doubt, and not sure if an item is design development or an actual change, the team was required to treat the item as a change. (Refer to Change Management in the Fog, page 226)

How to Close the Stage

One of the PMT members stayed after the issuance of the Project Authorization Document to close the stage. The other team members either left the project or continued to the next stage, depending on their roles and the organizational needs.

Closing the stage started as soon as the project was authorized (or terminated) and, once again, followed the organizational guidelines. The closure included performing a reconciliation of actual performance to plan, documenting the lessons learned, and updating the organizational records, along with other vital information.

Stage Management Team

The stage management team already included the project director and a couple of project management specialists. We added one more team member at the start of stage detailed planning to help with the management of the various resources. Table 26.1 outlines the primary responsibilities of the various team members.

Stage Detailed Planning Team

Here we needed a few non-project management subject matter experts (SMEs), representing various areas of expertise, to develop the Stage Detailed Plan. The personnel selected for this effort continued

Table 26.1 Project Management Team, Roles and Responsibilities, Feasibility Stage

Title	Responsibilities
Project Director	Responsible for: Overall management and control Lead stakeholders engagement Lead communication management
PM Specialist 1	Responsible for: Planning and estimating Project control Procurement Risk management
PM Specialist 2	Responsible for: Quality management, Safety, health, and environmental management Sustainability Stage close out

on the team during the development of the Feasibility Study as team leaders. These resources, along with the project management team (PMT), formed a stage planning team. These included experts in the following domains:

1. Market research
2. Aviation and airports
3. Marine and ports
4. Financial planning
5. Economy and commercial
6. Urban planning
7. Logistics and supply chain
8. Higher education
9. Sustainability
10. Legal
11. Medical/industrial medicine
12. City services and management

Initial Project Life Cycle

At this point, the team followed the CAMMP™ MP model and project lifecycle, since CAMMP™ and CAMMP™ MP had been adopted by the NPMA.

Process Gate 2: SMP Approval

☒ Approved ☐ Revise ☐ Hold ☐ Not Approved

Explanation of Decision

The project executive approved the Stage Management Plan without any modifications and requested that the project director continue the process and develop the Stage Detailed Plan.

26.3.3 Plan Details

Author Note

This section covers the Stage Detailed Plan for the Feasibility Stage. Although at this stage, the plan does not require the level of effort of some of the later stages, it still requires significant effort. Consequently, one cannot include all the details here. What is presented is just a brief and a fraction of what is needed, but it does cover the essential concepts.

Mobilize the Stage Planning Team

With the approval of the Stage Management Plan, it was time to identify and mobilize the various subject matter experts (SMEs) that would develop the Stage Detailed Plan. The SMEs would also lead the development of the Feasibility Study, each in his or her area of expertise. The planning team and SMEs consisted of internal resources. Therefore, the project executive and project director required two weeks for mobilizing the team; all personnel were re-assigned and mobilized to work on the ICMP. Thus, in addition to the three project management staff, the team now included various SMEs. During the detailed planning and execution of the stage work, these SMEs could seek support from their ministries or functional departments, on an as-needed basis.

The Stage Detailed Plan

Phase and Stage

Discovery Phase, Feasibility Stage

Scope of Work and WBS

The first steps in the detailed plan were to collect all the requirements for the stage, which the team did. At this point, the requirements were limited to those specific to the Feasibility Study, and the primary sources of information at this time were the documents produced to date. The team decided to meet with the originator of the idea (the strategic planning authority) and the Prime Minister's Office to explore for any expectations that might not have been captured in the initial documents. All items were clear and documented. However, during the meeting, the Prime Minister's Office emphasized the need to ensure that this project must be profitable but can accept a low return on investment, say, 10%. The low yield was conditional if the team could achieve 50% of the labor force (during the project development) to be nationals, and 80% of those who will operate and maintain the facilities also to be nationals. Although this expectation was not a scope-specific item, we document it here for use while performing the various activities and stages.

Based on the above, the scope of this stage was the study and analysis of numerous factors to determine soundness and validation of the business case. The team also needed to ascertain the feasibility of the project and assure that the team could deliver it successfully and realize the expected benefits. This would be necessary to gain the project executive's and higher management's decision on whether to proceed with the project or not.

The study must focus on all elements per the CAMMP™ model* as either as one integrated megaproject or as separate packages, such as airport and port, independently. The project executive decided to treat all parts as one integrated project, with a focus on the individual elements where necessary.

With the Feasibility Study, there was a Class 1 Estimate of cost and time (cost and time considerations). Further, there was a financial model that showed various estimated revenues, expenditures, profit, return on investment, breakeven point, among other factors (financial considerations). For projects like the ICMP, there was an environmental impact assessment, which was part of the sustainability considerations package.

Define Quality

For the Feasibility Study, the team referred to the written organizational standard to find the requirements for conducting the study, including the process, content, level of details, and other requirements. Also, the team had to comply with international and local standards.

At least once during the stage, the Business Excellence Department performed quality compliance reviews to determine that the team did comply with all standards and policies to ensure the quality of the process and the final project's product, inclusive of all components.

One of the assistant project managers was responsible for managing quality during the stage. Her job was to ensure that the team comply with the standard, while the ultimate responsibilities stayed with the project manager.

Health, Safety, and Environment (HSE)

For the Feasibility Stage, the HSE considerations were not of primary importance beyond the usual safe office practices, reducing waste, and offering a healthy working environment per the organizational policies and guidelines.

Sustainability

Similar to HSE for the stage, sustainability was not a primary consideration. However, during the development of the Feasibility Study, we needed to consider sustainability aspects for the project.

Schedule

From the stage breakdown structure, the PMT developed the high-level schedule for the Feasibility Study and the overall stage schedule, using analogous estimating techniques and expert judgment. Here are the scheduling notes:

- The Feasibility Study schedule is presented in Figure 26.1.
- The duration is roughly four months.
- The bar chart is generated from the network diagram, which reflects the logical sequence.
- The relationship is shown in this image, along with the critical path.

The team also developed the schedule for the whole stage, which gave a duration of about six weeks longer than what was specified in the Stage Authorization Document—five weeks if we exclude the closing process. The project executive accepted the revised date as the official date but requested to maintain the end of July date as a target. Once the consultant was selected, they developed their internal plan that will lead to updating this schedule.

---

* Discussed in Part A.

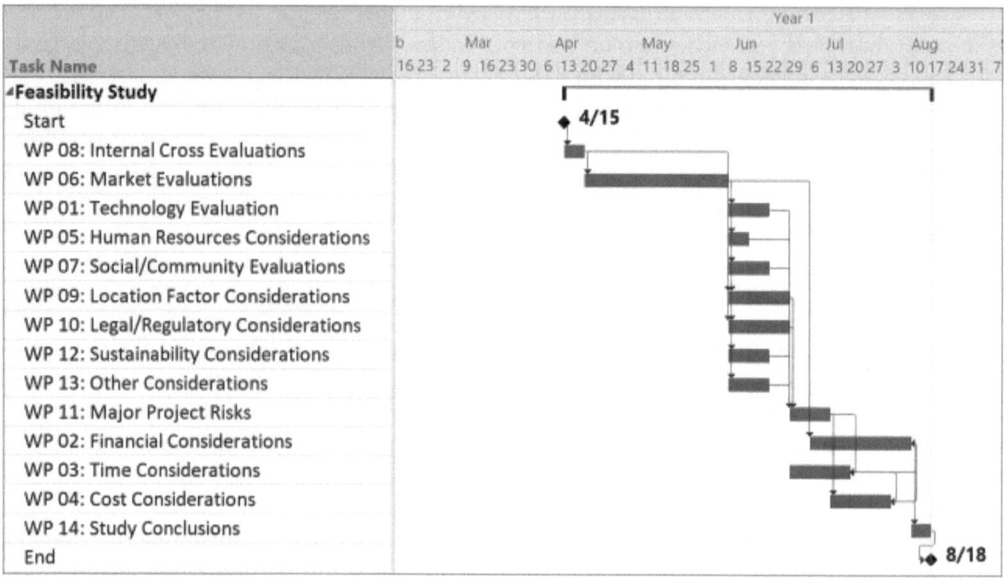

Figure 26.1 The Study Schedule

Cost Estimate

The cost of the Feasibility Stage, inclusive of the Feasibility Study, was roughly estimated at $2 million in the Stage Authorization Document. At this time, the PMT could update the estimate. The following is a list of the estimating basis (assumptions) and the high-level breakdown of the cost components for this stage. Because the scope was mostly people driven, the team decided to use a time-based estimating technique, durations of team members on the project, rather than work package estimating. This technique was not as detailed as work package estimating, but it fit the purpose.

- The resources durations were approximations based on the stage schedule, which is not shown here.
- The cost per month for internal resources was based on employees' costs.
- The cost per month for consultant's personnel was an assumption based on preliminary market research.
- The overhead cost was a rough estimate of the costs for items related to the team members performing their work. This did not include management overhead, which the team excluded at this stage.
- The miscellaneous allowances were for travel and other costs, which were associated with internal resources only. For consultant's resources, the assumption was that these costs would be built into the rates.
- The organizational guidelines called for 10% contingency with the Stage Detailed Plan. However, due to the high uncertainty on this project, the team decided to include 15%.

Notice that the estimated total cost for the Feasibility Stage was $2,246,640, which was about 12.3% over the amount listed in the Stage Authorization Document, and this was inclusive of the 15% contingency allowance. Similar to scheduling, the project executive accepted this deviation.

Table 26.2 Cost Estimate, Feasibility Stage

Item/Area	Duration	Cost/Month	People	Subtotal
PMT Members	8 months	$10,000	3	$240,000
SMEs	6 months	$8,000	12	$576,000
Consultants	4 months	$16,000[a]	15[b]	$960,000
Overhead[c]	8/6 months	$1,000[d]	3/12	$96,000
Misc. Allowances	@10% of Internal Costs			$81,600
Subtotal				$1,953,600
Contingency	@15% of the Subtotal			$293,040
Total Cost				$2,246,640

[a] Average per person for the members of the external consultant firm. In this estimate, the prices are based on assumptions and prior experience. Later in the stage, the consultant fees could be updated to reflect proposed costs.

[b] Equivalent to full-time positions, there could be more people involved for shorter periods of time or part-time.

[c] Offices, computers, supplies, and miscellaneous items.

[d] Per PMT–SME member, per month.

- **Cost Notice 1**: The project executive cost was not included above because, per organizational guidelines, the project executive cost was not charged to the project during this early stage, especially because her involvement was part-time. However, once the project is approved, the project executive will be dedicated on a full-time basis and will be included in the final cost.
- **Cost Notice 2**: The common practice in the region is that, for government projects, the internal costs were not allocated to the project cost. However, the recommended practice was that these expenses should be part of the project capital costs. Further, this project included multiple investors. Therefore, it was necessary to cover all costs.

Time-Phased Budget

The overall proposed budget was in line with the estimate but included a management reserve. Therefore, the proposed budget was $2.5 million. The time-phased budget, excluding management reserve, is shown in Table 26.3. All numbers are in US$.

Table 26.3 Time-Phased Budget, Feasibility Stage

Item/Area	Jan	Feb	Mar	Apr	May	Jun	Jul	Aug	Sep	Total
PMT Members	20,000	30,000	30,000	30,000	30,000	30,000	30,000	30,000	10,000	**240,000**
SMEs	-	-	96,000	96,000	96,000	96,000	96,000	96,000	-	**576,000**
Consultants	-	-	-	120,000	240,000	240,000	240,000	120,000	-	**960,000**
Overheads	2,000	3,000	15,000	15,000	15,000	15,000	15,000	15,000	1,000	**96,000**
Misc. Allowances	2,000	3,000	12,600	12,600	12,600	12,600	12,600	12,600	1,000	**81,600**
Contingency	3,600	5,400	23,040	41,040	59,040	59,040	59,040	41,040	1,800	**293,040**
Total Cost	**27,600**	**41,400**	**176,640**	**314,640**	**452,640**	**452,640**	**452,640**	**314,640**	**13,800**	**2,246,640**

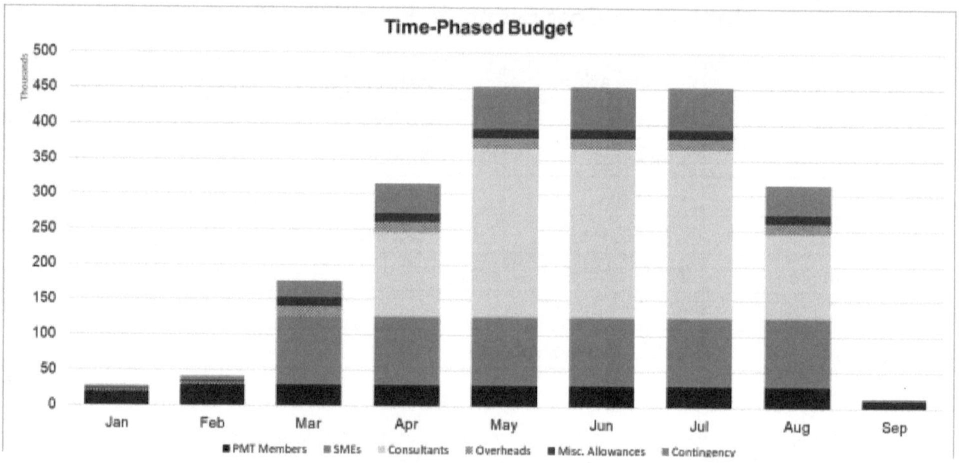

Figure 26.2 Time-Phased Budget for the Feasibility Stage

In addition to Table 26.3, Figure 26.2 graphically presents the time-phased budget. Each of the colors in the stacked histograms represents an account or type of cost, such as PMT, SME, or consultant cost.

Human Resources

The section is about the human resources roles and staffing requirements for implementing the stage work. First, the stage management team was already in place and consisted of three people: the project director and two project management specialists. These were enough to manage the stage. The stage planning team, including the above and the SMEs, were already in place. Once the Stage Detailed Plan was finalized and approved, these SMEs helped the PMT during the implementation of the stage work. While implementing, the SMEs functioned as internal advisors and points of contact to the consultants, each within his or her specialty. No other internal personnel joined the team, although there might have been a need for sporadic support from various government resources.

The roles and responsibilities were defined earlier in the Stage Management Plan and above. The team did not develop a RACI chart, and the agreement was that they would let the consultant develop one, including all personnel. The staffing plan was indirectly presented as part of the cost estimate.

The rest of the resources, who performed the stage work, were outsourced from the consultant firm. The consultant's management was responsible for staffing to accomplish their work. However, the project manager reserved the right to approve or reject the consultant's personnel. Once assigned, the consultant could not remove a team member without prior approval of the project manager.

Stakeholders

The stakeholders were defined in the Stage Authorization Document, and included:

- SPA personnel and management
- NPMA personnel and management
- Prime Minister's Office and staff
- Project executive, project director, business director, and feasibility team
- Other ministries
- Feasibility Study consultants

The project director assumed the responsibility for stakeholders' engagement to ensure obtaining their timely input to the project, exploring for their expectations defining their requirements, and agreeing to objectives. It was crucial to maintain the engagement for the Project Authorization Document and would be vital in the next stage, before Stage Gate 3. The project executive supported the project director in this area, especially with other government agencies outside the reach of the project manager.

Communications

- The PMT had brief meetings daily.
- The project team met weekly.
- The project director published brief bi-weekly reports to the project and monthly reports to the Prime Minister's Office and the concerned authorities.
- The team followed the company filing system for all paper and digital files.
- All communications were in digital format, except contractual documents.
- Urgent matters would be identified and resolved immediately.
- During the stage, issues and action item logs would be maintained.
- Open issues, clarifications, and similar items must be resolved as soon as possible.

The stage reports included a summary of issues and their resolutions, progress, risks, and resolutions, in addition to any lessons learned. Further, the report consisted of an overview of accomplishment and planned activities for the upcoming period.

Risk Management

The team met a few times to identify, analyze, and respond to risks. Risks were at two levels: project and stage level. The project level risks were part of the Feasibility Study. The stage risks are addressed in this section.

The following is a partial list of risks:

1. The government did not have enough expertise to perform a proper Feasibility Study of this large and complex project and may produce a Feasibility Study that does not reflect reality.
2. Specific individuals could bias the outcome of the study for or against the project, due to whatever hidden agendas that may exist.
3. Some rough studies existed in the ministry of transport for some of the components considered here, in particular for a port and an airport. If we could use these studies, it could save time.
4. There was also a study of the Ministry of Housing for a new community, a small city, which would be useful and save us time if we could use.
5. International consultants could be more expensive than estimated, causing us more cost pressure.
6. There could be a delay in obtaining the necessary information, which may delay the study.

The above are the highest priority risks that the team identified and analyzed.* Risks 1, 2, 5, and 6 are threats, whereas risks 3 and 4 are opportunities.

To respond to risks, the team decided the following:

1. For the first risk, the decision was assumed from the start, which was to outsource to a reputable consultant. A reliable consultant can also help mitigate the other risks, because they do not have a conflict of interest to bias the outcome. This risk treatment led to a secondary risk, which was

* The probability and impact for each risk are not shown here.

to ensure the proper selection of the consultants by being diligent about the selection process and not falling into the lowest-bid trap.

2. Opportunities 3 and 4 are similar. Because the team had a representative from the Ministry of Transport and the particular agencies, it was possible to find these studies and the people who developed them, so they could be consulted. It would be a great help if the team could use these studies with minimal modifications.
3. The cost of consultants: The procurement strategy must include a few candidates to propose on this project. The studies that exist can also be referenced in the request for proposal so the bidders can adjust their prices to compensate for the work already done.
4. To mitigate the last threat, it was necessary to ensure that all stakeholders involved understand the strategic importance of this initiative and are also aware of the schedule, so they know what is expected of them. Further, the team asked the Prime Minister's Office for an instruction letter to be sent to all government agencies for their corporations.

The project director, working with the project team, led the implementation of the above action plans. It was also necessary to seek the support of the project executive for risk number 6 and obtain the letter from the Prime Minister's Office.

Define Procurement

As outlined earlier, the team needed to outsource the development of the Feasibility Study and start to conduct research on the potential market for a supplier of such services. It was determined that it might be challenging to find one consultancy to perform all of the work, especially with the diversity of the project components.

The above is another risk that was not identified earlier. Therefore, the procurement strategy was to identify as many consultants with expertise in feasibility studies as possible and ask them to bid on all of the work, or only on some components. This would make the selection process a challenge but might be the best, if not the only, choice.

The team then developed the scope of work and selection criteria. The decision was to select a reimbursable-cost-plus-incentive contract. The team also included an initial list of potential consultants, which they obtained from the other ministries where the studies mentioned in the risk area were available.

Other Considerations

The team agreed that there were no other considerations to evaluate at this time and decided to finalize the Stage Detailed Plan and submit it to the sponsor for review.

Process Gate 3: SDP Approval

☐ Approved ☒ Revise ☐ Hold ☐ Not Approved

Explanation of Decision

During the review of the Stage Detailed Plan with the project executive, a few points were raised and required revisions to the plan.

1. The cost was over the Stage Authorization Document amount, and the project executive requested a re-evaluation. The team did re-evaluate and concluded that it would be best not to change the cost due to the high uncertainty. The project executive accepted the situation.

2. The project executive also raised a concern about the schedule being about five or six weeks longer than anticipated and requested a re-evaluation. The team did re-evaluate and discovered an error that had made the situation worse—they had not included time for procurement and consultant selection. This was even more challenging because of the risk of not getting an expert that could do the whole Feasibility Study. On the other hand, there was a potential reduction in time as a result of the existing studies, but the team decided not to take credit for that. This situation resulted in two additional weeks, with project authorization on September 18 and projected close on September 25. Consequently, a target of the end of July was not realistic anymore, and a goal of the end of August for issuing the PAD was agreed to.

With the above, the Stage Detailed Plan was re-submitted to the project executive and approved with the above modifications.

Process Gate 3: SDP Approval (2nd Review)

☒ Approved ☐ Revise ☐ Hold ☐ Not Approved

26.3.4 Implement Stage Work

Resources

With the approved detailed plan, the project director and team proceeded with the implementation work. However, the first thing to do was to mobilize the project (stage) team that would perform the Feasibility Study. As stated in the Stage Management Plan and Stage Detailed Plan, the overall feasibility team consisted of:

1. The project management team, including the project director. These were already onboard.
2. The internal subject matter experts who developed the Stage Detailed Plan and would work with the consultants' personnel to develop the Feasibility Study. These were also onboard.
3. The consultants' personnel, who would be acquired through procurement and formal contracts as specified in the Stage Detailed Plan.

Procuring Consultant Services

One of the project management specialists was responsible for procurement. To continue with the work started with a detailed plan, she conducted more research to identify as many potential bidders as desired. The bid slate was finalized, selection criteria were done, and request for proposals was ready to go when the team was re-evaluating the cost and schedule for the stage. Therefore, they were issued on the day of Process Gate 3 (PG3) approval. Consultants were given one week to deliver their proposals.

There were eight companies on the bid slate, and six of them submitted proposals. Three were able to submit for all work; the other three submitted for some of the components, which meant all three would have to be selected to cover the full scope. Consequently, the evaluation team had four choices: the three that could perform the whole thing and the combination of the other three as the fourth option. During the evaluation, one was disqualified for a variety of reasons. Another proposal was priced too low and deemed a threat. That left the team with two choices:

1. Option 1: The provider that could do all of the Feasibility Study work
2. Option 2: The combined option

Option 2 was about 10% lower in cost than the first choice. However, Option 1 could offer a better-integrated study and less coordination because all of the work would be within one organization. Therefore, once the team considered the other factors and total cost (including internal costs), the decision rested on the provider that could do the whole study: Option 1.

Cost and Schedule Update

Here, it is a situation of good news–bad news. The good news: the winning bid was $880,000, which was $80,000 below the estimate. However, because the cost estimate was time sensitive, most of this cost saving would be lost due to the extended schedule. The bad news is the schedule. The winning consultant and most bidders advised that the estimate of four months was not realistic for the Feasibility Study of this large and complex project. Their requested time was from six to 12 months. After negotiation, it was agreed that the winning bidder, who had asked for seven months, would update the schedule but with a six months' duration.

The consultant submitted a revised schedule, which included some revisions to the logical relationship between the major deliverables, because they were able to subdivide the 14 deliverables of the Feasibility Study further into more realistic work packages. Figure 26.3 is the revised Feasibility Study schedule, with the breakdown of a couple of the major deliverables.* As a result of this deviation, the expected completion date for the Feasibility Study was October 26, with completion of the Feasibility Stage now set for November 20 of Year 1.

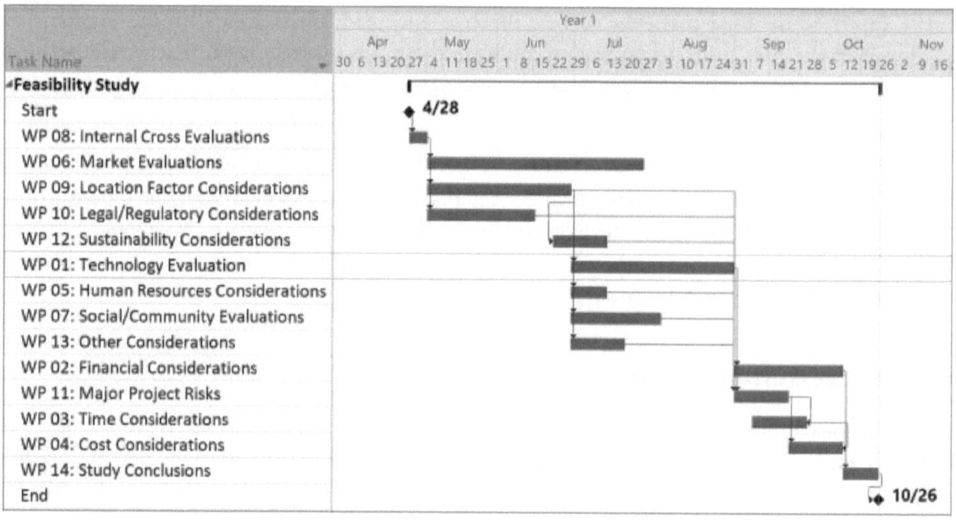

Figure 26.3 The Revised Feasibility Schedule

The Path Forward

Due to the significant changes in schedule, it was necessary to revisit the decision to go ahead. Consequently, the project executive was informed of the outcome of the procurement process and the revised cost and schedule situation. The project director and project executive agreed that this was

* Obviously, this is not a complete schedule.

logical and acceptable, despite the significant deviation from the Stage Authorization Document, which was due to the lack of experience in megaprojects.

The project executive informed the Prime Minister's Office, and they accepted the less-than-ideal situation and authorized the project director to award the contract for the Feasibility Study and start the work. However, the agreement included incentives for cost and schedule as an opportunity for expediting the process and reducing cost. The consultant and the team was made to understand that expediting the work did not mean taking short cuts or sacrificing quality.

26.3.5 Control Stage Work

The focus of this section is on control during the Feasibility Stage.

The Steps of Project Control

In general, control can follow these steps:

1. Monitor actual performance to determine what has been accomplished, at what cost, time, and resource usage, among other factors
2. Compare the actual performance against the plan to determine if there are deviations from the plan
3. Analyze and assess the situation, calculate performance metrics, and identify what actions might be necessary
4. Using actual performance to date, predict future performance for the remaining work
5. Produce the forecast, taking into consideration actual plus predicted performance. The forecast included other factors, including contingency management

In parallel to all of the above, change was managed when necessary.

Performance

As discussed earlier, during the Plan Details and Implement Stage work sections of this chapter, there were a few deviations that had taken place already. These included the cost and time variances between the Stage Detailed Plan and Stage Authorization Document. During the review at Process Gate 3, the team had discovered the scheduling error (missing allocating time for procurement, discussed above). What was not discussed earlier is project performance during the implementation work, which is next. Just a reminder, this is a simulation!

1. The team already highlighted cost and time differences once the team started the work, due to the outcome of the procurement—that is, the cost and schedule differences between the Stage Detailed Plan and the winning bids.
2. While working on the Feasibility Study, the team was able to save time due to the studies that existed in the other ministries. These studies resulted in cost savings because the contract was cost plus incentive contract.
3. There were other cost savings due to better-than-planned productivity of consultant's personnel.
4. There were other cost savings, because some of the work was completed by the internal resources (the government SMEs) rather than the consultant.
5. A change resulted in a few days of savings.

6. As a result of the above and other factors, the team was able to deliver the work ahead of the approved schedule. The Feasibility Study was completed on September 15 of Year 1 instead of October 26—a six weeks' savings.

Management of Change

During this stage, there was one change. The team that was researching for a location identified a military airport that was not utilized efficiently. Discussions with the Ministry of Defense determined that this airport could be converted to civilian use. Because this location was an excellent fit with the ICMP, the team initiated a change to request using this site. This change saved a few days from the Feasibility Stage duration but had high cost and time savings for the future stages, especially construction cost.

Forecast

The budget for this stage was about $2.5 million with the management reserve. Due to the above performance conditions, the team was able to complete the stage work at $2.2 million, which was closer to the SAD amount. This amount was reflected in the Class 1 Estimate for the total project cost.

26.3.6 Close Stage Work

The Stage Close-Out Report is presented below.

Performance Reconciliation

Most of the reconciliation items are related to the cost and schedule. These and other items are summarized below.

1. Pre-Implementation
   a. Cost and time difference between SDP and SAD*
b. Errors in missing procurement time in the schedule
3. During Implementation
   a. Finding the military airport and the ability to use it on the project
b. Better than planned productivity
c. Using internal government resources to help with conducting some of the feasibility work

The above is just a brief illustration of what one can consider for reconciliation. It is preferred that the above include actual numbers, such as (a) missing procurement in schedule led to two weeks' cost impact, (b) using internal resources resulted in $50,000 savings.

Lessons Learned

The main lessons learned were:

* Here, the author is only including the headlines; in a real project's report more details will be provided.

1. Underestimating the schedule: seven months initially (complete by the end of July), whereas the revised schedule required 10 months, and the actual schedule was 8.5 months. Need to improve planning and scheduling effort and improve organizational performance.
2. Internal resources can handle some of the work and should not have outsourced everything.
3. The cost-plus-incentive contract was a right choice because it allowed us to shift work efficiently plus save time.

Others

We are not listing all of the details of the other actions in the closeout report.

26.4 Stage Deliverables and Stage Gate

There are two stage deliverables in this stage: the Feasibility Study Report and the Project Authorization Document.

26.4.1 Feasibility Study Report

Introduction

The project Feasibility Study Report* is presented below. However, as noted earlier, this is a highly truncated study and only shows vital points to address rather than a real and comprehensive study of a large and complex project such as the ICMP. One can view the Feasibility Study as the first significant risk management exercise, because if not done (or not done well), the organization was likely to be taking a considerable risk with potentially adverse consequences, if not total failure, for project objectives.

Phase, Stage

Discovery Phase, Feasibility Stage

Internal Cross Evaluations

The primary concern here is that we did not want multiple organizations (ministries in this case) or divisions within ministries to be pursuing similar projects, or even components of this megaproject.

In response to a request from the Prime Minister's Office, all ministries clarified that they were not pursuing such a project. However, a few ministries stated that they had done similar work but only to specific components, such as airport and port, which the team already knew. The ministries were instructed not to pursue similar work because the SPA and NPMA in the Prime Minister's Office were jointly working on this megaproject. The ministries were also asked to keep this information confidential.

Based on information obtained during stage planning, it was recognized that some studies were performed for some of the components of the ICMP (port, airport) when those items were considered as projects that were under study. The team collected all of that information and evaluated them with the consultant. There were three studies:

---

* The structure and details were presented in Part A of the book, Section 19.2.3 (on page 173).

1. Study of an airport that is mostly for passenger traffic, which is useful but not a total fit because in the ICMP, the airport will be mainly for freight shipment with passenger traffic as a secondary requirement.
2. The port study was an excellent match. It was recent and at the same location where the ICMP project is expected. This was one of the constraints listed earlier.
3. The community (city) study was also recent and a good match because it incorporated most of the necessary facilities that would be required by the ICMP, although the ICMP city would be more prominent.

Market Evaluations

The consultant subdivided the market evaluations work into five work packages:

1. Study potential competitors
2. Study potential demand
3. Study market pricing
4. Evaluate market risks
5. Finalize market evaluation

Scheduling Note: The first two studies could be done in parallel, as Figure 26.3 illustrates. Once these studies were completed, the other three would be sequential in the order shown.

What is important here from a project management perspective was to ensure that the team had the right level of expertise and the work was adequately reviewed by leaders in this domain. Once the project director and project executive had confidence in the team, then it was a matter of monitoring performance and awaiting the final output—the Market Evaluation Study.

Author Note

The author is not an expert in market research, and any attempt to simulate this here would be futile. Further, this is technical, specialized work that is performed by subject matter experts and is not project management work. Therefore, we will continue to focus on the project management effort, sequence, and process. Yet, we cannot ignore the other "product"-related efforts. Therefore, we will list them and touch on them, but what is included must not be considered as a validated expertise.

The evaluation team concluded that the market evaluation justified the project, the demand was high, the location was an enabler due to access via rail links to neighboring countries and airport for neighboring countries and beyond. However, the leading competitors and market risk came from a similar logistics city in another country. However, the ICMP location offered better land connections, better integration with the other facilities, and lower cost. The team concluded that they could achieve a 30% market share within five years of starting operations. The existing facilities infrastructure was at about 8% market share.

Location Factor Considerations

The location had one constraint, which is the location of the port, because there was already a port in the region, and the government was planning to expand it. Therefore, it was a direct fit into the ICMP.

A nearby city, the largest in the region, would be able to house the workforce. However, the team considered developing an area near the city that would facilitate land and rail lines and keep them outside the congested city. Further, a new city provides modern facilities in term of housing, hospitals, schools, and colleges. The team identified a location for the airport that was far enough from the airport of the nearby city not to be a commercial threat. This place had excellent access to rail routes, which could easily be linked. The shaping stage (both parts) included the work to perform site surveys to validate that the selected locations were suitable.

Legal and Regulatory Considerations

There were many laws and international standards to comply with. These included permitting from federal and local government, security, and various other actions. More information is outside the scope of this book.

Sustainability Considerations

The team performed a preliminary sustainability assessment to determine the most sustainable actions for this project in terms of impact on people, planet, and prosperity. The details are outside the scope of this book.*

Technology Evaluations

One of the primary concerns of this item in the Feasibility Study was to identify any potential and significant technology challenges. Examples of these challenges would be the use of new or unproven technologies which could add cost and time threat to the project. The project executive instructions were to use the latest technologies in all aspects of the project: *"We must have modern and world-scale facilities,"* was what the Prime Minister had ordered. However, the use of the latest technology excluded any items that were still in the testing or experimental phase. With the above instructions, the team subdivided the technology evaluations into five significant packages:

1. Airport technology
2. Seaport technology
3. Railroads technology
4. Technology for all other areas, such as the city, land transport, educational institutions
5. Finalize technology evaluation, including identifying potential synergies among the various packages

The consultant assigned a dedicated team for each of these work packages. The teams performed their research and assessment and submitted their report with the recommendations in line with the sponsor's instructions.

Human Resource Considerations

Human resources considerations in the Feasibility Study were mostly about availability, capability, and expertise. Further, it considered whether the organization could do the project with internal or external resources. The focus, at that time, was in the early stages, because for the later stages, the team would update the staffing requirements once the Project Management and Project Detailed Plans were done.

It was clear that most actions during the development and delivery phases would require global resources, but the team had to endeavor to achieve the Prime Minister's Office instructions for obtaining about 50% of the workforce locally. For operations, the target is 80% local; therefore, during the Definition Stage, the team had to identify all of the required skills for operations activities (post-construction) and hire and train those resources. This would become part of the Operational Readiness Stages.

Social and Community Evaluations

This would be similar to market evaluations, but, in this case, it was specific to the social and community needs and requirements, which would apply to government and not-for-profit projects.

* We realize we might be overdoing it with these reminders; we will continue to include in this chapter and will minimize in future chapters. However, it is critical not to limit the learning to what we are including here.

The area chosen for the project is an area rich with natural resources, but which had been underdeveloped due to years of neglect and poor economic infrastructure conditions. This area has a high potential for development and growth, and this project would be leading enabler. Although one of the primary drivers for this project was a return on investment, the strategic imperative was that it must be an active contributor to economic growth and a facilitator for community development and the social welfare of the whole country, particularly this underdeveloped region.

This section of the feasibility could include studies of the workforce, education, social needs, among numerous other social and community factors. This was necessary to identify how the project utilizes what exists, determines gaps, and how the ICMP contributes to the region.

The 50% and 80% localization target was applicable here as well—it was a requirement shared in both of these areas.

Other Considerations

Other topics considered in the Feasibility Study were:

1. Political stability
2. Terrorism and regional conflicts
3. Tax incentives for direct investment

Major Project Risks

In the Stage Detailed Plan, the focus was on *stage* risks, and now the team shifted focus to significant *project* risks—risks to the overall project, where some of them could be resolved in this stage, while others could be postponed to future stages, once there was enough information to address them. Typical significant project risks* included:

1. Finding enough investors willing to take on such initiative, in a country in which instability is still an ongoing threat.
2. Dropping oil prices, which would hinder the ability of the government to fund its portion of the ICMP.
3. Increasing oil prices, which would encourage the government to take on a higher percentage of ownership and reduce international investment.
4. Finding wealthy nationals with international businesses that would be willing to repatriate some of their wealth back into their home country. Investors could provide funds through direct investment or purchasing project bonds.

Although it is not listed here, the team had to assess each of the above risks and many others, and then develop response strategies. Things such as political instability are hard to manage, but investors' risks can be mitigated by providing tax incentives and other financial security measures.

Class 1 Estimate: Schedule

The Class 1 Estimate† was a rough order of magnitude estimate because of the lack of details, especially in the future stages. Therefore, at this point, one could only use the analogous estimating technique

* In this book, risks would be threats and opportunities.

† The estimate classification and numbering system is per CAMMP™ and might not align with your organizational practice or professional references and guides.

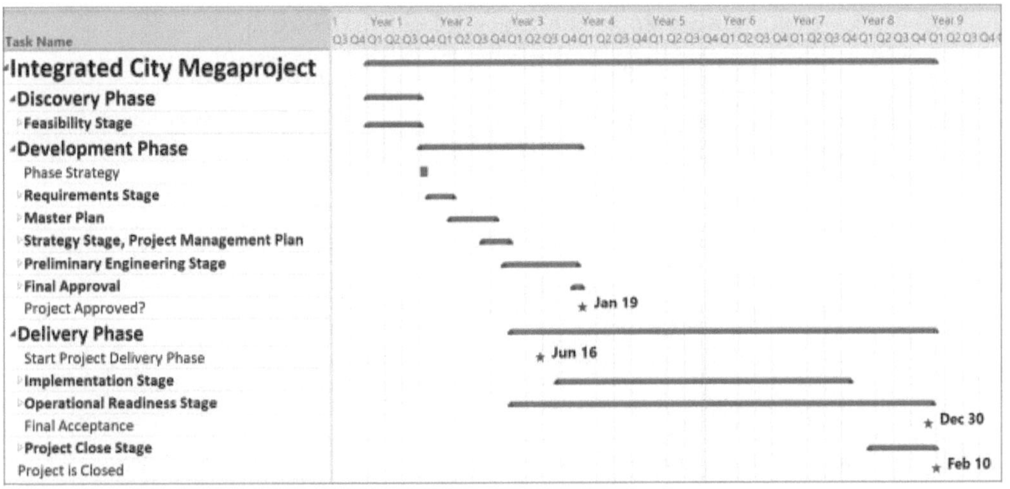

Figure 26.4 The Integrated City Project, Executive Summary Schedule

at the highest level. With this situation in mind, the project management team developed the overall schedule, with an understanding that it would be updated during the later stages.

The current schedule shows the project duration is just short of eight years, about three years for the first two phases and less than five years for project delivery.* It is noted that this schedule reflects the completion of all components, whereas some of the parts could be delivered earlier. The sequence of completion would be defined initially in the Class 2 Estimate and finalized in the Class 3 Estimate, just before final approval. Figure 26.4 presents the Executive Summary Schedule. This figure illustrates the duration of each phase and stage, along with the critical milestones. The figure also highlights the dates for starting the Delivery Phase, Final Acceptance, and Project Close.

Class 1 Estimate: Cost

Similar to scheduling, the team used analogous estimating techniques for the upcoming phases and the forecast cost of the first phase, which was about to complete. The project cost is summarized below. Please note the following concerning these costs.

The actual estimate was performed with more details than are shown here.

1. Other costs, such as escalation costs, cost of money (interest rates and cost of borrowing), anticipated currency fluctuations cost, overheads, and similar items are part of the estimated cost listed for each component.
2. Because the basis of these costs is historical records, they were updated to today's value based on historical inflation rates.
3. Project management costs are built into the individual component cost.
4. Contingency for each step is included in the estimated cost for that line item, and the percent used is 25%, which is per the organizational guidelines for contingency at the Feasibility Stage.

* I originally wrote Part B in 2015, before the literature review and research that I have shared in Part A. Interestingly enough, the schedule information I am sharing is close enough to the average megaproject in the IPA database.

Table 26.4 The Class 1 Cost Estimate

Phase	Estimated Cost
Discovery Phase (Feasibility)	2,200,000
Requirements Stage	1,000,000
Strategy Stage (Master Plan)	1,500,000
Definition Stage (Preliminary Engineering)	90,000,000
Implementation Stage (Engineering)	400,000,000
Implementation Stage (Construction)	5,000,000,000
Operational Readiness Stages	350,000,000
Close Stage	1,000,000
Subtotal All Stages	5,845,700,000
Management Reserve (@25%)	1,461,425,000
Overall Estimated Project Cost	7,307,125,000
Proposed Budget (round off)	7,300,000,000

5. Working capital, operations, and maintenance costs are not included here. Those are included in the financial model.
6. Management reserve is included as an independent line item and estimated at 25% for this early stage of the project, per organizational guideline and similar to the contingency reserve.

Financial Considerations

The financial consideration section is in two parts;

1. Part 1 is the financial model, including costs, revenues, cash flow, expected payback period, and estimated profitability calculations.
2. Part 2 is for funding and investment, which would be a work package that requires significant effort.

Once again, the details of the above are outside the scope of this book.

Feasibility Study Conclusion and Recommendation

Based on the Feasibility Study and all of its packages, the conclusion was that this project was feasible despite the numerous challenges, on the conditions that:

- The government could provide 30% through direct investment.
- 45% through BOT/BOOT or similar mechanism.
- 25% from the government reserve, which is possible based on $40 per barrel for oil at current production levels.

26.4.2 Stage Gate 2: Project Initial Approval

☒ Approved ☐ Revise ☐ Hold ☐ Not Approved

Explanation of Decision

The Stage Gate 2 approval was at the Prime Minister's Office and included numerous high-level officers from the relevant ministries.

The project executive discussed the Feasibility Study Report with the project director, business director, and project team to fully understand what was being submitted. Once satisfied, the project executive raised the study to the Prime Minister's Office and requested the stage gate review.

After reviewing the study (and after the recommendation from SPA and NPMA), the Prime Minister's Office agreed that the megaproject was feasible and understood the conditions that made it viable. The Prime Minister decided to approve the project and requested that the project executive issue the Project Authorization Document per the agreed terms and organizational guidelines. The reviewers also agreed to form a steering committee consisting of:

1. The general managers of the planning divisions from the following ministries: Transportation, Housing, Health, Public Works, Commerce, Foreign Affairs, and Environment
2. The project executive, head of the strategic planning authority, head of national project management authority, and the executive manager of the Prime Minister's Office

26.4.3 Project Authorization Document

The Project Authorization Document was next. Please note that most of the information was copied from earlier documents.

Phase, Stage

Discovery Phase, Feasibility Stage

Product Description

The project was to develop an integrated logistics city. The city includes airport, seaport, railroads, roads and highways, housing community with adequate infrastructures, and educational and professional development centers. The development centers will focus on the skills needed to manage the port, airport, and other facilities.

Project Justification

This major center will be a vital economic stimulus, driving direct international investment, contributing to the national income, offering excellent employment opportunities for thousands of people, and facilitating social and community development.

Location

The ICMP was planned for an underdeveloped region with a great potential in this country, as outlined in the Feasibility Study. The exact locations of some of the facilities will be defined at the Master Plan stage.

Duration

Based on the schedule (Figure 26.4), the project was estimated to be about eight years. Expected funding (Final Approval) was set for January of Year 4; Final Acceptance in December of Year 8; and project close in February of Year 9.

Project Total Cost

The estimated total cost of the project, including the management reserve, was US$7.3 billion.

Project Director

The project director for the upcoming phase remained the same. His authority was defined in the organization project management (OPM) system. This means that the project director would proceed with the project stage by stage, awaiting approvals at each stage gate before proceeding to the next stage.

Project Executive

The project executive for the upcoming phase also remained the same. Her authority was also per the OPM system. However, for this megaproject, the project executive will report to, and follow the guidance of, the steering committee.

Steering Committee

As instructed at the Stage Gate 2 reviews, a steering committee was formed for the project oversight. The committee membership was as outlined earlier. The steering committee met every other month during the project development phase and for all stage gate reviews. The project director and other team personnel could attend parts of the steering committee meetings to present any item of interest. For the process gates, the steering committee was to be informed, but the decisions were delegated to the project executive.

Assumptions

No explicit assumptions beyond what has been listed earlier.

Constraints

No specific constraints beyond what has been listed earlier.

Major Risks

Refer to the Feasibility Study, paragraph 26.4.1.

Stakeholders

The stakeholders included:

- Strategic Planning Authority personnel and management
- NPMA personnel and management
- Prime Minister's Office and staff
- The project executive
- The project director, business director, operational readiness director
- The steering committee
- The various teams, for the different project stages
- The involved ministries
- The impacted provinces
- The communities around the various sites
- Non-governmental organizations
- Suppliers and service providers

Project Success

The project followed the Four Dimensions of Project Success™.*

26.5 Chapter Summary

This is probably the most significant chapter in Part B of this book. Consequently, many of the topics that repeat throughout the stages which required explanations were presented for the first time here. In later chapters, many of the same issues will repeat but will be truncated and refer to this chapter.

* Refer to Section 17.5 (page 148).

Section III
The Development Phase

Chapter 27

Introduction to the Phase

27.1 Overview (A Refresher)

The Development Phase is the second phase of the project life cycle. The primary objective of this phase is to develop the concept of the project and expand the level of details to make it ready for the Delivery Phase. We typically accomplish this objective via three major steps:

1. Understanding the product, exploring the stakeholders' expectations and other requirements, and documenting them via a Project Requirements Document (PRD).
2. Establishing the strategies to deliver the product and managing the project, which ends up in a Project Management Plan (PMP).
3. Finalizing the scope and all other details, including implementation planning, which we would include in a Project Detailed Plan (PDP).

The above is per the standard CAMMP™ Model. CAMMP™ MP adds the Shaping Process. With the above major steps, we have the business decision and advanced funding with the second step and final funding with the third step.

27.2 Phase Strategy

Per the schedule of the Class 1 Estimate, this phase is about 27 months, which theoretically makes it equivalent to a large and complex project itself. Therefore, it requires special attention and an overall strategy. The project director, after consultation with the project executive, decided to take the time to develop a strategy for this phase. This was necessary because there are no models and no data in the national project management system for a project like this. They agreed to amend the project life cycle for this phase and keep the delivery phase as specified in the earlier work. The final project life cycle is a subject and would be finalized in the Strategy Stage. The project director committed that the extra time that will be used for this step, which was not in the Class 1 schedule, will be recovered during the phase.

27.3 Revised Model

The project director and project management team conducted research on megaprojects, their causes of failures, and critical success factors. They reviewed multiple case studies of project successes and failures, and they consulted with the government's National Project Management Authority (NPMA). As a result of all of the above, the project director decided, and the project executive approved, changes to the Development Phase. The decision was to add a Master Plan Stage between the Requirements Stage and the Strategy Stage. This split would help minimize the threats to the project, because it emphasizes the need to work each part correctly and to incorporate a stage gate at the end of each significant step. These were part of the facilities stream on the CAMMP™ MP project life cycle.

In addition to these stages, we still have the Shaping Stage, Part 2, happening in the governance stream in parallel to these three stages of the facilities stream.

27.4 Closing Remarks

With the above strategy and tailoring of the project life cycle, the team is ready to move ahead into the next stage, the Requirements Stage.

Chapter 28

Requirements Stage

28.1 Introduction

The Requirements Stage is the first stage post project authorization and would be the starting point for the Development Phase. The sequence is per the customized and adapted model, as discussed in the previous chapter. There is no need to discuss the stage sequence and stage management process, because we have discussed them in more than one place. The process deliverables will also clarify the sequence.

28.2 Process Deliverables and Process Gates

28.2.1 Authorize Stage Work

Stage Authorization Document

The truncated Stage Authorization Document is presented below.

Phase, Stage

Development Phase, Requirements Stage.

Instructions

The project director has been involved from the beginning and is already familiar with the project. Therefore, he was authorized to form a team to define the stakeholders' needs and expectations and the project requirements, and to produce the Project Requirements Document (PRD). The requirements were for all aspects: technical, social, community, business, and even financial. In the governance stream, the business team also started to research and identify potential individuals and organizations for investments and partnering. Ideally, the group of investors was defined and mobilized before the Project Management Plan. However, it would still have been acceptable if they mobilized before the Preliminary Engineering Stage.

Duration

Per the Class 1 Estimate, the duration of this stage was roughly five months.

Estimated Cost

Per the Class 1 Estimate, the cost for this stage was $1 million, excluding the management reserve.

Assumptions

Internal resources from the various ministries could do the work but continued to need external resources as advisors.

Constraints

None; only the location of port and airport as defined in the earlier stage.

Stakeholders

The stakeholders for this stage are per the Project Authorization Document.

Other Notes

None.

Process Gate 1

☒ Approved ☐ Revise ☐ Hold ☐ Not Approved

Nothing to add.

28.2.2 Plan Stage Management

Stage Management Plan

Although this was quite similar to the Stage Management Plan (SMP) of the previous stage, it was still addressed to ensure clarity and re-emphasize the concepts. For the other stages, the SMP was truncated to avoid repetitions. In those stages, the focus was on the differences or items unique to the stage.

Phase, Stage

Development Phase, Requirements Stage.

How to Develop the Stage Detailed Plan

The project management team and other project personnel, as defined below, formed the stage planning team (detailed). This team consisted initially of only internal resources and was able to develop the Stage Detailed Plan per the organizational guidelines. The Stage Detailed Plan included all of the functions and other elements, such as scope, quality, time, sustainability, and risk. Once again, the focus here was on the Requirements Stage, only.

How to Implement and Manage the Stage Work

The stage management team managed all personnel and project work from this point onward, all the way to stage closure. The final product of this stage work was the completed Project Requirements Document.

How to Control the Stage Work

The project management team monitored the ongoing stage work to identify potential defects, deviations from the plan, or any areas for improvements, and take the necessary action.

How to Manage Change During the Stage

The project management team followed the company change management system, which emphasized the need to review and assess all changes before approval and implementation.

How to Close the Stage

One of the PMT members stayed after Stage Gate 3 to close the stage, while the other team members either left the project or continued to the next stage, depending on their roles and organizational needs.

Stage Management Team

The stage management team already included the project director and two project management specialists from the earlier stage. The project director decided that the three-person team was enough for this stage as well because the volume of work was less than or about equal to the prior stage.

Stage Planning Team

A few non-project management subject matter experts were needed to represent various areas of expertise to develop the Stage Detailed Plan. The project director decided to keep the lead SMEs from the previous stage as team leaders for the work of this stage. These individuals had gained good experience working on the project so far, especially their daily interface with the international consultant. However, two of the 12 SMEs were replaced earlier for a variety of reasons. The SMEs on board were enough to develop the Stage Detailed Plan.

Project Life Cycle

A revision of the project life cycle was covered in the Phase Strategy chapter.

Process Gate 2: SMP Approval

☒ Approved ☐ Revise ☐ Hold ☐ Not Approved

Explanation of Decision

The project executive approved the Stage Management Plan without any modifications and requested that the project director continue the process and develop the Stage Detailed Plan.

28.2.3 Plan Stage Details

Stage Detailed Plan

Phase, Stage

Development Phase, Requirements Stage.

Scope of Work and WBS

The first step in the detailed planning effort was to collect all the **requirements for the stage**. The primary source of information at that time were the documents produced to date.

Further, because the same people were still involved, the initial requirements for the stage were readily understood. Based on the above, the team could define the scope of the stage. The primary scope of this stage was to produce a Project Requirements Document. Figure 28.1 illustrates the major area of research and work that the team conducted before they could finalize the PRD.

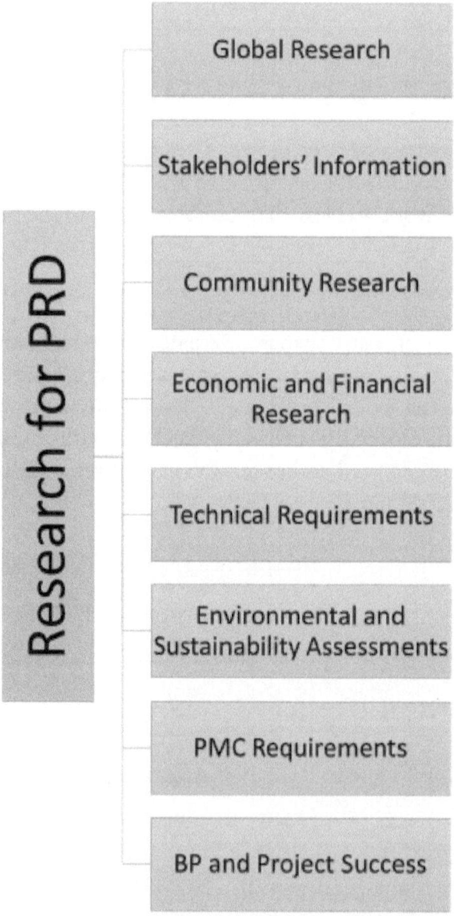

Figure 28.1 Research Required to Develop the Project Requirements Document

1. **Global research.** Researched similar projects from around the world, whether fully integrated cities or partial initiatives, such as port, airport, city, or urban development. The global research was necessary to capture ideas, success factors, lessons learned, and future trends.
2. **Stakeholders' information.** Explored for the expectations and requirements of the Prime Minister's Office, the steering committee, and the various ministries and organizations that would be impacted by this megaproject. This research was necessary to understand the vision for the project and its role in developing the country and community.
3. **Community research.** Held community meetings to understand the people needs and explore their expectations in term of jobs, community development, education, and other factors. This research also sought to identify what infrastructure existed and whether the project could build on what was available or have to start from scratch.
4. **Economic and financial research.** Met with economic and financial advisors to determine trends in the global economy and requirements of global investors. Also visited countries in which there were concentrations of national expatriates, especially those with the financial capabilities or expertise necessary for such a project.
5. **Technical requirements.** Identified all technical requirements for the various facilities, such as airport, port, railroads, city, and educational centers.
6. **Environment and sustainability.** Conducted environmental assessment and sustainability reviews to identify any requirements related to these areas.
7. **NPMA requirements.** Met with the National Project Management Authority (NPMA) to identify all of their expectations and requirements, especially in terms of complying with and improving the national project management system for the government. This type of project was not common and could be an excellent learning opportunity for years to come.
8. **Best practices and project success.** Best practices (BPs), as discussed in Part A of this book, were of great value and were considered for the Integrated City Megaproject (ICMP). These practices were vital to help enhance performance and deliver project success. This project used a few of the best practices as recommended by organizations such as the Construction Industry Institute (CII). The Requirements Stage was the right timing for starting using the CII Best Practices (CII, 2015) and defined the project success criteria. For project success, the project used the Four Dimensions of Project Success™ (SUKAD Group, 2015).

The above research and analysis led to the Project Requirements Document (PRD). Figure 28.2 illustrates the outline of the PRD per the standard model (Ajam, 2017). Technically, each of the items listed on this graphic can be a significant deliverable. For the ICMP, the PRD was a modified version of the outline illustrated in Figure 28.2.

Define Quality

For the PRD, please refer to organizational standard No X-Y-PRD-05 to find the requirements for conducting this work and how to ensure the quality of the deliverables. Because most of the work at this stage was still research and process oriented, quality management would be mostly about complying with the organizational project management policies. It was also about ensuring that the team conducted its work with care and endeavored to consider all aspects. The business excellence department participated in the reviews of the PRD pre-Stage Gate 3.

Health, Safety, Environment

For the Requirements Stage, health, safety, and environment (HSE) considerations were not of primary importance beyond the usual safe office practices, reducing waste, and offering a healthy working environment per the organizational policies and guidelines.

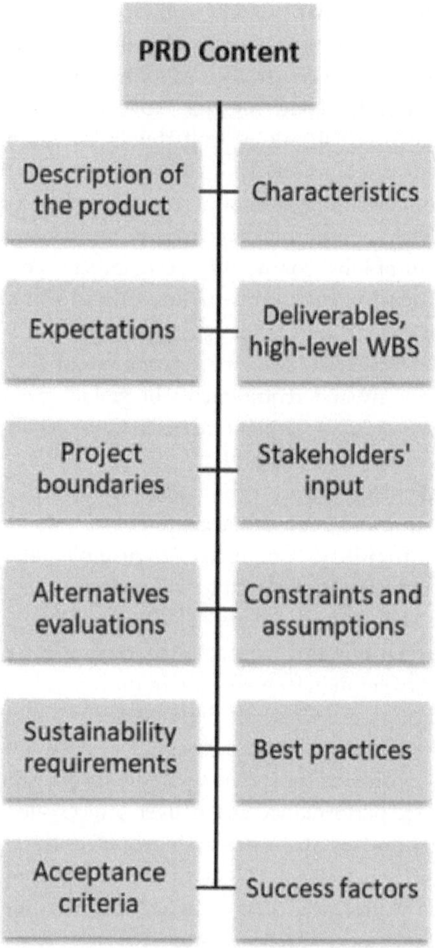

Figure 28.2 Outline of the Project Requirements Document

Sustainability

Similar to HSE, for the stage, sustainability was not a primary consideration. However, during the development of the Project Requirements Document, sustainability aspects were vital for the project, and it was one of the main areas of research in the scope of this stage.

Schedule

For reference, the schedule from the Stage Authorization Document (SAD) and Class 1 Estimate was 21 weeks, about five months. This was the estimate before the Stage Detailed Plan. At this point, the team had the stage breakdown structure (the research packages in Figure 28.1 and other items), and the project management team (PMT) developed the schedule for the project requirements work and the overall stage schedule. Because most of the team members were continuing from the earlier stage, that resulted in these distinct time benefits:

- The project director did not need time to identify and mobilize new resources.
- The resources were already familiar with the project, so they could expedite the stage processes and steps.

This favorable situation was estimated to save about four weeks from the earlier estimate. On the other hand, learning the lesson from the previous stage, the PMT added two weeks to the implementation time. The overall net result was two weeks' saving from the earlier schedule: 19 weeks' total duration. Therefore, the expectation was that the stage would be closed on March 25 of Year 2.

Cost Estimate

For reference, the cost, per the SAD and Class 1 Estimate, was $1 million, without the management reserve. In a similar fashion to the previous stage, the project management team estimated the cost of the stage work, which was still, mostly, people and time driven. They calculated the stage work to cost $900,000. The saving was due to a shorter duration and a few other factors, such as better cost estimate of resources that was formed in the previous stage.

Time-Phased Budget

Not included here because the concept was demonstrated in the earlier stage.

Human Resources

The section is about the human resources roles and staffing requirements for executing the stage work.

- First, the stage management team was already in place and consisted of the three people mentioned earlier.
- The stage planning team, including the above and the subject matter experts, were already in place.
- Once the Stage Detailed Plan was finalized and approved, these SMEs helped the PMT during the implementation of the stage work.
- While implementing, the SMEs were the advisors, each for his or her area of specialty. Some of them were assigned to lead a research package.

In addition to the above, the project team needed to add personnel from various ministries to help with the research work. For example, a team from the Social Affairs Ministry would lead the community research, a team from the Commerce Ministry would lead the economic and financial research, and so on.

Figure 28.3 is the organizational roles and responsibilities matrix, where the "R" indicates the team responsible (and accountable) for the work, and "C" represents the need to consult with that organization. This matrix is not detailed and is only at Level 1—the main areas of research for the PRD. Each of these teams developed a further breakdown that would include all roles and responsibilities, down to the individual level within the team or sub-team. In addition to the matrix in Figure 28.3, the project management team also developed the staffing plan, showing when to mobilize each sub-team and individual. The staffing plan was used to update the cost estimate and schedule where necessary. Since the team was growing, the staffing plan was more detailed and included the various areas of expertise required. The information consisted of the ministries and organizations that provided these resources.

Stakeholders

The stakeholders were already defined earlier, and in this case, it was more detailed. The team developed a plan to engage all stakeholders, including the community members that would be involved in the research work while developing the Project Requirements Document. In exploring for the expectations

Organizational Responsibilities Matrix

Area of Responsibility	Team						
	PMT	PT	MSA	MCT	MEA	PMC	PMO
Project Management	R						
Global Research		R		C			
Stakeholders' Information		R					
Community Research			R				
Economic and Financial				R			
Technical Requirements		R					
Environment & Sustainability					R		
PMC Requirements	R					C	
BP and Project Success	R						C
Integration of Packages	R						

MSA - is the Ministry of Social Affairs
MCT - Ministry of Commerce and Trade
MEA - Ministry of Environmental Affairs

PMC - Program Management Council
PMO - Prime Minister Office

Figure 28.3 Organizational Responsibility Matrix, Project Requirements Stage

of the various stakeholders, it was vital that "*Members of the owner's staff that should be involved in defining quality should include financing, engineering, operations, and maintenance personnel. This group must separate 'what owner needs' from 'what the owner wants'*" (Oberlender, 2014, p.7).

Communications

Because the work was still mostly internal, the team followed the same communication plan from the earlier stage, with modifications to the names and positions of the various stakeholders and their contact information.

Risk Management

The team performed a risk assessment for this stage, which is not shown here.

Define Procurement

The only procurement requirement for this stage was to amend the contract with the Feasibility Study consultant to include some of their consultants as advisors on a standby basis. The agreement was to use the same rates as agreed to in the contract. This condition was negotiated in the previous stage as a potential requirement.

Other Considerations

The team agreed that there were no other considerations to evaluate at this time and decided to finalize the Stage Detailed Plan and submit it to the project executive for reviews.

Process Gate 3: SDP Approval

☒ Approved ☐ Revise ☐ Hold ☐ Not Approved

Explanation of Decision

The project executive approved the Stage Detailed Plan after discussions with the project director and requesting minor changes to the staffing plan. Then the project executive notified the steering committee that the team was ready to proceed and develop the Project Requirements Document.

28.2.4 Implement Stage Work

With the approved plan, the project director and team proceeded with the implementation. However, the first thing to do was to mobilize the project team that would perform the project requirements work. The team started the implementation work with two major steps:

- Taking action on the SDP to mobilize the various team members from their organizations
- Amending the contract with the consultant firm for the advisors

28.2.5 Control Stage Work

Author Note

Because this is a simulation, we can make up stories about performance deviations, corrective and preventive actions, and changes, among other things. However, there is no need to make this book longer than it should be and we will not revert to storytelling. If there is anything unique to this stage, it will be addressed.

Performance

Performance deviations were applicable for all knowledge areas and functions.

- There could be deviations in scope, and scope management was to ensure that the team did the work, no more, no less.
- There could have been defects that the team would identify and do the necessary to repair or replace, as appropriate.
- There could be time or cost variances, which may or may not have required corrective actions.
- There could be changes in the human resources aspects, such as changes in personnel, delays in mobilizing people, lower or higher productivity than planned, among other factors.
- The same goes for all other areas.

In summary, let us say that, for a variety of reasons, work on this stage did not go well. It required significantly more effort than planned because of the difficulty of identifying all of the project requirements to ensure achieving the objectives.

Management of Change

For this stage, there were quite a few small changes that were necessary to meet the objectives. There was a significant change—the team—and the NPMA found that they did not have enough experience in best practices. However, because they did believe these were paramount and would reduce the overall total cost and improve profitability, they agreed to hire a specialized consultant to advise the team and lead the effort. This consultant would stay on the project in future stages if necessary.

Forecast

Same as before: based on the team's performance and change during the stage, the project management team updated the stage and project forecast and communicated to the appropriate people. Due to the above factors, the team was able to achieve the schedule, quality, and other planned requirements, but the cost grew significantly, from the estimated $900,000 to $1.4 million.

28.2.6 Stage Closing

Same as Stage Controlling, no need for storytelling here.

28.3 Stage Deliverable and Stage Gate

28.3.1 Project Requirements Document

The various teams were mobilized, each to perform its duties. They conducted the necessary research and reviews, leading to the Project Requirements Document summarized below. However, as noted before, this is a highly truncated study and only shows key points rather than a real and comprehensive study of a large and complex project like the ICMP.

Phase, Stage

Development Phase, Requirements Stage.

Description of the Product

The project was to develop an integrated logistics city to include airport, seaport, railroads, roads and highways, housing community with adequate infrastructures, and professional development centers. Once again, to avoid storytelling, the product description could be a lengthy document with details about the whole project and all of its components.

Characteristics

In this section, the team expanded on the description and included information that could better characterize the various facilities. The following are some notes on what this means.

- **Airport.** The primary airport purpose was cargo, with passenger traffic as a secondary objective, because there was another international airport in the region. The airport facilities needed to be modern, with the latest proven technologies that complied with leading industry practices. Automation would be utilized to an optimal level to facilitate efficient operations. As stated earlier, the project used the existing military airport and converted it to civilian use to meet ICMP requirements. The airport land area was smaller than required; therefore, the ICMP had to acquire more land around the airport, especially for longer runways. The existing building was also not large enough and needed renovation.

- **Port.** The port location was defined at the earlier stage. The team research identified the required modifications necessary for the project. Like the airport, more land was needed for two additional berths, other buildings, and automated equipment for offloading the ships and moving containers to the transition area before loading onto trucks or trains. At this time, the port did not have railroads connections, and these were to be added. There was one concern here—the water was not as deep as necessary, which led to exploring alternatives (presented in the Alternatives Section on page 300).
- **Railroads.** There was already a railroad route connecting the megaproject location, which was in the south of the country, with the central and northern regions. However, this did not reach the port and airport locations. Further, it was also not near the area planned for the community's residential areas. The project had to provide all of the interconnections to the new areas within the integrated city. The work included new routes, connections to existing routes, and other facilities such as a command and control center. The ICMP depended on another project that the government was developing, connecting the national railroads to the regional network into neighboring countries and beyond.
- **Roads and highways.** Similar to railroads, this element was a mix of utilizing existing infrastructure and the need to build what was missing. The highways were to include proper rest areas with service facilities, especially for the large trucks. Within the new city, modern traffic control would be implemented to facilitate the movement of all types of vehicles, with consideration for the higher proportion of truck traffic and the slow movements of loaded trucks. This was vital because of the hilly areas that exist around the region.
- **Community.** *Community* refers to the "city," residential areas and their needed infrastructure. The requirements in this part were to build new residential areas with ready access to the port, airport, and other centers, but also close enough to a major neighboring city. This requirement was related to the expectations as outlined below. The proximity allowed the residents of the new areas to utilize the facilities in this large city (some hospitals, international schools, universities, malls) and people from that city to use some of the modern facilities to be built within the new areas. During the Master Planning stage, more details would be developed.
- **Educational centers.** Similar to the community expectations and requirements, the team would not replicate what existed and embark on developing redundant facilities. The team evaluation concluded that there was no need for universities offering typical programs such as business and engineering, because enough universities and colleges offered these programs in the region. The need here was mostly to fill a gap in terms of specialized colleges and learning centers that focused on the domains required to run the new integrated city. There was an existing marine science college, but it needed to be modernized. Also, there were other needs such as vocational training centers, management and leadership development, project management, and others.

Expectations and Requirements

Most of the requirements were incorporated into the characteristics. More specific requirements for each of the particular ICMP components were documented in the Appendix (not included here). During the Feasibility Study, the team met with the originator of the idea, strategic planning authority, and the Prime Minister's Office to explore for any expectations that might not have been contained in the initial documents. All items were clear and documented. However, during the meeting, the Prime Minister's Office emphasized:

- The need to ensure that this project must be profitable but can accept a low return on investment (10%).
- The expectation that this project must help in reducing the unemployment rates, especially among professionals and university graduates.
- The aim to have 50% of the labor force, during the project development, be nationals.
- The aim to have 80% of those who would operate and maintain the facilities also be nationals.

Other expectations that were outlined earlier were to use the latest proven technology for this project in order not to have issues with experimental technologies or things that were still in the initial stages of use. One more expectation from the steering committee was to maximize the use of what existed, which was necessary to:

- Minimize the cost of the ICMP
- Enhance the capacity—utilization factors—of what existed, such as educational centers, residential areas, and others
- Not create over-capacity or redundant facilities

These were the most important expectations and high-level requirements that the team would need to clarify and comply with throughout the remainder of the project development and delivery.

Stakeholders' Input

In preparation for the Project Requirements Document, the various teams conducted extensive research to collect all of the stakeholders' inputs and incorporated them here. There were numerous meetings with groups and key decision makers to understand their vision, expectations, desires, and requirements.

Naturally, not everything could be implemented, and there were many conflicting requirements. The team endeavored to document all of the notes from these various engagements, study them, and incorporate what was feasible into the Project Requirements Document. At the upcoming stage gate, the team presented the PRD to ensure alignment.

Alternatives Evaluations

As we discussed in Part A of this book, it was best not to keep alternatives open beyond the conclusion of the Requirements Stage. We wanted to assess all options at this point before the team finalized the shaping stage and the Project Management Plan, which would also be the point of the business decision.

- **Airport.** In the earlier stage, the alternatives were whether to build a new airport or use the existing military airport, and the path was cleared to use what existed. No other options to study here.
- **Port.** The depth of water in the port allowed for certain size ships and ocean liners to dock there. This was enough for large ocean-going ships but not enough to accommodate some of the modern superliners. The team would have had to expand the channel and deepen the water to take on these humongous ships. The team had a decision to make: exclude these superliners or deepen the canal. The team evaluated the possibility of increasing the depth of the channel, but that was rejected because there were too many complications, including political concerns with neighboring countries.
- **Railroads.** One idea was to consider a dedicated rail route. This was quickly rejected as not necessary, as the value added was a fraction of the cost involved. In other words, the benefit/cost ratio could not justify this option.
- **Roads and highways.** Here there was a consideration to route the traffic through the neighboring large city, but the preference was for building a by-pass road. Although this would require

additional capital and initial costs, it was deemed necessary because it would reduce life-cycle costs and increase the economic value of the megaproject, owing to efficient traffic and movements of cargo and people.
- **Community.** No alternatives evaluated because the expectations sections clearly outlined how to proceed.
- **Educational Centers.** Same as the community—ruled by the expectations.

In closing this section, there were alternatives in terms of funding and whether these would be developed by the project or opened to direct investment. The team was open and addressed this point in the financial section of the Project Requirements Document.

Initial Product Breakdown Structure

From the various sections, the high-level product breakdown structure (PBS), Level 1, is clear and included in the Figure 28.4. This figure also includes Level 2 elements. This PBS does not include the project stages and the work of the stages. The PBS is a visualization of the final product, which would help the team in clarifying the requirements that ultimately led to building the facilities that the organization would operate to realize benefits. The Master Plan expanded on all of the above facilities.

Project Boundaries

The project incorporated all of the facilities mentioned, including the hospitals, schools, educational centers, and anything else outlined here. The project does include facilities (buildings, furnishing, and equipment) but does not include:

- Curriculum, staffing, and operations of schools (elementary to high school); these will be run by the ministry of education.
- Staffing and operations of clinics and hospitals; these will be run by the ministry of health.
- Any work that will be direct investment, although the ICMP management will coordinate this work.

It is worth noting that some schools or other facilities could be privatized as part of the ICMP direct investment scheme. In these cases, the investors would provide everything, including the facilities and their operations, in compliance with the government and responsible ministries' regulations.

Constraints

- **Airport.** The existing land area was limited but could be expanded via land acquisition, specifically to extend the length of runways.
- **Port.** The depth of water limited the size of ships docking, but that was accepted during the alternatives evaluations.
- **Railroads.** Location of existing primary routes.
- **Roads and highways.** None specific at this time.
- **Community.** Location must be near large neighboring city and readily accessible.
- **Educational centers.** None specific at this time.

Assumptions

Will be able to find investors willing to invest in the future of the country.

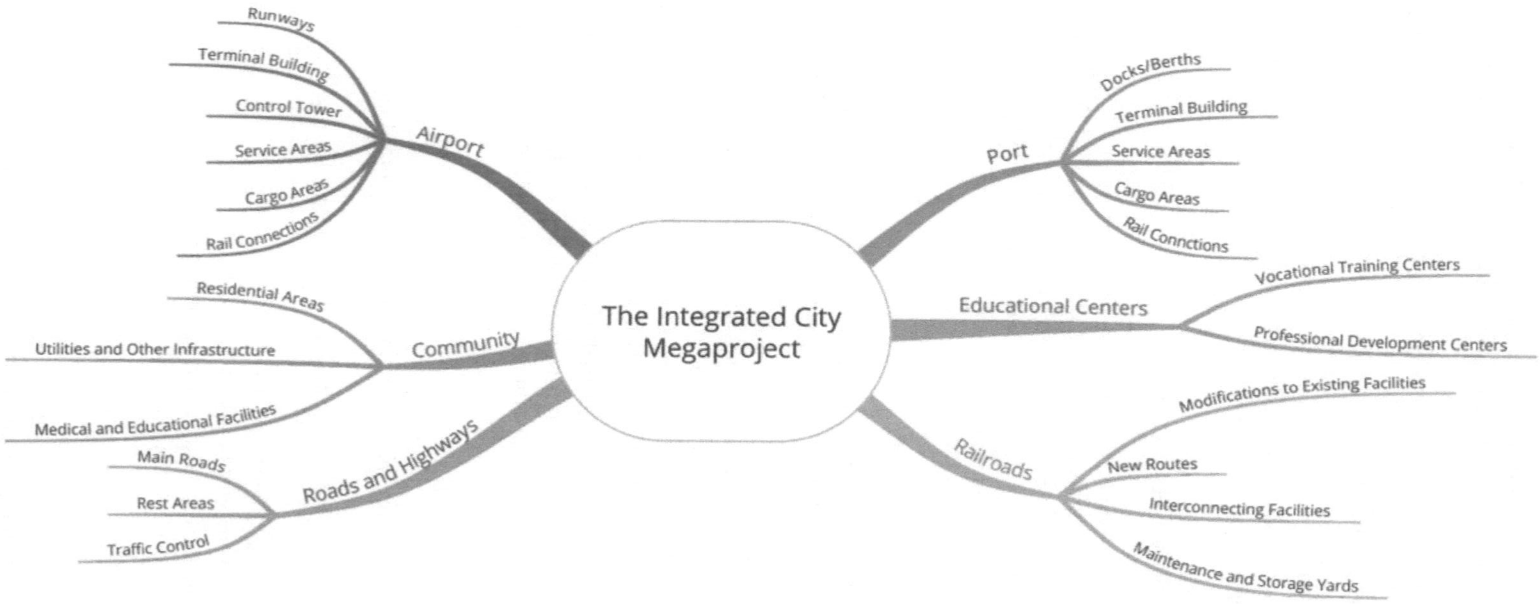

Figure 28.4 The Initial Product Breakdown Structure for the Integrated City Megaproject

Best Practices

> **Lean Design and Construction**
>
> Lean Design and Construction is a production management-based approach to project delivery—a new way to design and build capital facilities. Lean Construction extends from the objectives of a lean production system—maximize value and minimize waste—to specific techniques and applies them in a new project delivery process (Lean Construction Institute, 2015).

The project methodology that the team followed (CAMMP™ MP) and the government national project management system already incorporated best practices. Also, due to the large and complex classification of the ICMP, the team decided to supplement the organizational project management guidelines with the following best practices from CII (CII, 2015). It was the first time CII practices would be utilized extensively on a megaproject in this country. Some of these best practices would be implemented now, whereas others will be applied in later stages, where appropriate. It is also worth stressing that some of these best practices were utilized more than once during the project life cycle, each time with a different focus.*

- **Alignment.** This best practice was also a CAMMP™ MP featured practice. In addition to all of the stage gates and stage deliverables, the primary purpose of the Project Requirements Document and Stage Gate 3 is to ensure stakeholder alignment before reaching the business decision point.
- **Change management.** Some aspects of this best practice element have been addressed in the chapter on change management, and at this time, this is enough. The team had been monitoring changes during each stage. Further, once the team reached the preliminary engineering stage, change management became more critical, and the change management plan for that stage had to build on this best practice.
- **Constructability.** The team considered the constructability reviews at three different points on the project life cycle. First, during the Master Planning to ensure the proper and efficient layout of the various facilities. Second, during the preliminary engineering for more accurate reviews. Finally, at the detailed design stage to ensure that the designs could be built efficiently. This CII best practice is also related to Lean Construction Institute (2015).
- **Front-end planning.** In a simple explanation, the front-end planning best practice was also a CAMMP™ MP feature, in which the project life cycle approach and stage gate process provided the necessary discipline to develop the project concept properly before the project was fully funded and moved into the Delivery Phase. The team also followed CII guidelines, especially before Stage Gate 5, which was the final approval and funding for the project.
- **Lessons learned.** Covered in every closure of every stage and the project.
- **Materials management.** This was mostly applicable for the construction stage but had to be addressed earlier, during the preliminary engineering and detailed engineering stages.
- **Partnering.** The nature of the ICMP made it an excellent match to consider partnering with service providers with long-term service agreements.
- **Planning for start-up.** The CAMMP™ Operational Readiness Stages (three stages in the production stream) were excellent matches to this best practice.
- **Zero accidents techniques.** Part of the project safety management.

Sustainability Requirements

In order not to provide partially accurate information, a useful reference here would be the GPM® Organizational Model for Sustainable Change Delivery, from Green Project Management® (GPM®).†

* For more reading, please refer to CII site: https://www.construction-institute.org/scriptcontent/index.cfm

† http://www.greenprojectmanagement.org/index.php/the-psm3-sustainability-assessment.

Technical Requirements

This topic is too broad to cover here, and there was not enough expertise in technological aspects of the type of facilities that the ICMP would deliver. The subject matter experts developed each within their areas of expertise.

Community Requirements

This would include, among many others, things such as:

- The number of housing units required with a split between single or multiple uses
- The number of schools and what levels
- The space allocation for shopping and other common facilities
- Current unemployment rates
- Training and developmental requirements to maximize the use of existing local and regional resources before hiring expatriates

Economic and Financial Requirements

For ICMP, this was the outcome of the research that the team conducted. The main topics of concern here were things such as:

- Identifying potential investors, either for a cash investment to the overall project or direct investment in specific facilities
- Recent trends on investment schemes such as build–operate–transfer (BOT) and build–own–operate–transfer (BOOT), their advantages and disadvantages
- Changes to employment laws, if necessary, to encourage foreign investments
- Tax incentives, at least for a period to promote foreign investments
- Understanding the economic indicators such as expected growth rate in populations and commerce

NPMA Requirements

The NPMA requirements were:

- The project director and project management team would comply with all national project management system and guidelines.
- If any element of the NPMS was not applicable or required updates, the team would highlight and offer their recommendations; bypassing any guidelines without prior approval was not acceptable.
- From the time of completing the Project Management Plan onward, the team would include an NPMS manager who would be responsible for monitoring all of the work to identify areas of opportunities to enhance the national project management system, either by updating what existed or including new areas that were not in the system.
- The NPMS manager would be involved in all best practices sessions because they were new areas, and the government system lacked information in this vital area.
- The NPMA would approve all project management metrics to use on the project.

Acceptance Criteria

The acceptance criteria for this project were in multiple parts and for every component of the ICMP. The team developed standards for each of these components and summarized them in the Appendix. In general, acceptance was linked to the various products, and they were delivered in compliance with the different specifications, regulations, and laws. Any deviation had to be evaluated for impact before acceptance. Any work that did not meet the acceptance criteria would be redone until it reached a satisfactory level.

Success Factors

The team developed the criteria for each of The Four Dimensions of Project Success™ and the timing on when these would be assessed and who would be responsible (not included here).

28.3.2 Stage Gate 3: Stakeholders' Alignment

☒ Approved ☐ Revise ☐ Hold ☐ Not Approved

Explanation of Decision

The Stage Gate 3 approval was at the Prime Minister's Office and included numerous high-level officers from the relevant ministries. The project executive had insisted, and the Prime Minister agreed, that all essential stakeholders participate in these reviews, and no approval would be granted without their input. After a couple of trials, all stakeholders did end up attending. There were quite a few valid points raised, requiring more clarifications and re-work of some parts of the Project Requirements Document. The Project Requirements Document listed here was the final approved version, which would form the basis for the Master Plan. The Prime Minister instructed the steering committee to proceed to the next stage.

28.4 Chapter Summary

As you will have noticed, in term of process, every stage had to follow the six-stage management processes of Authorize, Plan Management, Plan Details, Implement, Control, and Close. These six processes have many process steps, which we mentioned in Part A of the book, and we will not detail them here due to the size and scope of the book. We think it is clear by now how we separate the management of each stage from the overall management of the project. The key differences between the stages are stage deliverables, which are the products of each stage. These stage deliverables are related to developing and delivering the megaproject, leading to the final product, the Integrated City Megaproject.

In this stage, the focus was on developing a Project Requirements Document that included the needs, expectations, and requirements of all stakeholders. In many situations, the team had to balance some conflicting demands. This was necessary because different stakeholders might have had different objectives, even if all worked for the same organization. The PRD served as the basis for the upcoming stages, once all stakeholders were aligned. The PRD was also necessary to establish the acceptance criteria for the various components and address project success measures.

Chapter 29

Master Plan Stage

29.1 Introduction and Stage Purpose

The Project Master Plan Stage is not in the CAMMP™ or CAMMP™ MP standard model; it was added owing to the nature of this project and the geographic spread. In smaller and simpler projects, this stage could be merged with the Requirements Stage. However, that would be per the organizational policies and systems.

It is crucial to explain the purpose of the stage since it is not in the standard model, and it is a stage that is specific to this type of projects. The primary purpose of this stage is to develop a master plan for the project and gain the approval of this plan at a stage gate. The Project Master Plan includes the locations of all facilities, their interconnections, and the block diagrams within each facility. The block diagrams are the rough placement of the major components of each facility. For example, the general layout of the airport (runways, passengers building, storage areas, and service areas). In addition to locations and block diagrams, the project team defined all significant elements and expanded on the project requirements.

29.2 Process Deliverables and Process Gates

29.2.1 Authorize Stage Work

Stage Authorization Document (truncated)

The project director has been involved from the beginning and is already familiar with the project. Therefore, he was authorized to continue working on the project and form a team to develop the Master Plan. The Project Requirements Document was the basis (input) for this work. The Master Plan would be comprehensive and detailed enough to allow the project management team to formalize a Project Management Plan. Also, the Master Plan was the technical foundation for the preliminary engineering stage.

Duration

Per the Class 1 Estimate,* the duration for this stage was just about seven months.

Estimated Cost

Per the Class 1 Estimate, the cost for this stage was combined with the Project Management Plan because, in the Feasibility Study, the team had considered those as one stage. Now they were split, and the cost had to be split. Based on the early assumptions, the combined stage estimate was $1.5 million, but the team allocated $8.5 million for this work, excluding the management reserve. The balance was allocated to the Project Management Plan development.

Stakeholders

The stakeholders for this stage were per the Project Authorization Document. The only difference was the additional consultants that worked with this stage and different internal resources with experience in master planning.

Process Gate 1

☒ Approved ☐ Revise ☐ Hold ☐ Not Approved

Nothing to add.

29.2.2 Plan Stage Management

Stage Management Plan

This is quite similar to the Stage Management Plan of the previous stages. Therefore, to avoid repetition, only new information will be covered. The truncated Stage Management Plan is presented next.

Stage Management Team

The project director decided that the three-people team already on board was enough for this stage as well, because the volume of project management work was about the same as the prior stage. However, during the stage, the project manager started thinking about the next stage, where there would be a need to mobilize many project management resources to develop the Project Management Plan.

Stage Planning Team

To develop the Stage Detailed Plan, the team from the previous stage continued, and they handled all master planning work except for the airport and port, for which there was not enough expertise in the government and so required hiring specialized consultants.

Process Gate 2: SMP Approval

☒ Approved ☐ Revise ☐ Hold ☐ Not Approved

* Until we had the Class 2 Estimate, the Class 1 Estimate would continue to be the control reference.

Explanation of Decision

The project executive approved the Stage Management Plan without any modifications.

Plan Stage Details

Stage Detailed Plan

The author is not an expert in urban or master planning since this is technical, non-project management work. Therefore, the Stage Detailed Plan and the Project Master Plan that are presented in this chapter may lack technical information and should not be used as an example for master planning. In general, the structure of the Stage Detailed Plan would be similar to other stages' detailed plans, and there's no need for repetition here. Will touch on project-management related topics.

Schedule

For reference, the schedule from the Stage Authorization Document (SAD) and Class 1 Estimate was 31 weeks, about seven months. However, like all stages and projects, the team developed an updated estimate for the stage based on the best available information at the time. This estimate could have ended up requiring more or less time than the SAD time estimate, which would have been perfectly acceptable, rational, and expected, as long as the variations were within a reasonable tolerance, which would be expected to be about 15% to 20%. Let us say that the stage duration estimate was mostly in line with the schedule.

Cost Estimate

For reference, the cost, per the SAD and Class 1 Estimate, was $850,000, without the management reserve. As with the schedule, the team developed an updated estimate for the stage based on the best available information at the time. This estimate could have ended up requiring more or less money than the SAD cost estimate. Based on the experience of the previous stage, the team did increase the forecast to $1.1 million. This estimate overrode the last estimate stage and also affected the overall project estimate.

Human Resources

At this time, the team developed the detailed human resource requirements for the stage work, including roles and responsibilities, organization chart, and staffing plan. These plans were for the internal resources and the work done by them. The project management team (PMT) continued to manage the overall work, including the contracted work. The consultant(s) developed their own human resource plans.

Define Procurement

The team utilized two specialized consultants, one for the Airport Master Plan, and one for the Port Master Plan. These were necessary because there was not enough expertise in the existing government organizations. For knowledge transfer and minimizing cost, the Prime Minister's Office requirement was to embed many of the Ministry of Transport personnel with the chosen consultants. There were seven individuals selected for these roles. The team decided to use

> **Copy–Paste–Modify**
>
> Due to technology, Google, and many other online resources, many students and professionals have joined the copy–paste culture. In theory, this can be an acceptable practice; however, in practice, some often copy and paste without proper analysis or consideration of the different contexts. This is why we like to stress the need to copy, paste, but also *modify* to accommodate the varying context and conditions. This is what we label *CPM*.

a cost-plus, reimbursable cost contract with incentives. This was similar to the type of contract used before for the feasibility consultant. The agreement also included a clause for maintaining the team and rates for two years, since it was possible that the winning consultants would be considered for additional work in the upcoming stages. The team finalized the selection criteria, which was copy–paste–modified from historical data and the organizational project management system.

Other Considerations

Most of the focus in the detailed plan above was on the Project Master Plan work. However, at this timing on the project life cycle, it was crucial to start working on identifying potential investors. The project executive appointed a commerce and finance manager and a couple of assistants to begin working the financial and incentive package that would be presented to prospective investors. These individuals reported to the business director.

Process Gate 3: SDP Approval

☒ Approved ☐ Revise ☐ Hold ☐ Not Approved

29.2.3 Implement Stage Work

Master Plan is a term often used in facilities projects, but which may be more applicable to real estate, urban planning, or similar projects in which there is a large area of land required for the project. The dictionary uses the following definition: *"To develop or improve (land, a community, a building complex, or the like) through a long-range plan that balances and harmonizes all elements"* (Dictionary.com, 2015). In general, a Master Plan could be detailed or semi-detailed, or what one would call a *block diagram.* In either case, the Master Plan had to show the overall plot layout for all the Integrated City Megaproject's (ICMP's) facilities, including the residential communities, railroads, and roads corridors, in addition to the port and airport.

Conduct Procurement*

The team finalized the scope of work for each of the packages, port, and airport. A team from the Ministry of Transport was selected to lead the procurement process and management of the contracts. The team collected available information from the organizational records and also conducted market research to identify the leading providers for these types of services. As a result, the team was able to identify four potential providers for the port work and six providers for the airport work.

One of the conditions in the pre-selection was regional experience. The team sent out the request for proposals for these organizations and allowed three weeks for the proposal work. The request included the requirement for potential future work for the preliminary engineering stage. The project director appointed a bid evaluation and review team for managing the selection process. The proposals arrived as required, and the team went through the evaluations; they had allocated 80% for the technical and management expertise for this type of work and only 20% for pricing. However, the team also made it clear to all bidders that the selections were not based on the lowest bids.

* I am including a bit of story-telling here to help visualize how to avoid some of the major challenges that typically face megaprojects, which we discussed in Part A.

For the port project, the result was that the lowest bidder was also the most qualified, which was a uniquely ideal situation. The team was concerned because this was a Chinese company. However, after the proper due diligence and negotiation, the team awarded the contract to this provider. They had the right experience and price.

For the airport, it was a tougher selection process. The first round of reviews resulted in two bidders emerging as leaders. One was better technically, whereas the other had a better price, which created a dilemma for the team. The project director and project executive were consulted, and the decision was to use the "lowest conditioned bid" approach. This method adjusts the bid pricing due to factors the team considers relevant and critical. That brought the price of the lowest bidder higher than the other company. In the end, the team decided to go with the higher-priced provider because of their extensive experience in the country, and they were currently working on another airport project in the region. After negotiation, the vendor agreed to adjust their price down by about 5%, and if they won the subsequent preliminary engineering work, they would add another 5%, retroactively. Cost and schedule-wise, both contracts were within the estimated tolerances.

29.2.4 Control Stage Work

Control Within the Stage

This topic has been covered extensively already and there is no need to repeat.

Project Control

In this stage, project control is against the last Control Reference Point, which would be the Project Authorization Document and the Class 1 Estimate.

29.2.5 Close Stage

The stage close-out report would be similar to what we have covered already.

29.3 Stage Deliverables and Stage Gate

For this stage, the plan was to have two significant deliverables: the main one—the master plan—but also a commercial and financial plan. This second plan could be part of the shaping process and stage but is included here for discussion purposes.

29.3.1 The Project Master Plan

The Master Plan included many schematics and drawings, starting with showing the location of the Integrated City on the map of the country and southern region. This would be the first level. Next, there would be second-tier schematics showing the location of the main blocks, which were: the port, the airport, and the community. Notice the use of the word "main," since no details were available. The third level showed the block diagrams for the central facilities; as in the city, they displayed the location of single-family housing, multi-family housing, singles housing, common areas, parks, educational centers, and clinics/hospitals, but no more details. Further detail would be at level 4. Please note that levels 1 to 4 mentioned above are only for illustration; they might be different, depending on the organizational context.

29.3.2 The Commercial and Financial Plan

In general, what was required here was to address all commercial and financial aspects to solicit and acquire investors. This was for individuals or organizations, and whether they invested cash in the general funds or direct investment into specific components or partial components of the ICMP. To achieve the above objective, it is worth reflecting back on the earlier work, including the Feasibility Study, which concluded that this project was feasible despite the numerous challenges. The conclusion was on the conditions that:

- The government would provide 30% through direct cash investment, funds that come from investors through cash through bonds, or similar platforms.
- Another 45% was expected through BOT/BOOT or similar mechanisms.
- The remaining 25% would be from the government reserve, which was possible based on $40 per barrel for oil at current production levels.

In other words, up to 75% of the required funds needed to come from outside the government funds, from cash or institutional investors. Consequently, the team had to develop an investment program. Figure 29.1 shows the various packages that composed the components of this program. During this stage, the team finalized all of these packages and put them into action in the next stage, to mobilize the investors before initiating the Preliminary Engineering Stage.

The packages were:

- Tax incentives package
- Local employment programs
- Investment options
- Commercial models
- Investor selection process

We won't go into detail here and will leave it to the reader's imagination.

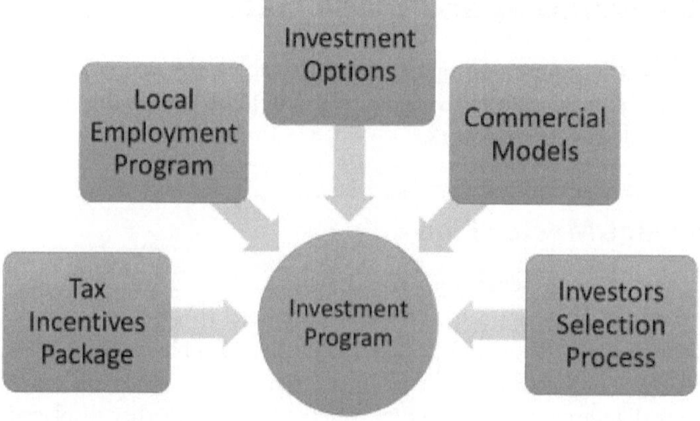

Figure 29.1 Commercial and Financial Plan, Investment Program

29.3.3 Stage Gate 4: Master Plan Approval

☒ Approved ☐ Revise ☐ Hold ☐ Not Approved

Explanation of Decision

The stage gate approval is at the Prime Minister's Office and includes all of the relevant stakeholders. The plan was approved after a few modifications, and the decision was made to continue with the Project Development Phase and develop the Project Management Plan. With the Project Management Plan, the team also developed the Class 2 Estimate.

29.4 Chapter Closing Comments

We continue to remind the reader that our focus is on the project life cycle, the tailored methodological approach. We are using concepts, such as master planning, to maintain a realistic simulation. We are also touching on stage management. In other words, I think it is becoming clearer and clearer that the CAMMP™ tailored approach is a product delivery and project management model. We progress from idea and Project Brief, all the way to closure and beyond.

As teams move along the project life cycle, you should be noticing that, so far, most of the work is happening along the facilities stream but is also tightly integrated with the governance (business stream) and to a lesser degree (so far) with the production stream. You should even be noticing the need to manage the overall work, at the project level, per the last defined control reference point. Also, concept to success means managing (including controlling and change management) one stage at a time.

Chapter 30

Strategy Stage

30.1 Introduction and Stage Purpose

The Project Management Plan (PMP) was the main deliverable of the Strategy Stage. Per the customized and adapted model for the ICMP, the Strategy Stage started after the completion of the Requirements Stage, but it could have started before completing the Master Plan. There were four major actions that we will highlight here:

> **Total Cost**
>
> Total installed cost, total estimated cost, or just total cost are all terms that refer to the expected cost of the project upon completion.
>
> Depending on organizational context, this cost could be the cost from idea to closure, or cost of delivery only (from point of funding), or maybe from project authorization to completion, excluding cost of the discovery phase.

1. CAMMP™ MP for the ICMP included an initial strategy assessment with the Feasibility Study. Then in the Strategy Stage, there were many things happening, including updating the various strategies, which would lead to the PMP, inclusive of the Class 2 Estimate. In every stage, we developed a Stage Management Plan. Those are the plans for managing a given stage. Now, it was time to develop the management plan for the whole project; or maybe a more accurate term would be the *management of the remainder of the project,* which in terms of the cost could be around 97% of the capital cost.* After the completion of the PMP, the effort and expenditures increased significantly—hence the need for a proper PMP at this point in the project life cycle. Although the primary focus at this time was the development of the management plan, there were other stage deliverables and objectives.

* Up to the end of the Strategy Stage, the total cost expended could be 1% to 3% of the total estimate cost.

2. As we said, the PMP included the Class 2 Estimate, which would be more detailed and accurate than the earlier Class 1 Estimate. This estimate was critical to update the project feasibility and financial situation but still not comprehensive and accurate enough to justify full funding of the project.

3. Another purpose would be to request *advance funding* with the stage gate approval. This advance funding would be necessary to cover the cost from that point until final approval. To reflect on the real world, the expenditures to this point on the project life cycle were low—maybe less than 1% of the estimated total cost. However, the next stage was the Project Preliminary Engineering Stage, in which expenditures would dramatically increase, potentially reaching as much as 3% or 5% of total estimated cost, assuming no early purchasing of long lead items.

4. The final objective of this stage, which was unique to this type of project, was that the commerce and financial team would also work on the investment program toward mobilizing the principal investors before the next stage. Although this item is listed as the fourth item, in reality, this started as soon as the stage was initiated. In other words, not all four major deliverables were sequential. Keep in mind, if an organization adopts the CAMMP™ MP tailored approach, this last objective will be part of the Shaping Stage.

In the end, all of these deliverables, with the Master Plan and the Project Requirements Document, came together for a critical decision: the business decision, as Merrow (2011) suggested and we covered in Part A.

In the end, all of these deliverables, with the Master Plan and the Project Requirements Document, would come together for a critical decision: the business decision.

30.2 Special Message

The PMP needed to be comprehensive and detailed enough to allow the Project Management Team (PMT) to plan the upcoming stages and develop the Class 2 Estimate, per the organizational guidelines. Although this plan would be detailed, it would not replace the Stage Management Plans for the future stages. This PMP, although with some details, would be a higher-level plan than the stage plans, which would be more detailed and specific to those stages.

30.3 Stage Sequence

Same as before.

30.4 Process Deliverables and Process Gates

Same as before, we will only highlight a few points, with a bit of storytelling time.

30.4.1 Authorizing the Stage

Per the Class 1 Estimate, the duration for this stage was just about five months.

Per the Class 1 Estimate, the cost for this stage was combined with the Master Plan because, in the Feasibility Study, the team had considered those as one stage. Now they were split, and the cost had to be split. Based on the early assumptions, $650,000 was the amount allocated to this work, excluding the management reserve. However, in the previous stage, it was determined a best practices consultant was needed, and the level of effort was underestimated. The project executive authorized $1 million, noting that during planning, the revised stage cost would be calculated, and the project executive anticipated it to be even more than $1 million.

30.4.2 Plan Stage Management

Same as before.

30.4.3 Plan Stage Details

Please take a moment to reflect on Part A, the Strategy Stage section for information about the details of the PMP, which is the output of this stage. If you recall, the output had four components: the strategies section, the how-to section, the class 2 section, and the people aspects section. All of those are now for the whole project. Therefore, the Stage Detailed Plan will highlight the detailed plan for developing the PMP. We do know that the plan could be confusing. However, it is important to realize that although the output of this stage (the Project Management Plan) has the word *plan* in the name, the work to get there is a stage that would require more than $1 million and many months, hence the need to apply the Stage Management Processes even for this stage.

Schedule

Like all stages and projects, the team developed an updated estimate for the stage based on the best available information at that time. This estimate could have ended up requiring more or less time than the Stage Authorization Document (SAD) time estimate, which is perfectly acceptable, normal, and expected, as long as the variations are within a reasonable tolerance, which would be 15% to 20% at that time in the project life cycle. *For reference, the schedule from the SAD and Class 1 Estimate was about five months.*

Cost Estimate

The cost per the Class 1 Estimate was $650,000, but the project executive raised it to $1 million in the SAD. The team developed the stage estimate, and as expected, the estimated cost was $1.3 million. This was only for the "traditional" work, which was the Management Plan, the Class 2 Estimate, and advance funding, but did not include the money needed for the commercial and financial team, which was estimated at $900,000. The expected requirement for this account (commercial and financial work) was $3 million, but only $900,000 was required for this stage, and $2.1 million for the next stage.

30.5 Stage Deliverables and Stage Gate

30.5.1 *Project Management Plan*

Introduction

For the purpose of the discussion, we have separated the Class 2 discussion from the PMP—in other words, treating it as a separate deliverable. We will continue to truncate these deliverables.

Strategies Section

Project Management Strategy

The management strategy is simple: follow the organizational policies and guidelines. However, due to the size and complexity of the project, the government did not have the expertise or resources required to lead the delivery of such a large and complex project into successful completion. Consequently, the PMT recommended, and the steering committee approved, hiring a project management services company to help lead the effort. However, the decision was to use an integrated team concept, in which project management personnel from the government and the consultant would work together as one team—a full integration.

Project Procurement Strategy

There are many significant components in this project, such as the airport and port. Therefore, the strategy is per the following guidelines:

- The project work would be subdivided into multiple contracts.
- Where possible, the team would use engineering and procurements (E&P) contracts on a cost-plus basis and construction (C) contracts on a fixed-price basis.
- In some situations, it was possible to use EPC (engineering, procurement, and construction) contracts to minimize the coordination and errors between engineering and construction,

We are only highlighting a few key points here. In the real world, there would be significantly more details for each of the components and packages.

"How to" Sections

How to Define the Project

This first item is specific to how the team would develop the Project Detailed Plan, using the CAMMP™ MP standard term, or **how to develop the preliminary engineering packages** (using the ICMP case).

The answer!

The project hired different specialized design firms to handle the various preliminary engineering packages for the ICMP components, except those that would be done through build–own–operate–transfer (BOOT). The investment team identified potential BOOT investors who would be determined by the end of this stage. In other words, BOOT investors would handle their own preliminary engineering, whereas specialized consultants would handle the remaining work. All of the work, regardless of whether contracted or BOOT, would follow conventional international design standards, where applicable.

How to Manage the Project

The project management strategy was highlighted earlier. To elaborate on the strategy, the project management services company was mobilized at the start of the next stage with a small team, then expanded their team as necessary once the project was fully approved and ready to go into full implementation mode with the award of contracts. However, the team would be integrated with personnel from the government.

One of the first activities that the project management consultant did is to review the government NPMS to identify any gaps that needed to be filled. Further, they identified any inconsistencies between the government and the consultant systems that they would have to bridge. In all of these reviews and improvement efforts, the NPMS manager was directly involved and functioned as the liaison between the integrated project management team and the NPMA. Once the alignment was complete, processes and systems updated, the integrated project management team followed this updated NPMS. The integrated team led the procurement effort to the next stage.

> **NPMS Manager**
>
> The NPMS Manager is a position specific to the project. This manager is an employee of the National Project Management Authority (NPMA), and the mandates of this project-specific role are:
>
> - To be involved in the development of new project management processes or update of existing ones.
> - To capture the learning from best practices implementation to incorporate into the national project management system (NPMS).
> - To act as the lead liaison between the project and the NPMA.

How to Control the Project

The integrated project management team (IPMT) included dedicated resources for project control, including all aspects and not limited to cost and schedule. Besides cost and schedule, the IPMT was responsible for controlling quality, safety, risk, and all other functions. Control was an ongoing effort, meaning daily monitoring and controlling at different levels. Contractors were responsible for the control of their contracts' work but maintained a direct reporting relationship with the IPMT. Obviously, the depth of control and level of involvement by the IPMT was a function of the type of contracts used for the various components.

How to Handle Change on the Project

Although the project was not entirely funded yet, a strict change management system was enforced during the Preliminary Engineering Stage and post final approval. The Preliminary Engineering Stage was a critical stage, because it was a transition stage from the Master Plan and requirements to detailed engineering. At this time the team expanded the scope details due to design development and progressive elaboration. Yet, the scope details were still not very clear; they were still foggy, although the fog was clearing.

Due to these conditions, the Preliminary Engineering Stage was fertile ground for **scope creep**; the IPMT needed to be diligent in controlling the scope and preventing such scope creep. If a scope item were to be added, it would need to go through the project change management system. The steering committee functioned as a change control board for any objectives change; these changes to the project authorization and objectives were established during the Feasibility Study.

Other changes that did not alter the overall objectives were handled by a Change Control Board (CCB) within the IPMT, led by the project services manager, with participants from engineering, construction, operations, planning, and cost control. Also, change approval, for all changes, were per two approval steps. When a change was requested, the CCB could reject the change or approve it for the evaluation only, unless the change was related to HSE (health–safety–environment) or to cover a clear error that would prevent the proper operation of an ICMP component. If the change was approved at the first step, it would be evaluated from all angles, and the impact presented to CCB. At that time, the CCB could make a final decision. In addition to all of the above, the team would utilize, to the maximum extent possible, the Change Management Best Practice from the Construction Industry Institute (CII, 1994 and 2015), since this was one of the requirements identified in the Project Requirements Document.

How to Get Ready for Operations

Similar to project definition, the operational readiness for the BOOT component would be handled by those investment companies in collaboration with government personnel. For the other work, the ICMP project mobilized an operational readiness team, starting with the preliminary engineering stage. This was headed by a director-level position reporting to the project executive and working closely with the project director. All of the ICMP components had similar facilities that already existed within the country, such as ports, airports, railroads, educational centers, and other infrastructure facilities. Therefore, the project team utilized these organizations to help get ready for operating the various ICMP components. The structure of this team is not shown here. In reality, this would consist of the numerous teams, each responsible for an element or a functional area (such as electrical, civil, controls systems).

How to Close the Project

A small team stayed post completion of the construction and handover to finalize the project close out. Also, a closeout manager was assigned to the project from the start of the next stage onward. The closeout manager began by collecting the closeout report from the stages completed to date, identifying any gaps and addressing them as the manager developed the requirements for closeout for each upcoming stage. This was necessary to facilitate the final project close out. Also, the closeout manager identified a few requirements that the project management team could follow in their daily monitoring and control that would also facilitate the final closure. The project metrics were also essential items to utilize.

It could be logical to have the NPMS manager take on the role of the close-out manager, or at least be part of the closeout team.

People Sections

Project Governance Structure

Due to the size of the project, management, proper governance, and control structure was needed. Project governance would be accomplished at two levels: Level 1, the high-level reporting and steering committee; Level 2, direct project oversight with the executive committee. Figure 30.1 presents the project governance and reporting structure. Per this image, the steering committee reports directly to the Prime Minister's Office. There is also a link to the NPMA and Strategic Planning Authority. The executive committee is also shown here, reporting to the steering committee. The executive committee would be responsible for all major decisions that did not have to go to the steering committee. It is important to note here, that once the project was approved, the steering committee membership could change to include top decision makers and possibly a few ministers. A minister would head the steering committee.

Next, Figure 30.2 shows the structure of the executive committee, which consists of the project executive, who is its leader. The committee also includes the various directors on the project. These are the project director, the business director, and operations director. This committee may be extended later to include the directors/sponsors from the BOOT components. Figure 30.3 is a sample high-level structure of the project leadership, the facilities stream under the project director.

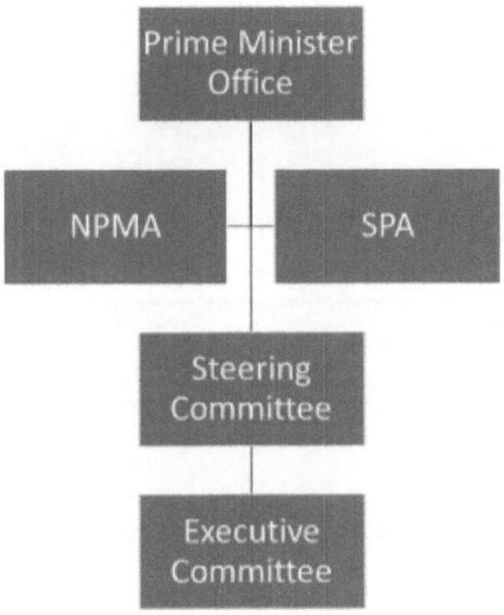

Figure 30.1 The ICMP Governance Structure

Project Management Team

We can continue to expand on each of the above branches, but we believe other resources would be more useful references. Both Oberlender (2014) and Merrow (2011) provide good information on this topic. We only show a possible structure of the facilities team to help the reader visualize the integrated project management team. The integration is demonstrated by identifying if the holder of a given position is from the government or the project management services company.

30.5.2 Class 2 Estimate

Cost Estimate

The project progressed quite well over the project life cycle since the Project Brief. During the Feasibility Study, the project management team developed the Class 1 Estimate based on the feasibility assumptions and the low level of project details at that time. Since then, the project team had developed the

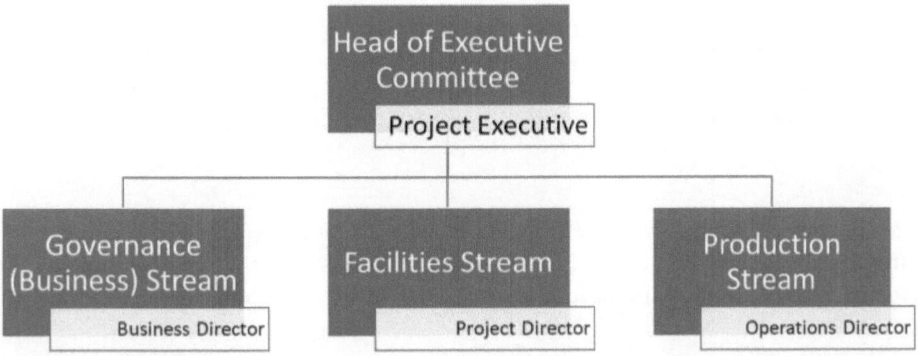

Figure 30.2 Executive Committee

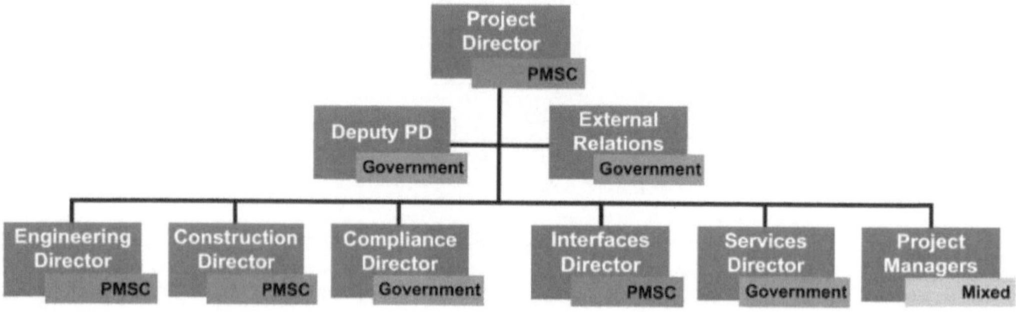

Figure 30.3 ICMP Integrated Project Management Team, Leadership Team

Project Requirements Document and the Project Master Plan. Also, project management planning was ongoing and finalized in parallel to this estimate. In other words, significant work has been done to allow the team to produce the Class 2 Estimate. The project management team's estimating group provided the estimate shown in Table 30.1.

Table 30.1 Comparisons of the First Two Estimates

Item/Area	Class 1	Class 2
Discovery Phase (Feasibility)	2.20	2.20
Requirements Stage	1.00	1.40
Strategy Stage & Master Plan	1.50	-
Master Plan Stage	-	1.20
Strategy Stage	-	2.20
Preliminary Engineering Stage	90.00	85.00
Project Management Services	-	140.00
Implementation Stage (Engineering)	400.00	382.50
Implementation Stage (Construction)	5,000.00	5,328.00
Operational Readiness Stage	350.00	300.00
Close Stage	1.00	1.00
Subtotal All Stages	**5,846**	**6,244**
Management Reserve	1,461	1,249
Total Cost	**7,307**	**7,492**

Please note that the Class 2 Estimate included actual costs for the work already completed. For the upcoming stage, we show the estimated cost but updated it based on the latest information. The key points to highlight at this time are the following:

- Per the estimating guidelines, contingency and management reserves should have been brought down to 15%, each. However, the project director recommended, and the project executive agreed, to keep these at 20% due to the size of the project and the lack of experience in similar work.

- Like the Class 1 Estimate, the contingency reserve is included in each line item, whereas the management reserve is kept as a separate line item.

Please note, the subtotal costs increased as expected, due to more details, but most of this growth was compensated for with the lower contingency reserve. The net change was about a 7% increase, as shown in the subtotal line in Table 30.1. Then, lowering the management reserve reduced the total cost growth to about 3% higher than the Class 1 Estimate (bottom line in Table 30.1). Part of this increased cost was due to the addition of the project management services consultant and other costs. It is worth noting that, excluding best practices estimated savings, the total cost of the project would have been estimated at about $8 billion, but due to the expected savings from the applied best practices,* the estimate was close to $7.5 billion.

Schedule

Without going through the details, let us say the schedule was also updated and is shown in Figure 30.4. This schedule is about three months shorter than the Class 1 Schedule (see Fig. 26.4 on page 279).

30.5.3 Advance Funding

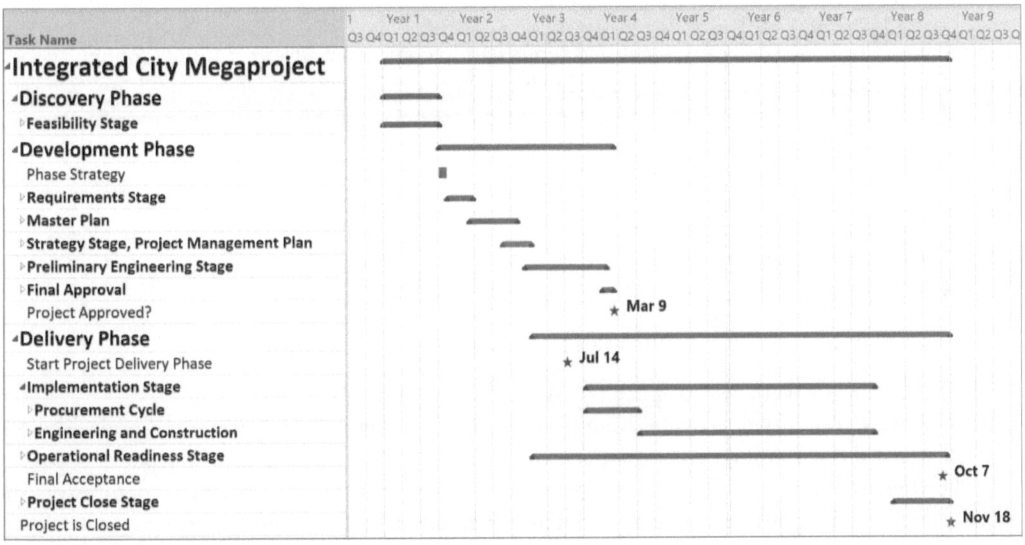

Figure 30.4 ICMP Class 2 Estimate, Summary Schedule

The advance funding was required to sustain the work until the final approval of the project at the end of the preliminary engineering stage. This was necessary since the government guidelines did not allow expenditures for projects beyond this point without advance funding, primarily because the amount of money far exceeded typical operating budgets. The amount required would be to cover:

* If you recall, earlier on we discussed that this megaproject would be using best practices. The story here is telling us that these best practices have produced a $0.5 billion in estimated savings, or about 6%.

1. All costs spent to date, including this stage
2. The estimated costs for the project preliminary engineering stage
3. The estimated costs for the procurement work, which would be in parallel to the preliminary engineering work
4. The estimated costs for the identified long lead items

The organizational guidelines allowed an authorization of up to 3% of the estimated total cost of the project (excluding management reserve), which in this case, it would be about $180 million. The team estimated needs were about $130 million; therefore, $180 million would be more than enough.

30.5.4 Investment Program

As outlined earlier, what was required here was to implement the commercial and financial plan developed in the previous stage, with the ultimate objective of mobilizing investors. To achieve this objective, it is worth reflecting back on the earlier work, including the Feasibility Study. The Feasibility Study concluded that this project was feasible despite the numerous challenges. The conclusion was on the conditions that:

- The government could provide 30% through direct cash investment.
- There would be 45% through BOT/BOOT or similar mechanism.
- Also there would be 25% from the government reserve, which was possible based on $60 per barrel for oil at current production levels.

In other words, up to 75% of the required funds had to come from investors if the economic conditions did not change. After discussions with the Prime Minister's Office, the above conditions were still applicable, and that set the targets for the team. The commercial and financial plan required:

1. Tax incentives package
2. Local employment programs
3. Investment options
4. Commercial models
5. Investors selection process

Let us say that the team implemented this plan according to the following:

1. The team was able to get tax incentives from the government for 20 years.
2. They were committed to establishing a professional development program to minimize the cost of recruitment, training, and expatriate workforce.
3. Further, the team prepared various investment packages to attract different types of investors and included a small business administration to promote start-up and entrepreneurship initiatives.
4. They created various commercial models for investors to select from.
5. They finalized the investors' selection criteria and open the doors for investment through international conferences and events.

The net result was that the commercial and financial team were able to mobilize a few major investors at this time, the rest to be done in the next stage. One major investor took on the airport component as a BOOT (Build–Own–Operate–Transfer). Another investor committed to the same scheme to build three power plants, one each for the city, airport, and port. The power plants had spare capacities that could be sold to the power company or other local users.

30.5.5 Stage Gate 4: Project Management Plan Approval

☒ Approved ☐ Revise ☐ Hold ☐ Not Approved

Explanation of Decision

The PMP and Class 2 Estimate were reviewed and approved after a couple of iterations. With a few investors already committing, the Prime Minister's Office authorized 3% of the total estimated cost to allow the project to move to the next stage.

30.6 Chapter Closing Comments

As you will have noticed, this was a critical stage, so let me summarize the vital actions that have been achieved as of this point in time:

1. The project's requirements and master plan have been defined and used for the PMP.
2. The Shaping Stage, in the governance stream, has been completed.
3. The Class 2 Estimate was finalized.
4. Using the above information, the business team was able to update the business case and feasibility of the project and decided to recommend continuing with the ICMP.
5. The above allowed the steering committee to accept the teams' recommendation and finalize the business decision about proceeding and gained approval from the Prime Minister's Office.
6. The Prime Minister's Office also authorized the necessary advanced funding to cover the preliminary engineering work.
7. With approval to proceed, the team could now shift the project control reference point from the Project Authorization Document and Class 1 Estimate to the PMP and Class 2 Estimate.

Chapter 31

Preliminary Engineering Stage

31.1 Introduction and Purpose

The Preliminary Engineering Stage is the last stage in the Development Phase. This stage is in the standard CAMMP™ and CAMMP™ MP models, but under the name Definition Stage. In some domains, organizations treat this stage as the final stage of the front end and focus on "execution" planning. The output in the CAMMP™ model is a Project Detailed Plan (PDP); however, in this case, the PDP would include the preliminary engineering design. For ICMP, we had five major components (objectives). These are:

1. Develop and finalize the preliminary engineering packages for the various components. In this stage, the consultants performed the preliminary engineering design for the various facilities, such as architectural design for the building, process design for power plants and other utilities, and major equipment design for port and airport, among numerous other activities.
2. Another objective of this stage was to continue to implement the commercial and investment program plan and mobilize other principal investors—a continuation of the previous stage.
3. The third objective was to develop the Class 3 Estimate, which would become the control estimate during the Delivery Phase all the way to completion. This would be the official project budget once final approval was obtained.
4. In parallel to the above, the operational readiness team developed a detailed plan for the operational readiness activities and initial operations.
5. The final objective was to update the Feasibility Study and obtain final project funding if the project was still feasible and would be continuing.

31.2 Stage Sequence

Because the sequence and process were identical to the previous stages, no need to repeat. The only difference is the implementation work.

31.3 Process Deliverables and Process Gates

Same as Stage Sequence, no need to repeat, just remember, they are needed. Also, as we proceeded along with the project life cycle, this was the "biggest" stage so far in terms of effort, duration, and cost. Therefore, the stage management planning effort was significant. The size and complexity of this stage on its own could be classified as a large and complex project—close to $100 million if we include the reserves.

31.4 Stage Deliverables and Stage Gate

31.4.1 Preliminary Engineering Packages

For more information about this stage, please refer to Part A. We are making the assumption that our readers have a good idea about this type of work. For those who might not have 100% clarity, we include a brief here.

Project definition (as we use in CAMMP™) is about developing a detailed plan for a given project. A thorough plan in the CAMMP™ context is about the details required for all aspects, such as the comprehensive scope, detailed quality management, detailed schedule, and the other project management functions. Looking at project definition differently, it would be the last stage before the final funding and before project delivery. In other words, delivery depended on this work and it is used as the control reference: control for all aspects, mostly scope but also time, cost, and quality. Further, this work forms the basis for the first dimension of project success—product delivery success (refer to Chapter 17 on project success).

Bringing the above concepts to the ICMP or a capital-intensive project, project definition would be equivalent to preliminary engineering, also known as FEL-3 in the IPA world (Merrow, 2011). Others use the terms *Front-End Engineering Design* or *Package, FEED* or *FEEP,* respectively. The preliminary engineering work can vary from one type of project to another. For example:

1. In real estate development, this stage work would be mostly about architectural design work. Preliminary engineering would be about expanding the Master Plan and providing the architectural design details.
2. In airport design, this could be finalizing the layout for the airport facilities, along with the industrial and architectural designs for the various parts. The design includes storage, loading and offloading system, and similar items.
3. In petroleum and process facilities, this is often called Front-End Engineering Design (FEED) and would be to finalize the plant layout, inclusive of all major equipment and major systems.

For the ICMP, there were multiple contractors working the different preliminary engineering packages for the various components. There was a package for the port and its associated facilities, another for the airport and its related facilities, and the same for all other components.

31.4.2 Class 3 Estimate

Cost Estimate

By this stage, the project had progressed quite well, which allowed the team to produce the Class 3 Estimate, which would be a detailed estimate used for the updated Feasibility Study and financial model; once the project was approved, this estimate was converted into the budget, time-phased budget, and performance management baseline. These were used for control of the approval appoint to the end of the project. Since there were BOOT investors at this stage, they did their own estimates to incorporate into the overall project estimate.

There's no need to re-explain the various concepts that have been outlined before or are available through other project management resources. The Class 3 Estimate total estimated cost should be close to the Class 2 Estimate amount if the team had done their work appropriately and the organization had enough relevant historical data. As we explained in Part A, the scope was more detailed, which led to higher certainty and more accurate estimating, justifying a reduction in the contingency and management reserves. Therefore, even if the known cost in Class 3 Estimate would be higher than the Class 2 Estimate, the increase would be compensated via the reduction of the reserves. Ideally, the total estimated costs of Class 2 Estimate and the Class 3 Estimate should be close; however, estimating is not a precision business, so we expected some differences between the two estimates.

The Class 3 Estimate included all direct costs, indirect costs, project management costs, organizational overhead, escalation rates, technology allowances if required, and the cost of money for financing and borrowing. At the time of final approval, contingency and management reserves should have dropped to 10%, although based on performance trends in the megaproject's domain, we were justified in increasing this amount to 15%.

Schedule

Similar to the cost situation, the schedule was also updated with important details, and there were quite a few changes from before. The number of activities in the Class 3 Estimate (schedule) were significantly higher than the number of activities in the Class 2 Schedule. Similar to cost, the schedule included reserve time and other factors to account for the various risks that the project would face if not mitigated.

Author Note

Let us assume that, in the context of this book, the amounts of the Class 3 Estimate and the Class 2 Estimate were close enough. In real life, they should have been close (±10%)—again assuming the cost engineers were highly qualified, and the organization had enough and relevant historical data and excellent project management processes.

31.4.3 Updated Feasibility and Final Funding

Based on the above, the team updated the Feasibility Study at this time, including all components, where applicable. Based on the revised market conditions and project costs and time, the financial team re-assessed whether the project was still feasible or not. Then they finalized the funding request for final approval. The application for final approval included the total cost broken down to highlight the share of each of the various investors.

31.4.4 Operational Readiness Planning

As we explained earlier, in the CAMMP™ standard model, this work would be included in the project detailed plan and led by the project director, working with personnel from the operational divisions.

However, for CAMMP™ MP, we have a parallel stage dedicated to this scope. Therefore, in the real world, this is an independent stage that is part of the production stream and led by the operational readiness director. However, to minimize the repetition, we include this item here.

The objective of the operational readiness planning was to start to plan for the upcoming stages that would be happening in parallel to the engineering and construction work; these would be the Operational Readiness Stage, and once a given facility was handed over, that would trigger the start of the Initial Operations Stage. The ICMP was a complex initiative with many components. Each of these components could be a standalone megaproject and requiring its own dedicated team. However, a centralized, coordinated effort could enhance the efficiency and effectiveness of this critical work. We had highlighted in Part A more than one case study of megaprojects completing their facilities but being unable to move into operations because the operational readiness work was not complete.

31.4.5 Investment Program

This is a continuation of the prior stage work, and all had to be finalized before the next stage gate, when the final approval was expected.

Stage Gate 5: Final Approval

☒ Approved ☐ Revise ☐ Hold ☐ Not Approved

Explanation of Decision

Final project approval was granted after all investors and government agreed to the various items, including cost sharing, profit sharing, investment, timeline, and all other factors. This approval could take weeks or even months. In the government, this is definitely a decision for the council of ministers. In private companies, this would be a board of directors' decision.

31.5 Chapter/Section Closing Comments

In Part A, we had mentioned that simplistically, we can divide projects into front end and execution. With this stage complete and approval granted, the front end was complete, and the project reached the final approval point and was sanctioned to proceed into execution, or what we call the Delivery Phase. On a project like this, it could take a few years to reach this point. In the schedule that we showed earlier, we allocated 38 months from the start of the Concept Stage to final approval. Merrow (2011) stated that the average project in the IPA database required about 23 months for the equivalent of the CAMMP™ Development Phase. Adding the Discovery Phase, what we presented would be a realistic simulation. Per Merrow, the upcoming work would be about 43 months for the average project. For the ICMP, we showed this work to be more than 50 months due to the size and spread of the project. If we expand the duration to include all the way to closure, then we are showing more than 65 months.

With this, we close this stage and the work on the Development Phase. Next, we will discuss the Delivery Phase.

Section IV
The Delivery Phase

Chapter 32

Introduction to the Phase

32.1 Overview

The Delivery Phase was the third and final phase of the CAMMP™ project life cycle. It was about delivering the project per the Project Detailed Plan. In the case of the Integrated City Megaproject, project delivery was in line with the preliminary engineering packages and implementation plan. The Delivery Phase encompassed implementation, operational readiness, initial operations, and close stages. For CAMMP™ MP, we had shared the project life cycle earlier in Part A and at the start of Part B.

32.2 Purpose of Phase

During this phase, the project director and the integrated project management team managed and controlled the implementation work of the project to complete the work packages that had been defined during the earlier stages—in particular, per the preliminary engineering packages. For the ICMP, implementation was about performing the engineering, procurement, and construction work up to the point of handover (turnover) of the product (components) to the operating organizations. Some work, typically fix-ups, would remain beyond the handover point.

Further, in parallel to the implementation, the operational readiness team prepared and performed all tasks to accept the transfer of custody from the project personnel at the handover point. With the transfer of custody, the operating organization started up (commissioned) the new facilities, then moved into a period of "initial operations" until the final acceptance point.

As part of Delivery Phase, CAMMP™ mandates including project closure as a stage within the phase. This is necessary since, from the performing organization's perspective, the project is not delivered until it is properly closed, even though a client might have already accepted the work. The client, in this context, was the government through the operating organizations and relevant authorities.

32.3 Author Note

In terms of the project management process, the upcoming stages within this phase are identical to the earlier stages. Therefore, we will not repeat the discussions on the various stage management processes. The author will only add comments where necessary to explain what is relevant to this simulation. In terms of the Implement Process in each stage, these are purely technical activities and not project management specific. For example, in engineering, the Implement Process is about carrying out the engineering work, and in operational readiness, the Implement Process is to perform the various activities that are necessary to accept the custody transfer. We trust this truncated work is still clear enough for the reader to understand the core objectives of this book, as we had outlined them earlier.

Chapter 33

Implementation Stages

33.1 Introduction and Stages Purpose

In the CAMMP™ standard model, implementation is one stage. However, in the CAMMP™ MP, we typically split this stage into an Engineering and Procurement Stage and a Construction Stage. We will be discussing these two stages together, but as we said before, the discussion will be extremely truncated.

The purpose of the Engineering and Procurement Stage was to perform all detailed engineering work for the various components and purchase the necessary equipment and engineered materials. Construction, on the other hand, included awarding sub-contracts, performing the construction work and pre-commissioning, and ultimately delivering the "products": the physical facilities. Construction included purchasing of bulks and other consumables. With this approach, procurement was technically split between the two stages.

In general, the sequence started with engineering. Once engineering had reached a reasonable level of progress, procurement began. Finish-to-start principles advocate that construction can begin after engineering is complete. However, on large and complex projects, especially capital projects and mega-projects, this approach would add significant time to the overall project duration. Therefore, it is quite common for these projects (and ICMP) to fast track, by starting construction as early as possible after the start of engineering. Usually, as soon as engineering achieve about 50% progress, construction can begin. However, as we highlighted in Part A, one of the problems that cause megaproject failure, as we saw in the North American Project case study, is starting construction too soon.

To deal with this problem, CAMMP™ MP re-introduced a stage gate with a focus on "readiness for construction" to trigger construction start. The above sounds like we were are treating ALL engineering and ALL construction as one block. In reality, we had various components in the ICMP. Further, for each of these components, we split engineering and construction into packages and systems that could be sequential or overlapping.

33.2 Process Deliverables and Process Gates

Since the implementation work was outsourced, authorizing these stages coincided with the contracting work and contract awards. However, keep in mind that not all work was done under contracts or with one contractor—hence the need for interface management. Further, the stage management was in two parts: the high level, with the owner team and project management services company, and the detailed, by each contractor. Comprehensive planning was left to the contractors, since this was mostly about the work that they would be implementing. The Control Process was like management planning, at two levels. Finally, the closing was also in two parts: the contractors doing their internal closing, and the owner team doing theirs.

33.3 Stage Deliverables and Stage Gates

33.3.1 *Detailed Engineering*

Discussion of engineering work is outside the scope of this book. Nonetheless, engineering work was a sub-stage, which meant the contractors handled their own project management, detailed at their level. At the same time, they were working and complying with the overall Project Management Plan that was led by the integrated project management team (IPMT). At the end of every engineering effort, whether all packages were released together, which was unlikely, or in partial releases (package by package), there needed to be a stage gate, which was to ensure that engineering had progressed enough to allow moving into construction. In cost-plus contracts, the project owner would have total control over this stage gate, whereas in the fixed-price EPC contracts, the project owner would not have full control. Therefore, for every ICMP component, the preferred contracting approach was to split the EP from C, which gave the project owner more control over the process and decision on readiness for construction.

33.3.2 *Construction*

Construction was similar to engineering. During the latter part of the construction work, each contractor performed their testing and pre-commissioning work. This is typically required pre-handover since, at the handover, the custody of the facility was to shift to the operating organization.

33.3.3 *Operational Readiness*

In parallel to the implementation work, the operational readiness activities, the pre-handover work, was ongoing and was needed before the next stage gate. The next chapter is dedicated to the operational stages.

33.3.4 *Stage Gate: Ready for Handover*

Similar to the gate between engineering and construction, this stage gate was not one stage gate for all work. It was a series of similar stage gates, one for each ICMP component, and even a system within these components. The principal purpose of this stage gate was to determine if all the work, construction, and operational readiness activities were completed, in a satisfactory manner, to allow for performing the handover review, which led to initial acceptance and transfer of custody.

33.4 Success Factors

Before we close this chapter, I would like to reflect on the literature review. Oberlender (2014) offers a good deal of information about engineering and construction. Similarly, the Merrow (2011) also provides excellent information. I will share a few pointers here from Merrow, from pages 297–298. *"Excellent construction management may be the difference between success and failure."* The critical success factors for effective construction management are:

- A modern safety system
- Sequencing of design and materials delivery
- Materials management
- Hiring of craft
- Productivity measurement and problem identification
- Construction management staffing
- Construction management systems

The Construction Industry Institute (CII) also offers best practices related to the implementation work, and some of them align with the previous set of bullets from Merrow. For example, materials management is a CII best practice, safety is another, and of course, there are Project Definition Rating Index (PDRI) assessment, constructability reviews, and change management.

33.4.1 Chapter Closing Comments

In terms of effort, resources, time, and cost, the implementation work was the most involved. However, this chapter was one of the shortest in this book. This is another way of saying, whether the project is small and simple or large and complex, and whether a stage is a few weeks long or a few years, from a project management perspective, the similarities are high. The main difference is the level of effort and the type of product related work.

A project such as the ICMP required hundreds of engineers working for about two years and thousands of construction resources working for many years. Without a doubt, this could have been a nightmare in terms of coordination and control; hence the need for real leadership. Just management was not good enough for the Integrated City Megaproject. As we said before, its complexity was one of the reasons we use the term "leading" in the title of this book.

Let us visualize the following:

A megaproject can be like starting a company that will grow organically in the first three to five years, then experience a huge and rapid growth, with staffing levels (direct and indirect) reaching the size of many large companies, with thousands of people who work and report to different organizations, often with different organizational policies and cultures. Therefore, despite having a centralized, integrated project team, such a team would not even come close in terms of power and authority as it would be in a mid- to large-size company.

The workforce, at all levels, would most likely be a global workforce with a range of nationalities, cultures, religions, and ethnic groups, which is another tremendous challenge. On one megaproject, we had all construction safety signs in four languages, and still, most of the labor force did not speak one of those languages. Therefore, we made sure that each foreman of every construction crew spoke one of the four project languages and the language of the crew.

In the Asian Project, mentioned in Chapter 7, there was a camp for some of the laborers on the island. In this situation, even the food had to cater to their different nationalities. During Ramadan, the project had to accommodate the fasting of Muslim workers.

If the megaproject leadership team does not pay attention, even to this level of details, the project could be disrupted with ethnic conflicts, strikes, and even riots.

Chapter 34

Operational Readiness Stages

34.1 Introduction and Stages Purpose

The Operational Readiness Stages are in the production stream of CAMMP™ MP. One stage was happening in parallel to preliminary engineering, which led to a detailed plan for the operational readiness activities. Next, in parallel to implementation, was the actual completion of the operational readiness work packages in preparation for the handover. Finally, there was the Initial Operations Stage. We are discussing these three stages together in another extremely truncated chapter.

The main focus of these stages was to transition from project development and product delivery into full operations as the team worked toward realizing the anticipated benefits. Of course, this transition work can be significant, as one will see below.

34.2 Stage Processes

So you do not forget, even for these Operational Readiness Stages, we still need the stage management processes of Authorize, Plan Management, Plan Details, Implement, Control, and Close for each of the Operational Readiness Stages. Since this work was led by the operations director, it seemed good to either have a deputy who had project management experience or borrow a resource from the National Project Management Authority (NPMA).

34.3 Deliverables and Gates

34.3.1 Operational Readiness Detailed Plan

We mentioned the detailed plan in the introduction. We also discussed it in earlier chapters in both Part A and B. Therefore, no need for repetition.

34.3.2 Pre-Handover Work

The pre-handover is a period on the project life cycle that typically starts with implementation and ends with the start of the handover review. The pre-handover works include various deliverables. For ICMP, these were major deliverables, requiring significant effort and time. These included, but were not limited to, the following:

- Putting in place policies if needed
- Developing operational procedures for the overall facility and all of its parts
- Developing maintenance procedures
- Recruiting the various operation's personnel
- Developing and delivering the necessary training for the operation's staff
- Performing testing and pre-commissioning work

Most of this work was by the operational readiness team assisted by integrated project management team (IPMT) and EP/C contractors. It is common for the pre-commissioning work to be performed by the contractor that is constructing the facility.

34.3.3 Stage Gate

In the CAMMP™ MP standard model, this is Stage Gate 7 (SG7). It is the stage gate requiring the completion of **construction activities and the pre-handover deliverables**. Approvals at this stage gate would officially allow the start of the handover reviews. As stated earlier, for a project like the Integrated City Megaproject, this stage gate consisted of multiple gates, one for each of the ICMP components and sub-components.

34.3.4 Handover Reviews

The purpose of the handover reviews is for the various relevant stakeholders to work together and review the work completed. This is necessary to verify that all work is done per standards, specifications, and other requirements. These reviews would be of documents and work on-site, reviews of quality control records, and final inspection of the completed facilities. This work should lead to initial acceptance (also called *provisional acceptance*). Initial acceptance is when the client (operation personnel) agrees that the deliverables are virtually complete and can accept the work per the acceptance criteria defined early in the project. The term *initial acceptance* is used because it is not the final acceptance.

Typically, at the end of the handover reviews, the operational readiness team will issue completion certificates as the team accepts to take over the custody of the product. This does not mean everything is perfect, but all things that are required for the start-up and commissioning are in place. Initial acceptance can be granted even where there are minor items that are not totally in line with what is required if these things are not necessary for the start-up. In this case, the team will create an "exception items list," or "snags list." Items on this list must be fixed and finalized before final acceptance can be granted.

34.3.5 Start-Up and Commissioning

Once the operational readiness team completed the transfer of custody, they started to commission and start up the facilities. *Commissioning* and *start-up* are terms common in facilities-related projects, and they represent the activities that would transition the project from construction into operations.

During this period, all systems (electrical, mechanical, safety) were tested individually and collectively, to validate the product completion and commence initial operations or full operations. These activities were the responsibility of the operation and maintenance personnel from the project owner organization, assisted by the IPMT, contractors, or specialized service companies.

34.3.6 Initial Operations

The Initial Operations Stage is often required for a short period and would be like regular (full) operations, but at a lower capacity. This is often necessary to help operations personnel become familiar with all aspects of the new facility; it is a learning period. As stated here, initial operations is often necessary but is not always required.

For the ICMP, it was needed for most components. To be more specific, initial operations was needed for port, airport, and railroads but not for the housing community. The duration of this period could be weeks or months, depending on the nature of the project. For ICMP, the airport and port initial operations lasted a few months. Next, full (normal) operations started at the normal capacity or as needed. It can also be the case where the duration will extend up to a year from initial acceptance. However, the year timing is often linked to contractual warranty rather than actual operations or final acceptance of the product.

Once again, let us reflect on the literature review. One of the challenges documented by Merrow (2011) and one of the Independent Project Analysis (IPA) five success factors is related to achieving beneficial production within the first year of operation. However, many industrial projects face many problems, which could have originated in the front end or engineering. For the ICMP, most of the projects were non-industrial but could still experience challenges; perhaps the airport proved to be the most complex component.

34.3.7 Stage Gate: Final Acceptance

Similar to previous stage gates, this stage gate is not one gate. It is a series of similar gates, one for each ICMP component. At this stage gate, operation personnel completed the product validation, ensuring that all work was accepted in line with the acceptance criteria established early in the project. Although some might treat this point as the end of the Delivery Phase, the CAMMP™ approach mandates that the project is not fully delivered until it is properly closed. Time to close the project.

Chapter 35

Close Stage

35.1 Introduction and Stage Purpose

The Close Stage is the last stage on the CAMMP™ and CAMMP™ MP project life cycle. Consequently, it is the final stage for the Integrated City Megaproject. The main focus of this stage was on properly closing the project and included all of the related activities, such as gathering the lessons learned, updating employee files, documenting project performance metrics, and updating organizational records, among other activities. The stage and close-out report actually consisted of the following components, per Figure 35.1.

Figure 35.1 The Project Closeout Report

These could be independent reports for a typical project and for a megaproject like ICMP, each of these components is likely to be an independent assessment. We had covered the concepts for these components in Part A. In this chapter, we will limit the discussion to a brief refresher.

35.2 Stage Sequence

Although this is the Close Stage, it is still a stage, and on megaprojects, it is a significant stage. It might take months and many resources to properly close a project like the ICMP, especially with the government's interest in using this project as a learning opportunity for future projects; this is why the government insisted on a National Project Management Systems (NPMS) Manager to be part of the team. Therefore, this stage had to be authorized, planned, implemented, even controlled, and closed. Remember, closing this stage is actually closing the whole project.

35.3 Stage Deliverable and Stage Gate

The main stage deliverable was the closeout report and all of its relevant components. We also had a stage gate at the end to verify completion of the work and closure of the project work.

35.3.1 Performance Reconciliation Report

Performance reconciliation is about reconciling (documenting) the actual project performance versus the plan. This not limited to cost or schedule; it is for all functions. In a way, the reconciliation report is like a story, telling the reader (organizational management, PMO, NPMA) what happened on the project journey. The reconciliation report is not a list—it must offer detailed explanations, because the purpose of this report is to learn for future projects, identify areas of opportunities and gaps, and help the organization enhance its project management system. In other words, it can help identify gaps in the system or in practice. For example:

- In the area of scope, the story is about what changed: Did the team deliver more than was required or less, and why? Did the project experience scope changes, what were they, their impact, sources, reasons, justifications? Could they have been avoided? etc.
- For quality, how did the project perform? Were there any deviations or waivers from the standards and specifications? Did the team change the standards for the project or any of its components or elements?
- For human resources, there are many similar questions to answer, including the performance of individuals or groups, changes in key personnel and why, poor or outstanding performers, actual versus planned productivity deviations and why, among numerous other factors.
- The same applies to all other functions, including—but not limited to—safety, cost, time, procurement, and sustainability.

For the ICMP, the performance reconciliation covered every component of the megaproject—airport, port, educational facilities, residential area, etc. It also included every major contract and sub-contract.

35.3.2 Lessons Learned

Lessons learned is different from performance reconciliation, because the latter starts by comparing final status versus plan and documenting all of the differences as an opportunity to improve future planning. Lessons learned are also about learning for the future, but with a focus on specific lessons, good or bad. We captured lessons learned on an ongoing basis from the Concept Stage all the way to closure and in every stage. Therefore, at this point in the project life cycle, and with all of the lessons learned that the IPMT collected from the various stages, it was time to close. The only thing left was for the team to reflect on the full project and consider lessons learned from the project perspective instead of the limited focus on the lessons learned from the various stages.

35.3.3 Organizational Records

Organizations cannot improve their processes and project/product delivery approach without proper project closure. Organizations cannot raise their organizational project management maturity without continuously improving their systems through adequate and timely closing. Therefore, project closure requires ensuring that the valuable information from the close-out report is organized and added to the organizational records to update the knowledge management base and modify the existing Organizational Project Management System, where needed.

35.3.4 Measuring Project Success

We have discussed success more than one once, and in closing, we reflect back on the Four Dimensions of Project Success™.

- Product delivery success, for each of the sub-projects and ICMP components, were assessed shortly after final acceptance and in comparison with the project detailed plans and preliminary engineering packages.
- Project management success was assessed at this time, during project closure.

Both of the above were included in the closeout report. Also, the other two dimensions, project delivery success and objective success, could not be determined at this point in time. The third dimension could be measured by the end of the first year, after beneficial production; therefore, this was a task for the NPMA or the Strategic Planning Authority (SPA). The objective success would likely be assessed a few years later. Therefore, these would also be the responsibility of SPA/NPMA or other designated units.

35.3.5 Stage Gate: Project is Closed

This final gate is quite different from all of the other gates since there is nothing after it. The CAMMP™ methodological approach requires this stage gate to stress the point that the project is not closed until all project deliverables are submitted and approved, and the project close-out report is one of those deliverables, along with ensuring the update of the knowledge management base.

Chapter 36

Part B Closure

36.1 Introduction

It is time to close Part B and the simulation of the Integrated City Megaproject, the ICMP. We will close the simulation with a few sporadic topics that are not specific to a stage, but are the project-level topics that link the stages of the project life cycle.

36.2 Governance

We stated earlier in the book that governance is everywhere along the project life cycle, from the idea to closure. Here are a few points on how the CAMMP™ and CAMMP™ MP approach to governance is applied to the ICMP:

1. Governance starts with the documentation of the Project Brief and Stage Gate 1; SG1 is mostly about strategic alignment. Because the ICMP is strategic in nature, all of the steps are subject to the stage-gate process, starting with SG1.
2. Next, a proper and comprehensive Feasibility Study is required that addresses multiple aspects, from project location to sustainability. Therefore, Stage Gate 2 is a critical decision-making point because, even if the project is feasible, it might not be authorized, especially if another project might offer higher strategic value. For ICMP, the project was approved. One can assume that the government would have been considering more than one megaproject or major project, and the government would have to select some of them and reject or postpone others.
3. Even in the first stage and phase of the project, and before the project is authorized, proper project management is critical and utilized. This is reflected via the stage management process deliverables and their process gates. In other words, governance at project and stage levels. This CAMMP™ approach is different than other resources that consider this work to be pre-project and outside the project life cycle.

4. The breakdown of the project life cycle into discrete, well-defined stages, each with a specific purpose and a stage gate in the end, is a form of the stage-gate process, which emphasizes the proper governance, especially with the strong gates approach that we discussed earlier.
5. Repetition of the stage management processes in every stage is also vital to help ensure effective and efficient project management processes across the project life cycle.
6. Recognizing that the stage gates are vital, and their roles are to not advance the project without the proper sequence of the necessary developmental steps. This is necessary to optimize the outcome and minimize over-expenditure.
7. Requiring formal reviews and complying with the budgeting process when the expenditures are expected to multiply and would be more than typical operating budgets; this is the need for advance funding after the Project Management Plan.
8. Formal funding at the point of no return after the preliminary engineering and before embarking on engineering–procurement–construction work, which is a considerable cash requirement and the bulk of the investment.
9. Requiring reconciliation reports and lessons learned at the closeout of every stage and in the end with the project close.
10. Requiring change management at the stage and project level, even in the early stages of the project at which ambiguity is high—even when there are no contracts.
11. Using the change control board and two-step approvals process for changes as another form of governance to ensure eliminating unauthorized expenditures.
12. Making risk management an integral part of every decision, at the process gates, at the stage gates, when awarding contracts, and all else.
13. Following the project success dimensions, which start with setting the criteria for success early on (different than acceptance criteria) and measuring success in the end. Further, the four dimensions allow a clear assessment of success on multiple dimensions to maintain focus on specific areas of accountability. The four dimensions eliminate the illusion of success, which is a result of emotions, ego, and subjective rather than scientific thinking.

36.3 Approvals

There are different types of approvals along the project life cycle. The formal ones, at the lower working levels, are the process gates within the stage management process; those are the four process gates. The next level are those at the project levels, which are the stage gates, at the end of every stage. Three of those stage gates have a higher purpose, which means possible approvals at the highest levels; these are:

- The stage gates at the end of the feasibility, leading to initial approval
- The stage gate at the end of the Strategy Stage, approving the Project Management Plan and advanced funding
- The stage gate at the end of the Definition Stage, which leads to the final approval and sanctioned funds

36.4 Estimating and Reserves

Throughout this book, we have stated that on capital-intensive projects such as the ICMP, there are three distinct **project** estimates in addition to the more specific stage estimates. These estimates are completed at critical points along the project life cycle. Each of these estimates has a particular purpose, usually linked to a project approval point, as discussed in the previous part. These estimates are also quite different in their level of detail and accuracy. However, they all include the same estimating components.

Since project estimates cannot be more than an approximation of the expected total cost of a project, although accuracy varies from one estimate to another, there are unknowns. In a study of project history within an organization, one can notice a common fact: things will happen that were not included in the project plans (and estimates). These could be put into two main categories: known unknowns and unknown unknowns. We can classify the known-unknowns as risks, and we use a contingency reserve to cover them. However, the unknown-unknowns cannot be covered within the contingency, nor can we expect the project director to be responsible for them. Consequently, organizations include in their budget request a management reserve to cover these uncertainties.

36.5 Risk Management

For risk management, we just need to remind the reader to consider the threats and opportunities, along with stage-specific risks and project level risks. Often, professionals struggle with the concept of identifying opportunities; let us include a few examples here:

- Researching to find out if there are other studies of similar projects or partial studies for a given component, and using those studies when found
- Using an existing port and expanding it, instead of building a new port
- Similarly, the use of an old military base and converting it to civilian use
- Pursuing tax incentives to enhance the chance of getting investors to participate in the project
- Implementing educational and professional development programs to increase local employment and reduce dependence on expatriate workforce, reducing the project life cycle cost
- In construction, on projects like this, using modularization and pre-assembly as a high cost and time saver
- Using best practices as opportunities for cost and time saving, and possibly enhancing quality and improving the safety performance

36.6 Performance Management

36.6.1 General Comments

Performance management is part of project management and control. It is linked to governance, and it can have various dimensions. However, in the context of this book, the discussion is limited to project metrics, the key performance indicators that the Prime Minister's Office and the Integrated Project Management Team (IPMT) used throughout the project. These metrics are necessary to ensure that the project progresses on the right track and stays healthy. The ultimate objective is the project's successful delivery and realizing its expected benefits. Like almost everything else in this book and the CAMMP™ MP methodological approach, project performance metrics are also at various levels—in particular, the project and stage levels. What we will discuss below is a focus on project metrics. However, many of these metrics apply at the stage level.

36.6.2 Project versus Organization

In one aspect, it is vital to consider projects and their performance as standalone initiatives. This is necessary to ensure that the team deliver the project and manage success. However, projects, at least from the project owner's perspective, are part of the ongoing programs and portfolios. Consequently, one must view projects' performance in relation to the organizational projects.

Organizational project management systems (OPMS) must be built to be sustainable. Sustainable OPMS requires ongoing and continual improvement, which comes from the various planned and delivered projects. OPM maturity depends on this continuing feedback from operations and projects. An organization cannot achieve a higher level of mature project management practices without the proper performance assessment at all levels. Therefore, an essential part of any organizational project management system is establishing and updating the performance metrics used on projects. Most of what we just mentioned here comes from Part A, in which we discussed these points in detail, except metrics.

Based on this discussion, there are metrics to use on any given project. Further, at the organizational level, there will be other metrics that would report on the performance of all projects (and programs) within the organization; this is the OPMS performance. To elaborate on the project versus OPM performance, here are some examples:

1. On the project level, one can calculate the cost performance index for the project.* While the project is ongoing, this index can help us in knowing if the project is below or above budget. We can even drill down to control accounts and work packages to find out how are we doing in the various areas. However, if the organization also has an OPM Cost Performance Index for all completed projects in the organization, one can also tell if the ongoing project is performing better or worse than the organizational average and historical performance.
2. A project change management index can also tell management how the team is doing on the ongoing project and how it is doing in comparison to historical norms.
3. The same applies to schedule performance, quality performance, safety performance, and all other functions and aspects.
4. We can expand on the above and drill down to the stage level. An organization that delivers capital projects with engineering and construction stages can also compare current stage and project performance (Engineering Stage, for example) to historical norms and how other engineering stages from previous projects performed.
5. Projects and organizational metrics are vital for continual improvement because they can help an organization identify gaps and whether these gaps are system-wide or in a particular area. Identifying the gaps is the first step in enhancing performance and reaching a higher level of organizational project management maturity.

36.6.3 Sample Project

Some of the metrics on the ICMP could be managed by the various contractors for their own work. However, the Integrated Project Management Team must require at least some of those metrics to be shared with the contractors' reports. Other metrics will be managed directly by the integrated project management team.

* Keep in mind, averages do not work, so when we consider project metrics, a metric for the overall project, we cannot view that metric as an average, we still have to drill down to the work package levels and consider multiple metrics to determine a true performance or status.

36.6.4 Project Metrics

It is outside the scope of this book to cover metrics in detail, explaining what each is and including formulas on how to calculate; there are other valuable references for this level of detail. What is listed here is just for illustration purposes and to trigger the reader to think about the possibilities. The sections below show some of the metrics and performance indicators that can be used to support the measurement of project success: the four dimensions. The metrics are listed here, but the project success discussion is left to the next chapter.

One last comment: many of these indicators might be meaningless on their own and should be compared to organizational or industry benchmarks. For example, a 3% defects on the number of welds on aluminum pipes—is this good, bad, or even ugly? One cannot tell unless we know what the industry benchmark and organizational norms are. If organizational norms are 5%, then 3% is good. If the industry norm is 1%, then this is ugly.

Product Development Indicators

In general, most indicators listed here will be technical and related to the standards and specifications as outlined in the Project Requirements Document and updated in the Preliminary Engineering Stage.

- The number of waivers to specifications approved
- The number of waivers requested
- Percent of waivers approved
- The number of issues not resolved at handover
- The number of significant defects (or below optimal solutions) that had to be accepted
- Customer satisfaction rating at final acceptance
- The number of design errors
- Front-end loading, Project Definition Rating Index (PDRI)
- Percent of the total cost spent on fixing defects post initial acceptance and handover
- The number of lessons learned identified (related to product development)

Project Management Indicators

For these project management indicators—metrics—to be useful, one needs to reflect on the earlier discussion and view them in isolation (for the project) and in comparison to the organizational benchmark—the performance of other projects in the system.

- Variance at completion in terms of cost (within established limits)
- Variance at completion in terms of schedule (within established limits)
- Percent of hours, cost, time spent on rework
- Percent of hours, cost, time spent on changes
- Percent of hours, cost, time spent on issues
- Percent of hours, cost, time spent on occurred risks
- Percent of project team turnovers
- The number of project requests for clarifications issued
- The number of estimating errors
- Claims index

- Percent of work packages over or under the budget/schedule control limits
- Cost and schedule performance indices
- Project management index (project management cost as a percent of total estimated cost)
- Variances from the Class 1 to Class 2 Estimates and from the Class 2 to Class 3 Estimates
- The actual cost of contracts over budgeted costs and awarded values
- The number of risks identified and their split in terms of how they were grouped or prioritized
- The categories of changes
- The source of changes
- Number of lessons learned identified (related to project management)

General Project Indicators

These indicators and metrics are related to project authorization.

- Was the project delivered with product and project management success?
- Did the project achieve all of the set requirements as defined in the project authorization?
- The rating of customer satisfaction
- The rating of management satisfaction
- The rating of team satisfaction
- Safety, health, and environmental performance indices
- Sustainability index
- The number of lessons learned identified (general)
- Did the project achieve beneficial production within the established time frame?
- Did it achieve the desired capacity or quality of the products?

Project Objectives Indicators

- Did the project deliver the expected outcome?
- Did the organization realize the benefits, per the established timelines?
- Did the team identify issues with the Feasibility Study?
- What lessons learned were identified (related to the objectives)?

36.7 Chapter Closing Comments

In this chapter, we covered some of the general topics that would not be specific to a stage. In Part B, we included a simulation of a megaproject. We tried to be as brief as we could and truncated most, if not all, process and stage deliverables. We wanted to provide the simulation mostly to show how we integrate the project level with the stage level and to demonstrate that we cannot lead a megaproject concept to success without managing each stage of the project. Messing up a stage, especially one of the early stages in the project life cycle, can have severe consequential effects on the project performance and can be the leading factor that pre-determines the project success or failure.

All that is left is a chapter to close the book. We warn you, in this next chapter, we will pose many points to ponder and even challenge some of the current thinking and views of megaprojects. We will end with a few calls for action.

Chapter 37

Call for Action

37.1 Introduction

A book on such a magnificent topic cannot do it justice. The level of effort, complexity, number of stakeholders, and numerous other factors that are required for megaprojects is incredible. Personally, the learning that I have acquired from working on megaprojects is what still shapes me today. The experience has helped me become a life-long practitioner of project management and enabled me to enhance the level of my contribution to the community of practice. Writing this book has brought back many memories, good and bad; yet these memories, magnified by years of experience, is what prompted me to write on a critical topic that is costing organizations and countries trillions of dollars in lost value. I do realize that there is no perfect solution and we cannot fix all of the challenges, but it would be unprofessional and even unethical, in my opinion, to accept the status quo or accept the illusion of making a change. We, as in the project management community, practitioners, academics, and various organizations and thought leaders, have a great responsibility to make a difference, a real difference. The only questions are how, when, and where.

I do not have all of the answers. However, I do know that a tailored approach is definitely a big piece of a huge puzzle. The CAMMP™ for Megaprojects (CAMMP™ MP) is the tailored approach that we have shared and used for the simulation; it provides an excellent starting point and that will lead to framing a comprehensive solution. We need others to fill in the other pieces, and we know a few have already filled in some of them before us.

> **Building the Puzzle of Megaprojects**
>
> The way we see it is that a tailored approach would help us build the edges of a picture puzzle, which define the boundaries. Then, we will need expertise from the various domains, such as technology, business, market research, engineering, construction, cost, quality, and many other fields of study. These specialized areas help us complete a beautiful picture.

I must warn those who want a ready-solution-in-a-book, you will not find one. The full solution is what your organization will build and sustain, over years of practice and dedication. Build it one piece at a time. Embed the culture of change and excellence to sustain it, and you will be on your way to becoming a true megaprojects center of excellence, enabling you to lead the megaprojects concept to success.

Again, this book can be a starting point that will give you a tailored approach. Our focus has been on the project management and product delivery model at the high level: the project life cycle level and the stage management level. We touched on cost, schedule, risk, scope, and procurement as functions and how we must treat them within a stage and across the stages.

However, that is not enough, as we have stressed throughout this book. As you build your organizational project management function and system, you need to build the scope management piece, the cost management piece, and the rest of the project management functions. You also need to build the governance structure and the stage-gate process. You will have to go deeper into the technical, business, and production aspects. Further, you have to consider the streams that we discussed, which are essential for helping your organization shift from *project* management into *asset* management.

In closing this introduction, we must reflect back on vital concepts that we think warrant further study. We believe we have identified a few concerns in the way we study and assess projects and megaprojects. It was not possible for us to research or resolve all of these critical concepts. Therefore, we raise these concerns through points to ponder, which are the calls for action that we raise. We humbly ask other practitioners and academics to contribute.

Finally, we believe that megaprojects require clear leadership and many leaders to deliver success, to realize the benefits, and to unquestionably achieve the strategic objectives. We trust that *CAMMP™ for Megaprojects is a practical, tailored approach for leading megaprojects, concept to success*!

37.2 Points to Ponder

We started this book with a few points to consider. I believe we have been able to answer most, if not all, of those questions in this book, to the best of our abilities. In closing, we will add a few open questions. We do not have the answers to these questions, although we could speculate. However, we prefer to post these questions for the professional community to consider. Maybe one of these points could be ideal for further research.

37.2.1 Point 1: Incremental Development

Let us call this, for fun, "Agile for Megaprojects" (jokingly,* of course). However, what is not a joke is this thought: could we use the concept of incremental development for megaprojects? Technically, this would mean converting megaprojects into (mega)programs, with multiple projects that could be developed and delivered over an extended, maybe intermittent period, instead of continuous delivery and a big bang effect. For example, the Integrated City Megaproject is definitely fit to be a program. However, even when it is not evident that a megaproject is a program, can we modify it to become a program?

Not to confuse this concept with the Agile Manifesto† (Agile, 2019):

* The reason for saying "jokingly" is because in Agile, the idea is that the increment should be about working products development in a short time frame, two weeks to two months. Therefore, Agile is not something we can use in capital projects or megaprojects. Sure, capital projects can use agility and lean practices, but that is not Agile as it has been promoted with the Agile Manifesto as a foundational concept.

† http://agilemanifesto.org/

- We do realize that in this case, the increment would not be two weeks to two months, but more like three to five years, each.
- We also note that each increment will be a full project with its own project life cycle and product that will lead to value generation.
- Based on the above, this concept cannot apply to all projects, because some megaprojects have to have their components delivered to go into operations and production in an integrated way.

Let us take a dam megaproject for example. The research shows that "*. . . three out of every four large dams suffered a cost overrun . . . with actual cost 96% higher than estimated costs.*" (Ansar et al., 2017, p. 73). The schedule performance is not much better, and although the researcher did not have enough data on benefits, the data they had shown stated an average "*benefits shortfall of at least 11%.*"

With the above scenarios, would it be more beneficial to build a few smaller dams in series* in different locations? In theory, this could mean power generation would be distributed to a broader geography, possibly close to a consumption location? In all transparency, I do not know anything about large dams projects, so my question could be naïve—but is it?

Other examples could be considered for factories, chemical plants, pipelines, power generations, among other possibilities. I do realize the economy of scale concept that bigger might be cheaper on a per-unit basis, but since many megaprojects fail, and smaller but still significant projects have much better performance, then would the avoidance of the overrun not cover this concept of incremental development? A refresher, per the IPA data: only 35% of megaprojects are successful (Merrow, 2011), but 60% of capital projects are successful (Barshop, 2016), which might support this point. In closing this part, "bigger is better" might be true in a particular situation, but with the fragility of megaprojects could we consider "small is beautiful"?†

37.2.2 Point 2: Are We Using NPV And BCR Properly?

We do realize that NPV (net present value) is an evaluation of the project in today's money and considers future revenues and expenses. We use the NPV to compare projects and select the most promising ones. However, are we assessing the real, actual, post completion NPV? For facilities projects, the product's useful life could be 25 to 30 years; therefore, are organizations‡ (or assessment researchers) actually collecting this information, which could be highly proprietary, if it can be calculated?

A similar point is benefit–cost ratio (BCR). How are researchers calculating the actual benefits and costs over the life of the product? I know that capital costs are readily available, and maybe data from the first year or two of operations are also available; however, how about after that?§

Merrow (2011) clearly states that, for example, the costs incurred to fix production problem are hidden. The consequences of this are that the IPA data on assessed projects (cost growth and other factors) could be worse than predicted, which means failure rates are also higher than reported. On the other hand, benefits could also be higher, and in that case, the success rate is higher than reported. This is a point worth considering to reach a logical answer.

---

* Series does not mean back to back, but could be on the same waterway, separated by enough practical distance.

† This concept was championed by E. F. Schumacher in his 1973 work titled: *Small is Beautiful: A Study of Economics as if People Mattered* (Schumacher, 1973).

‡ I would assume that organizations (project owners) are doing this, but is this information shared openly with researchers?

§ I could be wrong on these statements, but in all the literature reviews that I have conducted, I have not read any references to actual life-cycle benefits and costs.

37.2.3 Point 3: Are We Assessing Project Success Properly?

In this book, we have shared how to assess project success. We shared the IPA model and SUKAD's Four Dimensions of Project Success™. We also know that many references use BCR and NPV as mathematical indicators. We know that NPV and BCR are useful in the early stages of the projects and would help organizations decide on the viability of the project. However, what is challenging us is what is happening post completion. What we have seen is that IPA, for example, assesses production in their final assessment, but that is only a short time after startup—equivalent to SUKAD's third dimension—but are they (or others) determining for the fourth dimension, the asset success?

If we reflect on this point and the previous one, the above leaves us with the question, *are our assessments of success and failure valid without access to production data* (operation, maintenance, revenues, etc.)? I must be a cynic here and ask, *are we exaggerating failure reporting to sell services?* Many of us practitioners and researchers* wonder why project owners continue to deliver more failure than success. Maybe they know something we do not, and they are more successful than what appears on the surface. So instead of reporting failure correctly, perhaps we are over-reporting it or under-reporting success. I do not have the answers to these questions, but I would speculate that a bit of truth could be on all sides of the issue. Therefore, I believe further research is necessary.

To close this point, it is interesting to highlight that despite the extensive literature review that I have conducted, I did not see much on assessments per a life-cycle perspective except a paper by Love et al. (2017). Maybe I missed it.

37.2.4 Point 4: The Role of Education

I hesitated in including this point for fear of inadvertently offending anyone. However, I think it is critical to state what I believe and know. In performing the literature review, I reflected on a comparison between the academic world on one side and the practice world on the other. What I have noticed is that many of the papers and references that I have read from an academic perspective gave me the impression of theory over real-world applications. I have no idea how many of the professors writing these papers have actual experience in the field of managing projects or megaprojects—I mean experience working as professionals and team members on megaprojects, not as consultants. I also reflected on the references written with the practice and applied concepts in mind, and I observed a notable difference. I know that we need both types of writing and research to further the learning in this dynamic and growing field. In other words, I am not promoting an either/or scenario.

Yet, I kept thinking about a study about MBA programs, mentioned earlier, and the status of education in general with a higher emphasis on theory than learning. I have met many bright, young professors, but I have to ask, in an applied domain like project management, and with the tremendous complexity of megaprojects, do we need to relook at the academic educational system, especially at the postgraduate level? Do we need more practitioners leading classes than pure academics? If not more, then should we embed these programs with practitioners to support the academics and possibly work collaboratively?

37.2.5 Point 5: How About Shared Services or Authority?

A few large companies in the private sector—conglomerates—have established shared services divisions, or even separate shared services companies. In this case, the shared services include project management

* For example, "Why Do Organizations Never Seem to Learn?" (Drummond, 2017).

to lead or handle capital projects. Another approach that we have observed is a centralized organizational project management function. I have not done any research or seen any research on whether this is a logical solution or a leading practice. Our observations are that this is something that should be studied further; however, our position is that this is a leading practice. We believe in centralizations and standardizations as long as we follow the tailoring principles and develop fit-for-purpose solutions, which would be at the organizational level and for similar types of projects and not on a project by project basis. These shared services organizations must ensure the great customer-centric approach and avoid the pitfalls of bureaucratic processes.

Can something like this, shared services, be applied to the public sector? Again, researching this topic is a considerable effort and is outside of the scope of this book. However, we cannot help but ponder, would a project management authority, a form of a shared service for government projects, work? To our knowledge, the UK has established the Infrastructure Projects Authority, and I believe Saudi Arabia has the National Project Management Organization. We are not sure if similar arrangements exist elsewhere or if the work of these organizations has been analyzed independently to demonstrate their effectiveness.

CII presents us "partnering" as a leading practice, but what we are discussing here goes beyond partnering between two organizations or so. What we are advocating is a National (State) Project Management Authority that could have two significant mandates. One is being an authority that would be responsible for the development and sustainability of standards, guides, processes, for all types of government projects (mega and non-mega, capital and non-capital). The other (optional) mandate could be to lead large projects and megaprojects for all entities, in lieu of or in partnership with project management consultancies (PMCs). By leading, we do not mean EPC (engineering–procurement–construction) only, but being involved concept to success. We must stress that one of the pitfalls could be the silos effect, but this can be avoided through the proper setup and collaborative culture.

37.2.6 Point 6: Contingency and The Management of Risks

One of the recommendations of the Grattan Institute report on Australian transport infrastructure is to hold contingency at the portfolio level (Terrill and Danks, 2016). In other words, the contingency reserve would be estimated and included in the budget, but then shifted to the portfolio level, so it can be shared across projects. This sounds like a logical step and could have practical applications from an administration perspective; however, does it contribute to project performance? I do not have an answer.

37.3 Call for Action

More than one of the references that we have reviewed compared megaprojects with the frog in hot water story.* In other words, as we work on megaprojects from concept to closure, things could be trending toward failure, but the organizational culture and human nature might not recognize the cycle of doom trend, or hope they could recover, or even ignore the signs altogether. When the damage is done and is visible, the sunk-cost syndrome might prevent the organizational management from cancelling the project, which, like the frog, it is heading to its death. Death in the capital project would be projects that result in a negative net present value, or "stranded asset" (Ansar et al., 2017).†

* If you do not know the story, Google it! Just kidding: the theory is that if you put a frog in water and heat it slowly, the frog's biological makeup does not recognize the gradual increase in heat and it stays put, to its death.

† *The Oxford Handbook of Megaproject Management*, p. 68.

With this scenario, I want to close this book by reflecting back on what we have presented. I am not going to list everything here. I only want to double and triple emphasize the need to build the organizational project management function, which requires building the organizational competence and capacity of the project owners' organizations. Remember, building the organizational project management function can happen through building the organizational project management system and organizational knowledge base. However, sustaining the function and growing the level of project management maturity require culture.

It is time to reflect on the image on the next page*; is your organizational culture holding you down or lifting you up? If you let circumstances answer this question, then expect to be going down a cycle of doom route. However, the good news is that the choice is yours, and you can adopt a winning mindset and build a competent culture that has the will, motivation, and capacity to lead megaprojects to success.

The clear message that we want to close this book with is continual learning. The project management community can benefit from answering the points to ponder we raised here. It is our opinion that we need to study megaprojects and include this topic as an area of focus in project management programs, especially those linked to engineering or construction schools.† We need the professional associations to give this topic more attention or maybe establish a dedicated association for capital and megaprojects. I know that personally, I will continue to explore, learn, and share. Join me on an exciting journey of learning and growth. It is our ethical obligation and social responsibility.

* The original concept for this image (without the balloon) came from my good friend and colleague Nah Wee Yang, founder and CEO of Knowledge Systems and Knowledge Method in Singapore. He drew the original concept as he was listening to my explanation during a workshop on organizational project management. I had my graphic design re-draw and add the balloon to incorporate a positive twist!

† We know that many universities, including my alma mater, UC Berkeley; the University of Texas in Austin; and others have project management programs, but are they giving the proper focus to the asset or product life cycle, or even the project life cycle, concept to success, or are they focusing mostly on engineering and construction?

Is your organizational culture pulling you DOWN or lifting you UP?

Appendix A

Additional Resources

To support this book, SUKAD Corp. will have a dedicated section on the SUKAD Portal, www.sukad.com, for this book and another one for *Project Management beyond Waterfall and Agile* (Ajam, 2017). These sections will be part of the SUKAD Knowledge & Innovation division (SKI).

In these sections, we include various references and additional resources, such as templates, flowcharts, and mind maps. These website sections will be:

1. *Project Management beyond Waterfall and Agile:* https://sukad.com/ski/books/project-management-beyond-waterfall-and-agile.
2. *Leading Megaprojects, A Tailored Approach:* https://sukad.com/ski/books/leading-megaprojects-a-tailored-approach.

Some of the content in these sections will be open to the public, others will be restricted to purchasers of the relevant book or accessed through a SUKAD Knowledge & Innovation subscription. In addition to these sections on the website, SUKAD Corp will be happy to give limited access to an online learning program (course) related to CAMMP™. This access is also restricted to purchasers of either book. This course is on the Quest to Mastering Project Management site, https://pmquest.sukad.com.

Appendix B

Sample Templates

In this appendix, we will share a few templates that will allow you to see the difference between stage management and project management. We will share the template for the Stage Management Plan and for the Project Management Plan. We will also include an example of a possible header for each template and the template for one of the stage gates.

Project General Information

This is an example of the header information for every template. This is modified for megaprojects.

Project Title:			
Project Number		Assigned Date	
Idea Generator		Organizational Unit	
Project Director		Project Executive	
Part of a Program?		Program Title	
Document Date		Document Version	

Stage Management Plan

Phase and Stage
How to Develop the Stage Detailed Plan
How to Implement and Manage the Stage Work
How to Control the Stage Work
How to Manage Change During the Stage
How to Close the Stage
Stage Management Team
Stage Planning Team

Project Management Plan

Strategy Section

Project Life Cycle Selection
Project Management Strategy
Delivery Strategy
Project Procurement Strategy
Best Practices Strategy
Sustainability Strategy

How to Section

How to Define the Project
How to Manage the Project
How to Control the Project
How to Manage Change on the Project
How to Get Ready for Operations
How to Close the Project

People Aspects Section

Project Management Team
Extended Project Team
Project Stakeholders
Project Communication Requirements

Class 2 Estimate Section

Project Schedule
Project Cost Estimate
PMB and Time-Phased Budget

Project Manager Comments

Stage Gate: Project Management Plan Approval

Summary of Prior Work

Project Management Plan Approval

Decision

☐ Approved ☐ Not Approved ☐ Revise ☐ Hold

Funding

Is there a need for advance funding at this stage? ☐ Yes ☐ No

If yes, funding approved? ☐ Yes ☐ No

Date: ______________________

Explanation of Decision

Management Other Comments or Instructions

Acronyms

AACE	The Association for the Advancement of Cost Engineers, International (AACE® International)
AC	*See ACWP*
ACWP	Actual cost of work performed, *same as AC*
ANSI	American National Standards Institute
AP	Asian project
BCR	Benefit–cost ratio
BCWP	Budgeted cost of work performed, *same as EV*
BCWS	Budgeted cost of work scheduled, *same as PV*
BOD	Board of directors
BOOT	Build–own–operate–transfer
BOT	Build–operate–transfer
BP	Best practices
C2C	Concept to closure
C2S	Concept to success
CAMMP™	Customizable and Adaptable Methodology for Managing Projects™
CAMMP™ MP	CAMMP™ for Megaprojects
CAPEX	Capital expenditures
CII	Construction Industry Institute

CM	Construction management
COR	Close-out report
CPI	Cost performance index
CPM	Copy–paste–modify
CPM	Critical path method
CPO	Chief project officer
CRP	Control reference point
CSF	Critical success factor
ELE	Extinction level event
EPC	Engineering—procurement—construction
EPT	Extended project team
EV	*See BCWP*
FEED	Front-end engineering design
FEL	Front-end loading
FS	Feasibility Study
FSR	Feasibility Study Report
GAPPS	Global Alliance for Projects Performance Standard
GCC	Gulf Cooperation Council
GPM®	Green Project Management
ICB	Individual Competence Baseline (IPMA)
ICMP	Integrated City Megaproject
IPA	Independent Project Analysis, Inc.
IPA	Infrastructure Project Authority (UK)
IPMA®	International Project Management Association
IPMT	Integrated project management team
IPT	Integrated project team
ISO	International Organization for Standards
KPI	Key performance indicator
MP	Megaproject

NAP	North American project
NGO	Non-governmental organization
NPMA	National Project Management Authority
NPMS	National Project Management System
NPO	Nonprofit organization
NPV	Net present value
OCB	Organizational Competence Baseline (IPMA)
OECD	Organization for Economic Co-operation and Development
OKB	Organizational knowledge base
OPEX	Operating expenditures
OPM	Organizational project management
OPMF	Organizational project management function
OPMM	Organizational project management maturity
OPMS	Organizational project management system
ORT	Operational readiness team
PAD	Project Authorization Document
PB	Project Brief
PBS	Product breakdown structure
PD	Project/program director
PDCA	Plan–do–check–act
PDP	Project Detailed Plan
PDRI	Project Definition Rating Index (by CII)
PEM	Project Excellence Model (IPMA)
PG	Process gate
PLC	Product life cycle/project life cycle
PM	Project manager
PM	Project management
PMBOK® Guide	*A Guide to the Project Management Body of Knowledge®*
PMC	Project management consultant

PMI®	Project Management Institute
PMM	Project management maturity
PMO	Project management office (could also be Program or Portfolio MO)
PMP	Project Management Plan
PMP®	Project Management Professional PMI certification
PMT	Project management team
PPM	Project portfolio management
PRD	Project Requirements Document
PRiSM™	Projects Integrating Sustainable Methods™ (GPM®)
PT	Project team
PV	*See BCWS*
RFP	Request for proposal
SAD	Stage Authorization Document
SDP	Stage Detailed Plan
SG	Stage gate
SMP	Stage Management Plan
SOW	Scope of work, statement of work
SPA	Strategic Planning Authority
SPI	Schedule Performance Index
SPM	Senior project manager
SUKAD	Success Uniqueness Knowledge Attitude Development[a]
UCB	University of California Berkeley
VIP	Value-improving practices
WBS	Work breakdown structure

[a] SUKAD is the name of the company co-founded by the author.

Bibliography

Agile. (2019). *Principles behind the Agile Manifesto.* Retrieved from Manifesto for Agile Software Development: http://agilemanifesto.org/principles.html

Ajam, M. A. (2014). *Redefining the Basics of Project Management.* Amioun, Al-Koura, Lebanon: SUKAD Multimedia.

Ajam, M. A. (2017). *Project Management beyond Waterfall and Agile.* Boca Raton, FL: CRC Press.

Ajam, M. A. (2018). Establish a National Project Management Authority and System. SUKAD LLC.

Ansar, A., Flyvbjerg, B., Budzier, A., and Lunn, D. (2017). Big Is Fragile (An Attempt at Theorizing Scale). In B. Flyvbjerg (Ed.), *The Oxford Handbook of Megaproject Management*, pp. 60-95. Oxford University Press.

Aschman, A. (2018, December 20). Why Capital Project Systems Succeed or Fail. Retrieved April 11, 2019, from Independent Project Analysis: https://www.ipaglobal.com/news/article/why-capital-project-systems-succeed-or-fail-2/

Barshop, P. (2016). *Capital Projects.* John Wiley & Sons, Inc.

Bertram, S. (2019, March 27). Retrieved from Quora: https://www.quora.com/Did-Einstein-really-define-insanity-as-doing-the-same-thing-over-and-over-again-and-expecting-different-results

Booch, G. (n.d.). "A fool with a tool is still a fool." Retrieved April 30, 2019, from https://www.phertogen.com/2016/09/29/a-fool-with-a-tool-is-still-a-fool/

Brockmann, C.-I. and Girmscheid, G.-I. (2007). Complexity of Megaprojects. *CIB World Building Congress*, pp. 219–230.

CII. (1994). *Quantitative Effects of Project Change.* Construction Industry Institute.

CII. (2015, February 7). CII Best Practices. Retrieved from Construction Industry Institute: https://www.construction-institute.org/Store/CII/Publication_Pages/bp.cfm?section=orders

CII. (2019, March 27). The Construction Industry Institute. Retrieved from https://www.construction-institute.org/

Clairmont, N. (2013, July 31). Those Who Do Not Learn History Are Doomed to Repeat It. Retrieved April 25, 2019, from Big Think: https://bigthink.com/the-proverbial-skeptic/those-who-do-not-learn-history-doomed-to-repeat-it-really

Clegg, S. R., Sankaran, S., Biesenthal, C., and Pollack, J. (2017). Power and Sensemaking in Megaprojects. In B. Flyvbjerg (Ed.), *The Oxford Handbook of Megaproject Management.* Oxford University Press.

Dalton, G. W., Thompson, P. H., and Price, R. L. (1977). The Four Stages of Professional Careers—A New Look at Performance by Professionals. *Organizational Dynamics,* 6(1): 1942. Retrieved from https://www.researchgate.net/publication/232538105_The_four_stages_of_professional_careers-_A_new_look_at_performance_by_professionals

Davies, A., Dodgson, M., and Gann, D. M. (2017). Innovation and Flexibility in Megaprojects. In B. Flyvbjerg (Ed.), *The Oxford Handbook of Megaproject Management.* Oxford University Press.

Defining Governance. (2015, February 17). Retrieved from Institute on Governance: http://iog.ca/defining-governance/

Dictionary.com. (n.d.). Master-planning. Retrieved February 11, 2015, from Dictionary.com: http://dictionary.reference.com/browse/master-planning

Drummond, H. (2017). Megaproject Escalation of Commitment. In B. Flyvbjerg (Ed.), *The Oxford Handbook of Megaproject Management.* Oxford University Press.

Englund, R. L. (2006). *Project Sponsorship.* San Francisco: Jossey-Bass, a Wiley Imprint.

Flyvbjerg, B. (2017). Did Megaproject Research Pioneer Behavioral Economics? (The Case of Albert O. Hirschman). In B. Flyvbjerg (Ed.), *The Oxford Handbook of Megaproject Management* (pp. 155–193). Oxford University Press.

Flyvbjerg, B. (2017). Introduction, The Iron Law of Megaproject Management. In B. Flyvbjerg (Ed.), *The Oxford Handbook of Megaproject Management* (pp. 1–18). Oxford University Press.

Flyvbjerg, B., Holm, M. S., and Buhl, S. (2002). Underestimating Costs in Public Works Projects: Error or Lie? *Journal of the American Planning Association*, 279–295.

Global Alliance for Project Performance Standards. (2015, February 1). Project Complexity. Retrieved from http://globalpmstandards.org/tools/complexity-rating/project-complexity/

GPM. (2012). The GPM Guide to Sustainability in Project Management. Green Project Management Global. Retrieved from https://greenprojectmanagement.org/gpm-standards/the-gpm-reference-guide-to-sustainability-in-project-management

Hirschman, A. O. (n.d.). The Principle of the Hiding Hand. *National Affairs.* Retrieved from https://www.nationalaffairs.com/storage/app/uploads/public/58e/1a4/a29/58e1a4a298835643416368.pdf

Ibbs, W. and Reginato, J. (2002). *Quantifying the Value of Project Management.* Project Management Institute.

IPMA. (n.d.a). Individual Competence Baseline. Retrieved from https://www.ipma.world/individuals/standard/

IPMA. (n.d.b). Organisational Competence Baseline. Retrieved from https://www.ipma.world/organisations/standard/

IPMA. (n.d.c). Project Excellence Baseline. Retrieved from https://www.ipma.world/projects/standard/

ISO 21500:2012. Available at https://www.iso.org/standard/50003.html

ISO 31000:2009. Risk Management, Principles and Guidelines. Available at https://www.iso.org/standard/43170.html

Lean. (2015). Lean Construction Institute. What Is Lean Design & Construction. Retrieved from http://www.leanconstruction.org/about-us/what-is-lean-construction/

Lenfle, S. and Loch, C. (2017). Has Megaproject Management Lost Its Way? In B. Flyvbjerg (Ed.), *The Oxford Handbook of Megaproject Management* (pp. 21–38). Oxford University Press.

Levitt, R. E. and Scott, W. R. (2014). Institutional Challenges and Solutions for Global Megaprojects. In B. Flyvbjerg (Ed.), *The Oxford Handbook of Megaproject Management* (pp. 96–117). Oxford University Press.

Love, P. and Ahiaga-Dagbui, D. (2018). Debunking Fake News in a Post-Truth Era: The Plausible Untruths of Cost Underestimation in Transport Infrastructure Projects. *Transportation Research,* Part A 113, 357–368. Elsevier.

Love, P., Ahiaga-Dagbui, D., Welde, M., and Odeck, J. (2017). Light Rail Transit Cost Performance: Opportunities for Futureproofing. *Transportation Research,* Part A, 27–39. Elsevier.

Love, P., Sing, M., Ika, L. A., and Newton, S. (2019). The Cost Performance of Transportation Projects: The Fallacy of the Planning Fallacy Account. *Transportation Research,* Part A, 1–20. Elsevier.

Merrow, E. W. (2011). *Industrial Megaprojects.* John Wiley & Sons, Inc.

Miller, R., Lessard, D., and Sakhrani, V. (2017). Megaprojects as Games of Innovation. In B. Flyvbjerg (Ed.), *The Oxford Handbook of Megaproject Management.* Oxford University Press.

Newton, I. (1776). "Standing on the shoulders of giants." Retrieved April 12, 2019, from https://www.phrases.org.uk/meanings/268025.html

Nieto-Rodriguez, A. (2014). Why Top Business Schools Don't Teach Project Management to Their MBAs? Retrieved April 11, 2019, from Antonio Nieto-Rodriguez: https://antonionietorodriguez.com/why-top-business-schools-dont-teach-project-management-to-their-mbas/

Oberlender, G. D. (2014). *Project Management for Engineering and Construction.* (Third edition). McGraw-Hill Education.

Oxford University. (2017). *The Oxford Handbook of Megaproject Management.* (B. Flyvbjerg, Ed.) Oxford, UK: Oxford University Press.

Parkinson's Law. (n.d.). Retrieved from https://www.dictionary.com/browse/work-expands-to-fill-the-time-available-for-its-completion

Pennsylvania State University. (2019, April 15). Predictability Limit: Scientists Find Bounds of Weather Forecasting. Retrieved April 20, 2019, from Phys.org: https://phys.org/news/2019-04-limit-scientists-bounds-weather.html

Pfeffer, J. and Sutton, R. I. (1999). *The Knowing-Doing Gap.* Harvard Business School Press.

Pollack, J., Biesenthal, C., Sankaran, S., and Clegg, S. (2018). Classics in Megaproject Management: A Structured Analysis of Three Major Works. *International Journal of Project Management*, 372–384.

PMI. (2013). *A Guide to the Project Management Body of Knowledge.* (Fifth edition). The Project Management Institute, Inc.

PMI (2016). *Construction Extension to the PMBOK® Guide.* Project Management Institute, Inc.

PMI. (2017). *A Guide to the Project Management Body of Knowledge.* (Sixth edition). The Project Management Institute, Inc.

PDRI. (2015, February 18). Project Definition Rating Index. Retrieved from the Construction Industry Institute: https://www.construction-institute.org/scriptcontent/more/rr113_11_more.cfm

Schumaker, E. F. (1973). *Small Is Beautiful: A Study of Economics as if People Mattered.* New York: Harper & Row. Retrieved from https://static1.squarespace.com/static/57b0be1c579fb3136382b62d/t/57c8e89337c5815308cb0d22/1472784531949/Schumacher-E.F.-Small-is-Beautiful.pdf

SUKAD Group. (2015, February 03). Introduction to Project Success. Retrieved from The SUKAD Way for Managing Projects™: http://sukadway.sukad.com/project-success-introduction

Terrill, M. and Danks, L. (2016). *Cost Overruns in Transport Infrastructure.* Grattan Institute.

Van Wee, B. and Priemus, H. (2017). Megaproject Decision Making and Management (Ethical and Political Issues). In B. Flyvbjerg (Ed.), *The Oxford Handbook of Megaproject Management.* Oxford University Press.

Vickerstaff, M. (2019, April 25). *Adapting Our Approach: Three Lessons Learned for Future Infrastructure Delivery.* Retrieved from Infrastructure Project Authority: https://ipa.blog.gov.uk/2019/04/25/adapting-our-approach-three-lessons-learned-for-future-infrastructure-delivery/

Wideman, R. M. (2015, February 17). The Four Components of Project Management Governance. Retrieved from Project Management Wisdom: http://www.maxwideman.com/papers/governance/components.htm

Wood, M. (2019, April 25). LNG Industry Heading into $200 Billion Capex Boom. Retrieved from LNG World News:https://www.lngworldnews.com/woodmac-lng-industry-heading-into-200-bln-capex-boom/

Index

A

Agile, 15, 18, 23, 28–30, 50–54, 98, 101, 103, 110, 120, 131, 132, 141, 144–149, 158, 170, 198–201, 205, 219, 225, 227, 354, 361
agility, 15, 28, 29, 354
Asian Project, 67–70, 73, 75, 99, 232, 337, 369
asset life cycle, 17, 26, 117
asset management, 4, 7–10, 18, 25, 26, 92, 117–119, 125, 149, 157, 160, 165, 199, 354
Australian projects, 59, 120
Authorize Process, 200–203, 219, 220, 253, 257–261

B

baseline, 46, 143, 145, 164, 219–222, 328, 370, 371
benefit–cost ratio, 31, 59, 117, 355, 369
benefits realization, 7, 116, 135, 152, 195, 197, 241
best practices, 35–37, 73–75, 95–98, 112, 120, 135, 145, 177, 183, 190, 235, 293, 298, 303, 304, 317, 319, 323, 337, 349, 365, 369

C

CAMMP™ MP, 8, 11, 18, 22, 29, 79, 138, 155–160, 163–167, 171, 176, 180, 187, 189, 198, 199, 205, 219, 239–247, 251, 254, 263, 285, 286, 303, 307, 315–318, 327, 329, 333, 335, 339, 340, 343, 347, 349, 353, 369
CAMMP™ Processes, 201, 202, 219
CAMMP™ Process Steps, 207
capital project, 24, 28, 29, 40, 45, 48, 109, 128, 132, 157–160, 165, 208, 357
case studies, 3–10, 34, 45–47, 67–69, 77–83, 103, 125, 145, 188, 212, 235, 286
change management, 12, 47, 50, 51, 62, 71–75, 81, 84, 91, 96, 97, 126, 165, 183, 184, 194, 201, 210, 223–227, 243, 262, 291, 303, 313, 319, 320, 337, 348, 350
Change Management in the Fog, 226, 262
CII. *See* Construction Industry Institute
close-out report, 151, 195–197, 202, 243, 274, 311, 343, 345, 369
Close Process, 195, 201, 202, 205, 255
Close Stage, 150, 161, 186, 192, 195, 196, 205, 274, 280, 311, 343, 344
competence, 10, 24, 46–51, 61–65, 70, 73, 76–80, 84, 91, 92, 103, 105, 112, 119, 120, 125–127, 136, 140, 145, 174, 176, 187, 214, 218, 224, 230–232, 260, 358, 370, 371
complexity of megaprojects, 5, 7, 10, 39, 46, 51, 84, 91, 126, 235, 257, 356
Concept Stage, 145, 158, 161, 162, 170, 171, 202, 203, 210, 251–257, 330, 345
concept to closure, 7, 9, 16, 20, 22, 39, 40, 213, 357, 369
concept to success, 7–11, 38, 80, 134, 143, 155–160, 167, 170, 199, 227, 235, 236, 241, 313, 352, 354, 357, 358, 369
Cone of Uncertainty, 210
constructability, 74, 97, 100, 183, 303, 337
Construction Industry Institute (CII), 7, 35, 64, 74, 95, 96, 135, 157, 177, 208, 293, 320, 337, 369

contingency, 108, 210, 211, 266, 267, 273, 281, 282, 322, 323, 329, 349, 357
contingency reserve, 211, 280, 323, 349, 357
contractor, 27, 37, 47–50, 69–79, 97, 99, 105, 109, 115, 134, 136, 148, 165, 177, 187, 192, 193, 208, 213, 216–219, 224, 231–233, 336, 340
Control Process, 201, 204, 219, 254, 258, 336
cost management, 206, 208, 354
cost overrun, 58, 59, 355
cost plus incentive, 219, 231, 273
cost reduction, 61, 64, 84, 91, 109, 117, 126
critical success factors, 10, 120, 121, 127, 131, 136, 140, 143, 146, 154, 199, 286, 337
culture of fixed price, 47, 48, 51, 84, 91, 126
culture of low bid, 47

D

Definition Stage, 35, 73, 129, 149, 150, 158, 162, 164, 180, 182, 186, 189, 203, 222, 226, 279, 282, 327, 348
Delivery Phase, 18, 36, 56, 98, 105, 150, 153, 160, 162, 170, 177, 180, 181, 184, 186, 189–192, 196, 216, 222, 243, 281, 287, 303, 327, 330–333, 341
Development Phase, 18, 56, 138, 160, 170, 171, 180, 181, 189, 191, 222, 223, 243, 282, 287–292, 298, 313, 327, 330
Discovery Phase, 15–18, 40, 62, 73, 80, 129, 138, 140, 158, 160, 164, 170, 171, 177, 181, 185, 203, 223, 235, 243, 249, 259, 261, 264, 275, 280, 281, 315, 330

F

Feasibility Stage, 158, 163, 171–173, 176–178, 188, 202, 204, 208, 210, 214, 251, 256–268, 272–275, 279, 281
Feasibility Study, 153, 164, 170–177, 188, 191, 209, 213, 214, 221, 222, 230, 235, 258–282, 296, 299, 308, 312, 315–321, 324, 327–329, 347, 352, 370
FEL Index, 35, 99, 136
final acceptance, 150, 192–195, 279, 281, 333, 340, 341, 345, 351
fixed price, 47, 48, 51, 64, 69, 72, 73, 84, 91, 107, 113, 126, 193, 216–218
Four Control Reference Points™, 143, 220
Four Dimensions of Project Success™, 30, 116, 135, 142, 145, 148, 149, 153, 158, 260, 283, 293, 305, 345, 356
front-end loading, 34, 35, 40, 62, 98, 113, 129, 136, 180, 212, 351, 370
front-end planning, 35, 36, 64, 74, 96, 98, 129, 139, 140, 157, 303

G

gates, 16, 20, 21, 76, 98, 100, 134, 144, 145, 157, 160, 163–171, 178, 192, 193, 199, 204, 215, 247, 253, 259, 261, 282, 289, 303, 307, 316, 328, 336, 339–341, 345–348, 363
governance, 4, 15, 20, 23, 24, 76, 112, 160–165, 171, 172, 176, 178, 185, 188–192, 226, 233, 234, 253, 254, 257, 288, 289, 313, 320, 321, 325, 347–349, 354

H

handover, 192–195, 215, 234, 320, 333, 336, 339, 340, 351

I

Implementation Stage, 35, 62, 85, 145, 158, 164, 173, 177, 181, 191, 223, 280
Implement Process, 201, 204, 220, 254, 258, 334
Independent Project Analysis (IPA), 7, 34–37, 49, 53–62, 65, 74–76, 95–101, 107, 108, 135, 136, 145, 148, 151–153, 157, 158, 163, 165, 177–181, 184, 187, 195, 212, 214, 219, 230, 233, 279, 328, 330, 341, 355, 356, 370
Industrial Megaprojects, 34, 37, 55–60, 64, 65, 79, 99, 107, 109, 125, 136, 178, 217, 218, 235
infrastructure projects, 35–37, 46, 55, 59, 60, 65, 92, 119, 120, 125, 139, 189, 192, 216, 235, 357
Initial Operations Stage, 194, 330, 339, 341
integrated teams, 97, 99, 137, 183, 233
Intermittent Project Reviews, 100
IPA. *See* Independent Project Analysis

L

leading practices, 73, 76, 91, 92, 95–100, 109, 112, 119, 120, 125, 126, 135, 136, 140, 255
Lean, 28, 29, 303, 354
lessons learned, 67, 98, 129, 135, 136, 177, 195–197, 202, 205, 260, 262, 269, 274, 293, 303, 343, 345, 348, 351, 352
lock-in effect, 138, 139
long-lead items, 164

M

management reserve, 70, 211, 267, 274, 280, 282, 290, 295, 308, 309, 317, 323, 324, 349
method, 19, 27–29, 76, 98, 120, 127, 128, 133, 143, 175, 176, 311, 358, 370
methodology, 8, 16, 27, 29, 127, 132, 135, 141, 144, 155, 195, 197, 239, 303, 369
modularization, 74, 97, 177, 349

N

National Project Management Authority, 119, 120, 127, 239, 251–254, 260–263, 268, 275, 281, 282, 288, 293, 297, 303, 319, 320, 339, 344, 345, 370
net present value, 31, 58, 96, 117, 136, 153, 179, 195, 355, 357, 371
North American Project, 67–70, 74, 99, 183, 192, 193, 212, 230, 232, 335, 370
NPMA. *See* National Project Management Authority

O

objectives success, 30, 67, 68, 116, 152, 197
Operational Readiness Stage, 98, 129, 186, 193, 194, 330
optimism bias, 65, 80, 120, 140, 172
organizational competence, 10, 24, 46, 61, 63, 84, 91, 92, 103, 112, 119, 125, 126, 140, 145, 234, 357, 371
organizational culture, 24, 56, 76, 93, 94, 357, 358
organizational project management function, 31, 46, 48, 65, 76, 81, 89, 92, 101–107, 110, 119, 125, 134, 142, 149, 176, 186, 205, 215, 217, 230, 354, 357, 358, 370
organizational project management maturity, 24, 31, 55, 145, 176, 196, 345, 350, 370
overrun, 36, 47, 51, 56–60, 99, 109, 140, 148, 151, 176, 211, 355

P

partnering, 97, 179, 180, 183, 213, 217, 240, 289, 304, 357
PDRI. *See* Project Definition Rating Index
Plan Details Process, 202, 204, 220, 247, 254, 258
Plan Management Process, 201, 204, 220, 254, 258
portfolio, 15, 23, 31, 97, 112, 142, 178, 181, 197, 232, 357, 372
portfolio management, 15, 23, 97, 112, 178, 181, 372
preliminary engineering, 26, 73, 100, 101, 158, 161, 177, 180–183, 187–191, 208, 209, 215, 218, 280, 289, 303, 307, 310–312, 316–320, 323–328, 333, 339, 345, 348, 351
Process Gate, 145, 204, 220, 254, 255, 258, 261, 263, 270–273, 290, 291, 297, 308, 310, 371
procurement management, 206, 216
product breakdown structure, 183, 299, 300, 369
product delivery success, 149, 328, 345
product life cycle, 17–20, 117, 148, 149, 160, 358, 370
product phase, 18
product scope, 16, 47, 137, 182, 183, 208–211
program, 4, 10, 13–17, 36, 39, 92, 96, 109–112, 115–120, 140, 141, 162, 230, 241, 312, 316, 324, 327, 330, 354, 361, 363, 371, 372
program life cycle, 16, 17
program management, 4, 10, 15, 39, 92
Project Authorization Document, 16, 152, 164, 171–174, 203, 221, 258, 262, 269, 275, 281, 290, 308, 311, 325, 370
Project Brief, 170–172, 208, 220–222, 226, 227, 251, 255–259, 262, 313, 320, 347, 370
project classification, 16, 29, 132, 146, 155
project control, 60, 71–75, 85, 97, 101, 143, 148, 187, 193, 199, 215, 219–225, 231–234, 258, 262, 263, 273, 311, 319, 325

Project Definition Rating Index, 35, 96, 98, 212, 337, 351, 370
project delivery success, 30, 67, 68, 116, 148, 151, 152, 178, 197, 345
Project Detailed Plan, 149, 151, 183, 189, 190, 222, 223, 226, 287, 318, 327, 329, 333, 371
project failure, 103, 232
Project Management beyond Waterfall and Agile, 18, 23, 29, 30, 50, 101, 110, 120, 131, 132, 141, 144–149, 158, 170, 198–201, 205, 219, 225, 227, 361
project management consultant, 48, 189, 319, 371
project management office, 97, 150, 372
Project Management Plan, 62, 150, 153, 162, 176, 184–186, 189, 200, 209, 220, 226, 261, 287, 289, 300, 304, 307, 308, 313–318, 325, 336, 348, 363–367, 372
project management success, 30, 31, 67, 75, 116, 148, 150, 154, 158, 197, 260, 345, 352
project owner, 9, 14, 26, 27, 35, 47, 48, 61, 64, 69–74, 77–79, 84, 85, 91, 97, 106, 115, 116, 126, 138, 148, 149, 165, 174–177, 180, 189, 206, 208, 212, 215, 216, 230–234, 241, 336, 341, 350
Project Phase, 18
Project Requirements Document, 182, 183, 189, 204, 208, 222, 226, 287–307, 316, 320, 351, 372
project scope, 15, 16, 61, 62, 77, 137, 182, 183, 207, 208, 223
project stage, 18, 19, 186, 243, 282
project success, 26, 30, 31, 62, 63, 67, 98, 99, 108, 112, 116, 133, 135, 142–145, 148–155, 158, 170, 178, 183, 195–199, 217, 225, 235, 260, 283, 293, 305, 327, 345, 348, 351, 352, 356
Project Success Score Card, 153
provisional acceptance, 194, 340

Q

quality management, 98, 206, 214, 263, 293, 328

R

Requirements Stage, 129, 158, 178, 181, 182, 204, 222, 280, 288–293, 296–300, 307, 315
risk management, 85, 98, 100, 128, 189–191, 206, 212–214, 263, 269, 275, 296, 348, 349

S

schedule management, 206, 212
scheduling, 34, 50, 61, 63, 72, 84, 91, 105, 126, 175, 205, 212, 265, 266, 273–276, 279
scope creep, 223, 226, 319
service provider, 14, 26, 148, 194, 208
Seven Elements of Project Management Maturity™, 23, 90, 110–113, 127, 142
Shaping Stage, 171–181, 185, 213, 222, 251, 256–258, 276, 288, 300, 316, 325
Stage Authorization Document, 145, 173, 201–204, 243, 254, 257–261, 265–270, 273, 289, 294, 307, 309, 317, 372
Stage Detailed Plan, 145, 173–176, 200–206, 219, 220, 243, 254, 258, 261–273, 278, 290–297, 308, 309, 317, 364, 372
stage gate, 15, 20, 81, 98–101, 128, 143, 145, 150, 158, 161–167, 171–174, 177, 181, 184, 185, 188, 191–194, 203, 213, 215, 251, 255–261, 269, 275, 280–282, 288, 290, 298, 300, 303–307, 310, 313, 316, 318, 325, 328, 330, 335, 336, 340, 341, 343–348, 367, 372
stage-gate process, 5, 7, 21, 58, 89, 96, 111, 128, 133, 134, 138, 347, 348, 354
Stage Management Plan, 145, 173, 200, 201, 204, 206, 220, 243, 254, 258, 261, 264, 268, 271, 290, 291, 308, 309, 315, 363, 364, 371
stage management process, 20, 145, 157, 163, 165, 184, 192, 195, 200, 206, 289, 347, 347
stakeholder management, 57
stakeholders, 6, 20, 24, 46, 54, 56, 61, 63, 84, 91, 126, 162, 180–184, 188, 201, 206, 226, 233, 235, 240, 255, 259–263, 268–270, 282, 287–290, 293–296, 300, 305, 308, 313, 340, 353, 366
Strategy Stage, 150, 176, 181–184, 190, 208, 222, 280, 287, 288, 315, 317, 348
sustainability, 41, 101, 112, 117, 136, 145, 174–177, 261–265, 277, 289, 293, 294, 303, 344, 347, 352, 357, 365

T

tailored approach, 5–12, 20, 22, 26–34, 37, 38, 56, 59, 74, 83, 91–94, 98, 101, 111, 121–128, 131, 140, 141, 154, 157–161, 165, 198, 199, 235, 239–243, 313, 315, 353, 354, 360
tailored method, 27, 28, 98, 120, 143, 176
The SUKAD Way, 141–143, 154
triple constraints, 61, 84, 91, 126, 137, 138

U

uncertainty, 46, 47, 51, 64, 84, 91, 113, 126, 179, 183, 208–213, 218, 266, 270

W

work breakdown structure, 183, 372